Free Nitroxyl Radicals

Free
Nitroxyl Radicals

E. G. Rozantsev

CHIEF, STABLE RADICALS LABORATORY
INSTITUTE OF CHEMICAL PHYSICS
ACADEMY OF SCIENCES OF THE USSR, MOSCOW

Translated from Russian by
B. J. Hazzard

Edited by
H. Ulrich

MANAGER, CHEMICAL RESEARCH
THE UPJOHN COMPANY
DONALD S. GILMORE RESEARCH LABORATORIES
NORTH HAVEN, CONNECTICUT

SPRINGER SCIENCE+BUSINESS MEDIA, LLC 1970

E. G. Rozantsev, a specialist in organic, physical, and biological chemistry, was born in 1931. A 1955 graduate of the Chemical Faculty of Moscow State University, he received the degree of Candidate of Chemistry in 1958. From 1959 to 1961 he directed a research group in the Institute of Synthetic Alcohols and Organic Products, simultaneously working in the Institute of Scientific Information, Academy of Sciences of the USSR. In 1961 he became director of a research group in the Institute of Chemical Physics and in 1965, having prepared a thesis on paramagnetic organonitrogen compounds, obtained the degree of Doctor of Chemistry. Since 1967 E. G. Rozantsev has been chief of the Stable Radicals Laboratory, Institute of Chemical Physics, Academy of Sciences of the USSR. He is the author of more than 100 scientific papers and holds a number of patents. His studies in general pertain to the synthesis, structure, and reactivity of stable organic radicals.

Library of Congress Catalog Card Number 69-12541

ISBN 978-1-4757-0712-0   ISBN 978-1-4757-0710-6 (eBook)
DOI 10.1007/978-1-4757-0710-6

The text of the author's manuscript was made available for translation under an agreement with Mezhdunarodnaya Kniga, the Soviet book export agency.

© 1970 Springer Science+Business Media New York
Originally published by Plenum Press, New York in 1970

FOREWORD

In the short time since the first nitroxyl radical was obtained in 1959, a new branch of chemical science has arisen and taken shape—the chemistry of stable nitroxyl radicals.

The present book was written by E. G. Rozantsev, one of the pioneers in this field and a prominent specialist on stable radicals at the Division of Chemical and Biological Processes of the Institute of Chemical Physics, Academy of Sciences of the USSR. His numerous papers have contributed greatly to the present situation, wherein nitroxyl radicals have acquired unusually wide popularity, including the fields of biology, medicine, chemistry, physics, biophysics, and instrument construction. A clear illustration of the astonishingly vigorous development of this new field of chemistry can be given by the enormous flow of information on the synthesis and use of nitroxyl radicals. There is no doubt that this monograph, which in part generalizes the results of many workers studying these radicals, will be received with interest by specialists working in this field.

The author has not attempted to give an exhaustive account of the material. His aim is to introduce the reader to this new field and to show the wide possibilities for using radicals in scientific experiment.

The voluminous bibliography, including many papers by the author himself which may not be well enough known to the American reader, will undoubtedly contribute to the usefulness of the monograph.

Considerable attention has been devoted in the book to new reactions of radicals without the involvement of the free valences. These reactions were discovered by Rozantsev in 1962 and have subsequently acted as a basis for the preparation of so-called "spin-labeled" compounds. The latter, thanks to the outstanding work of McConnell, have become a fine and reliable instrument for studies in molecular biology.

Extremely valuable is the final chapter of the monograph, which serves as an independent introduction to the preparative organic chemistry of individual nitroxyl radicals.

Thus, Rozantsev has succeeded in giving for the first time in a book of convenient size a concise and clear account of a large amount of material which had previously been extremely scattered. This book may be recommended in equal degree to chemists, physicists, and biologists interested in theoretical and applied questions of the chemistry of stable radicals.

Prof. N. M. Emanuél'
Member of the Academy of Sciences of the USSR

PREFACE

As developments proceed in chemistry and physics, in the chemical and fuel industries, in rocket technology, and in astronautics, processes involving free radicals are acquiring ever-increasing importance. Such processes include reactions of combustion and explosion, thermal cracking, and polymerization. Problems of the stabilization of gasolines, lubricating oils, polymers, and resins, and of the prevention of the oxidative spoilage of food products and drugs are also connected with free-radical processes. The biochemical processes of enzymatic oxidation and inhibition are apparently chain reactions in whose intermediate stages free radicals are formed.

A detailed study of the enormous number of complex chemical reactions in the liquid and solid phases, including photochemical, radiation-chemical, and biochemical reactions, shows that free-radical processes are extremely widespread.

As was the case half a century ago, free radicals continue to play a prominent part in the progress of the theoretical ideas of modern chemical science. A powerful driving factor in the development of this field of chemistry was the discovery and wide use of the method of electron paramagnetic resonance.

Free radicals are exceptionally useful compounds for solving numerous problems arising in the course of chemical, physical, and biological experiments.

With the discovery of the phenomenon of dynamic nuclear polarization, stable radicals with spectra having a simple hyperfine structure have come into use for the construction of high-sensitivity precession magnetometers, gradiometers, and maser-type quantum generators.

Enormous contributions to the chemistry of stable radicals have been made by the fundamental studies of M. Gomberg, H. Wieland, G. A. Razuvaev, S. Goldschmidt, K. Ziegler, and E. Miller, which may justifiably be regarded as classical. At the present time, this field is being intensively worked by such well-known specialists as K. Dimroth, A. Rassat, N. McConnel, K. Hausser, G. Russell, S. Hönig, Yu. Schmidt, A. Balaban, R. Kuhn, R. Normant, G.

Fischer, P. Frangopol, G. Schoffa, R. O. Matevosyan, A. Forrester, V. Griffiths, and N. Tsimmerman.

The vigorous development of the chemistry of stable free radicals has been going on at an ever-increasing rate since the 1950's. The abundance of experimental and theoretical work in this field makes it an urgent necessity for there to be a classification and generalization of the material in its individual branches.

This monograph deals exclusively with the chemistry of nitroxyl radicals. As a necessary introduction, the main classes of free radicals are enumerated and their relationship outlined. An attempt is then made to classify the material in accordance with the structural features of nitroxyl radicals. Chapters are then devoted to discussions of the structure, reactivity, methods of synthesis, and applications of nitroxides, nitroxyls, and polynitroxyls.

Because of limitations of space, questions involving quantum-chemical and kinetic investigations of nitroxyl radicals have been almost completely omitted; it has also been necessary to ignore π-complex formation and photochemistry, as well as the ultraviolet and infrared spectroscopy of radicals of this class.

I am extremely grateful to V. B. Stryukov for reading Chapter VII, and also to V. I. Suskina, R. I. Zhdanov, L. A. Krinitska, G. L. Grigoryan, and other colleagues of the Stable Radical Laboratory for invaluable assistance in the preparation of the manuscript.

I consider it a pleasant duty to express my gratitude to the scientific director of the department, Academician N. M. Émanuél', and also to Professors N. M. Chirkov and B. M. Mikhailov for a critical review of the manuscript and valuable advice and observations.

In conclusion, I offer a tribute of sincere gratitude and appreciation to the respectful memory of the late Prof. M. B. Neiman, with whom I had the opportunity to collaborate during the last three years of his life. His idea of the possibility of reactions of radicals without the involvement of the unpaired electron long inspired myself and my colleagues in our work by its freshness and by the perspectives which it opened up in chemistry. While he was not an organic chemist, Prof. Neiman had a lively interest in the work of synthetic chemists, frequently visited our small laboratory, shared our joy in our successes, and kept us in a state of constant creative enthusiasm.

CONTENTS

INTRODUCTION

1. Free Radicals

"Free radical" is the name generally given to a chemical moiety possessing an unsatisfied valence, i.e., an unpaired valence electron.*

The presence of a free valence imparts paramagnetism and high chemical reactivity to the radicals. The following reactions are the most typical for free radicals: (1) substitution, (2) addition and decomposition, (3) isomerization, (4) recombination, and (5) dismutation. It is an essential feature that all these reactions take place with the indispensable participation of the unpaired electron:

$$R\cdot + R_2-R_3 \longrightarrow R-R_2 + R_3\cdot \tag{1}$$

$$R\cdot + A_1{=}A_2 \rightleftarrows R-A_1-A_2\cdot \tag{2}$$

$$CH_3-CH_2-CH_2\cdot \longrightarrow CH_3-\dot{C}H-CH_3$$

$$R_1\cdot + R_2\cdot \longrightarrow R_1-R_2 \tag{4}$$

$$2R-CH_2-CH_2 \longrightarrow R-CH{=}CH_2 + R-CH_2-CH_3 \tag{5}$$

Of the chemical methods for detecting and identifying free radicals, the most important is the metallic-mirror method, which enables their concentration and stability in the gas phase to be determined, and Szwarc's method, based on the reactions of free radicals with toluene, giving bibenzyl.

Of the physical methods of investigation, the spectrophotometric (in the visible, UV, and IR regions of the spectrum), colorimetric, chemiluminescence, chromatographic, and mass-spectrometric methods have become widely used.

Investigations of the magnetic properties are specific for free radicals (Gouy's method). Among the most effective methods of investigation are gas microwave spectroscopy and electron paramagnetic resonance (EPR).

*All paramagnetic salts of the elements of the transition groups are excluded from the category of free radicals, since, in such type compounds, the unshared electrons are localized in inner unfilled atomic shells and have no direct relationship with the molecular bond [1].

In view of the fact that all free radicals are paramagnetic, the EPR method has proved to be particularly useful in the detection and the measurement of the concentrations of radicals. Of fundamental importance for determining their nature are the position of the resonance maximum and the structure of the EPR spectrum, which is expressed in the splitting of the resonance curve into a series of components, the number and relative intensities of which are frequently determined by the chemical structure of the radical.

The first free radicals with which man became acquainted were apparently the paramagnetic oxides of nitrogen: $\cdot$NO and $\cdot$NO$_2$. These extremely stable nitroxyl radicals are formed under natural conditions in lightning discharges, and they are produced on a large scale by the modern chemical industry [2].

In 1845, another extremely stable radical was obtained in solution, nitrosodisulfonate ion:

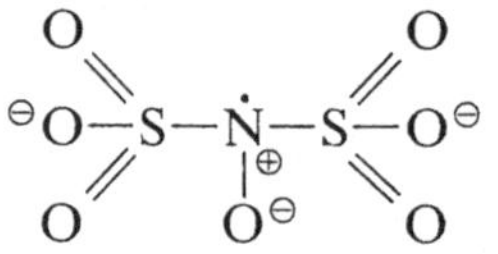

A free neutral radical with trivalent carbon was discovered in 1900 by Gomberg in an attempt to synthesize hexaphenylethane by the action of molecular silver on triphenylbromomethane [3].

In solutions, free triphenylmethyl exists in thermodynamic equilibrium with its dimer:

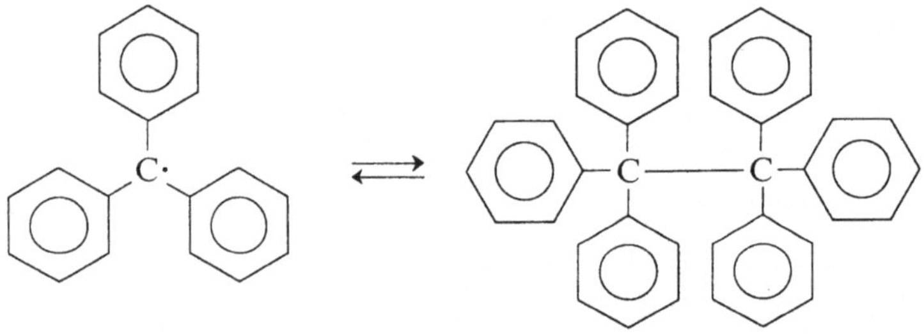

The yellow solutions of triphenylmethyl avidly absorb oxygen, iodine, and oxides of nitrogen and form the corresponding diamagnetic addition products. The degree of dissociation of hexaphenylethane in $0.1\,M$ benzene solution at 23°C is 2.4%, and it increases with rising temperature.

A number of analogs of triphenylmethyl are known to be more or less stable in solution, depending upon the nature of the aromatic substituents. The relative stability of triphenylmethyl and its analogs is explained by the delocalization of the unpaired electron in the system of aromatic bonds of the chemical moiety, in consequence of which, the electron is described by a wave function corresponding to a lower energy level. According to Pauling, triphenylmethyl is stabilized by the resonance of 44 limiting structures, some of which are

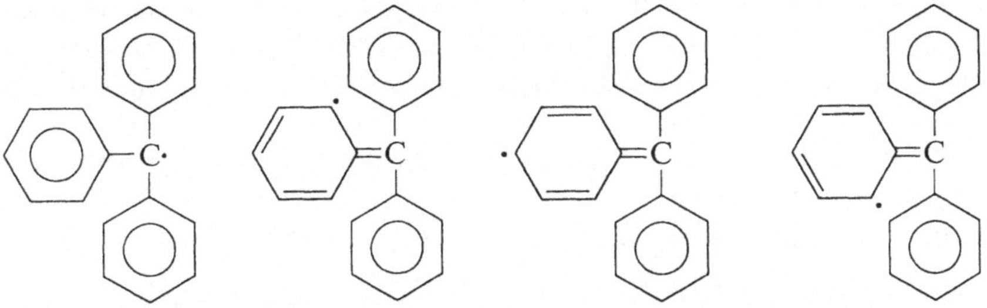

The benzyl radical can be regarded as the hybrid of five canonical valence structures [4]:

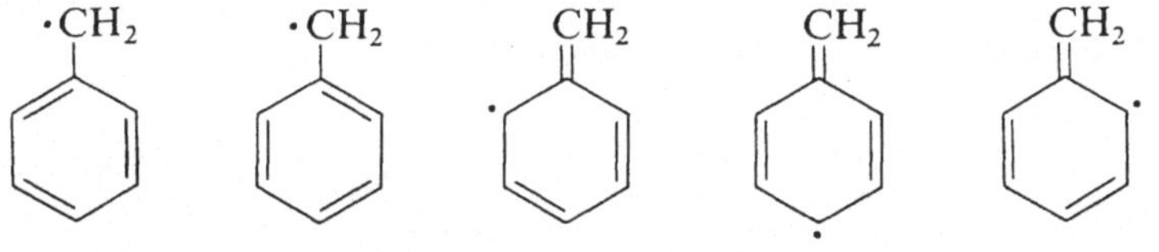

Consequently, benzyl is approximately 15 kcal/mole more stable than methyl, the half-life of which is 0.006 sec.

Free radicals formed in thermal dissociations are generally in equilibrium with the starting materials. Thus, when tetramethyl lead is heated, it dissociates reversibly into lead and methyl radicals [5]:

$$Pb(CH_3)_4 \rightleftharpoons Pb + 4 \cdot CH_3$$

The degree of dissociation of the molecules into radicals under otherwise identical conditions depends upon the bond strength of the substitute. Thus, at 200°C and atmospheric pressure, chlorine (with a heat of homolytic dissociation of 57.2 kcal/mole) is 35% dissociated, while at the same temperature iodine (with a heat of dissociation of 35.5 kcal/mole) is 95% dissociated [6].

Active free radicals can be obtained by the ultraviolet irradiation of certain compounds—for example, azo compounds:

$$R-N=N-R \xrightarrow{h\nu} 2R \cdot + N_2$$

or by the action of an electric discharge, in which the molecules are split into radicals under the influence of impacts of electrons or ions, or by the action of a high-temperature discharge. Penetrating radiation—for example β or γ rays—also causes the homolytic cleavage of molecules (radiolysis). Free radicals may also arise in the course of chemical reactions.

The development of spectroscopic and mass-spectroscopic methods has permitted the identification of a large number of active free radicals in various reactions, such as: BH, CH, NH, PH, SH, C_2, CN, ClO, HCO, NH_2, CF_2, PH_2, C_3, N_3, C_6H_5O, C_6H_5S, C_6H_5NH, CS, C_6H_5CO, $C_6H_5CH_2$, and C_6H_5. The radicals OH, H, O, and HO_2 have been detected in a hydrogen flame.

A class of free anion-radicals stable in the absence of oxygen and called metal ketyls was discovered in 1891 [7]. Metal ketyls are obtained by the action of alkali metals on nonenolizable ketones—for example,

$$C_6H_5-CO-C_6H_5 + H \longrightarrow C_6H_5-\dot{C}(OH)C_6H_5$$

Examples of radicals with divalent nitrogen are the unstable diphenyl-nitrogen, which is obtained by heating tetraphenylhydrazine in benzene at 80°C, and diphenylpicrylhydrazyl, which is stable in air and exists in the form of a radical in solution and in the solid state [8]:

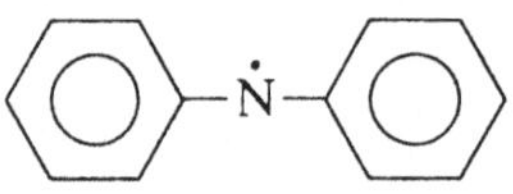

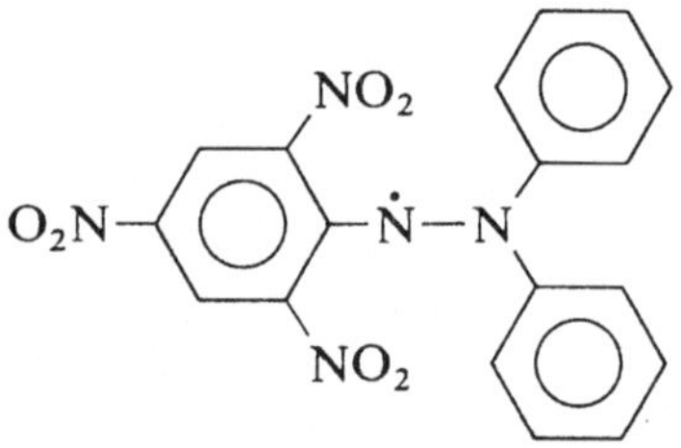

Examples of free radicals containing formally quatervalent nitrogen are paramagnetic aromatic derivatives of nitric oxide of the general formula:

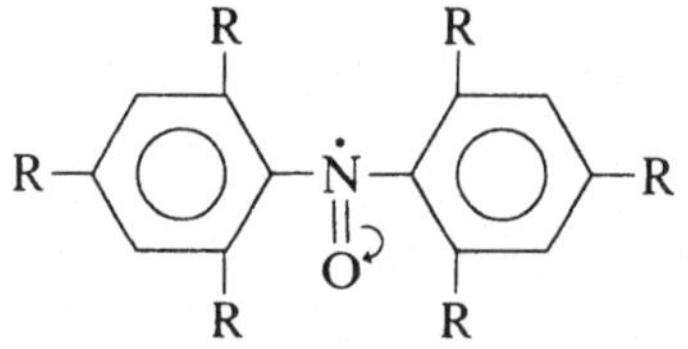

These compounds are characterized by a deep color which does not disappear even when the radicals are cooled to −60°C. Some of them can be kept for years under ordinary conditions—for example [9],

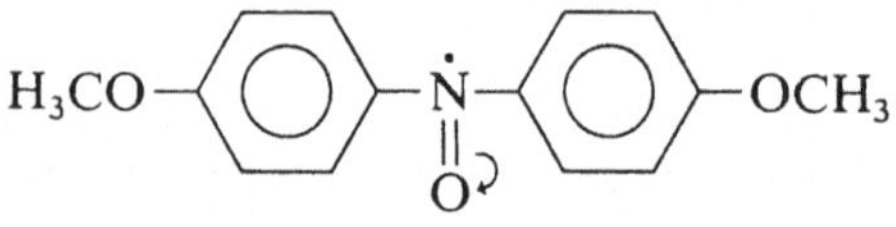

The intermediate products of the oxidation of various phenols belong to the large class of stable free radicals of univalent oxygen. The delocalization

of the unpaired electron leads to the situation that the canonical formula with a free valence on the oxygen is only one of many mesomeric structures of the radical. Compounds of this type possess various tinctorial and chemical activities depending on their structure: thus, the radical of "Ionol" is stable only in the absence of oxygen, while "Galvinoxyl" can be kept in air for several weeks [10]:

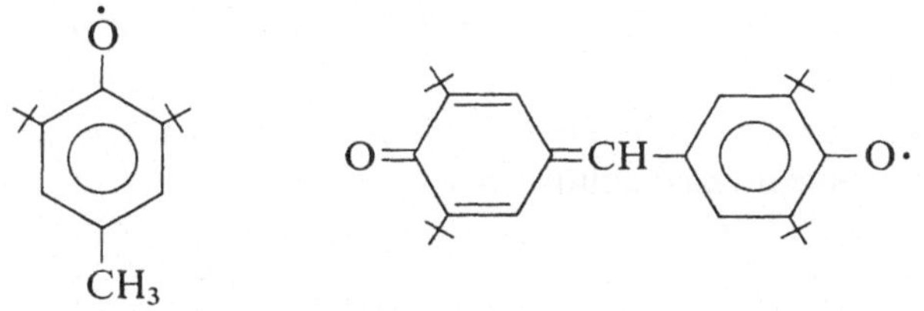

The delocalization of the unpaired electron is clearly shown in the EPR spectrum of the Ionol radical. Its spectrum consists of a quadruplet (splitting due to the protons of the methyl group) of triplets (additional splitting due to the *meta*-protons of the phenyl ring).

A special position is occupied by a family of recently discovered free radicals with a highly localized valence, the so-called free nitroxyls, which possess extremely high stability with regard to oxygen and to heating. In contrast to stable radicals known earlier, the stable nitroxyls do not possess a system of conjugated π-bonds—for example [11],

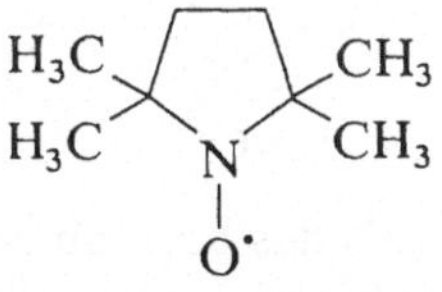

Radicals of this class dissolve in nonpolar and polar solvents with the formation of brightly colored solutions and give triplet EPR spectra.

One of the unusual properties of such radicals is their capacity for taking part in "nonradical" reactions without the participation of the free valence—for example [12],

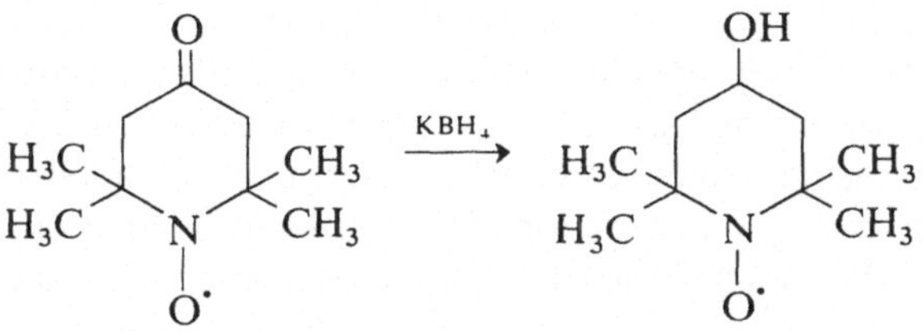

Negatively or positively charged free radicals, called anion- or cation-radicals, respectively, are formed in one-electron oxidation and reduction

processes [13]. This type of process takes place most readily in solvents favoring electron transfer. Thus, the action of sodium on a solution of naphthalene in tetrahydrofuran leads to the transfer of an electron from the sodium atom to the naphthalene molecule, with the formation of an anion-radical:

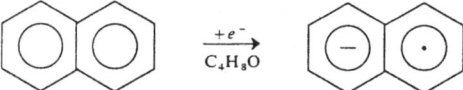

Correspondingly, a naphthalene cation-radical is formed when naphthalene is dissolved in concentrated sulfuric acid:

Ion-radicals of this type give EPR spectra showing the marked delocalization of the unpaired electron over the π-orbital system of the aromatic nuclei.

An independent group of anion-radicals are semiquinones, which are intermediates in the reduction of quinones of alkaline media [14]:

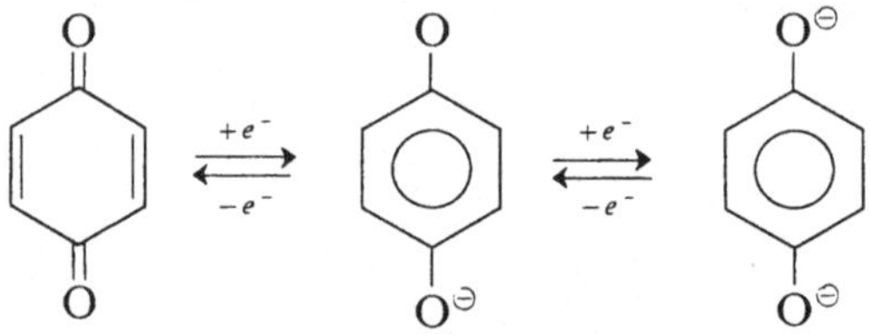

A number of cation-radicals has been obtained in acid solutions in the one-electron oxidation of various derivatives of paraphenylenediamine—for example, Wurster's red and blue [15]:

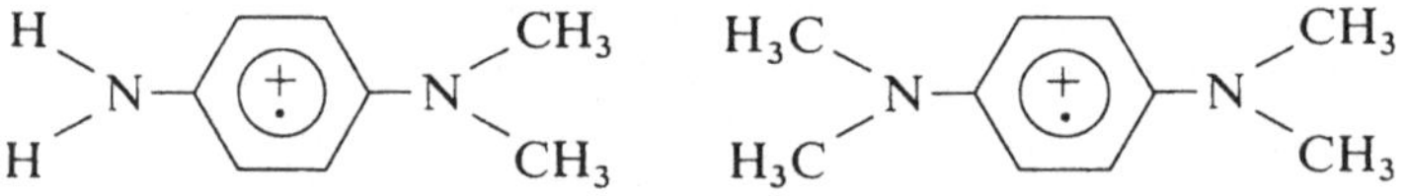

A special position in the chemistry of free radicals is occupied by compounds having not one but several free valences (unpaired valence electrons). A molecule possessing two unpaired valence electrons is called a biradical. If there is a strong spin–spin interaction of the unpaired electrons in the biradical, this state of the molecule is called the triplet state. The transition of a molecule from the ground (singlet) state to the biradical (triplet) state is accompanied by the absorption of energy (excitation of the molecule).

For example, when a molecule of ethylene is excited, a rearrangement of the spins of the π electrons takes place, their vectors becoming almost parallel with one another:

$$H_2C{=}CH_2 \quad \rightleftarrows \quad H_2\dot{C}{-}\dot{C}H_2$$

The excitation of the diamagnetic molecule of Chichibabin's hydrocarbon takes place analogously [16]:

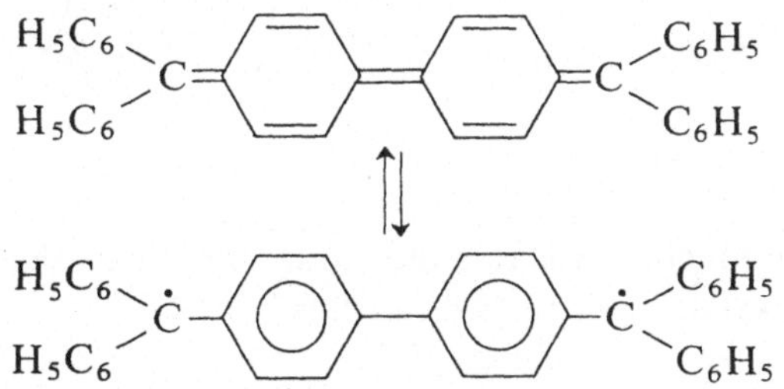

The proportion of excited molecules is determined by the magnitude $p\exp(-E/RT)$, where E is the excitation energy and p is the statistical weight of the given state, which, for the singlet and triplet states, is 1 and 3, respectively.

The greater the amount of conjugated π-electron bonds in a molecule, the easier do singlet–triplet transitions take place; for example, in the case of multinuclear condensed aromatic systems, the difference between the energies of the singlet and triplet states is almost zero.

The simplest substance with a biradical ground state is oxygen, because it possesses two unpaired electrons with almost parallel spins.

Examples of organic biradicals existing in thermodynamic equilibrium with their associates in solutions are Schlenk's hydrocarbon and chlorinated Chichibabin's hydrocarbon [17, 18]:

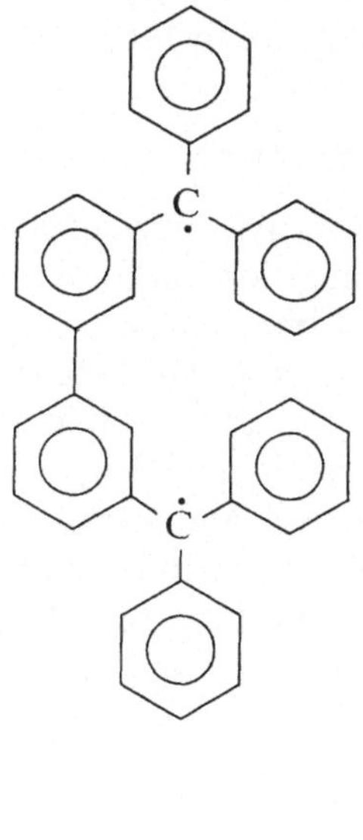

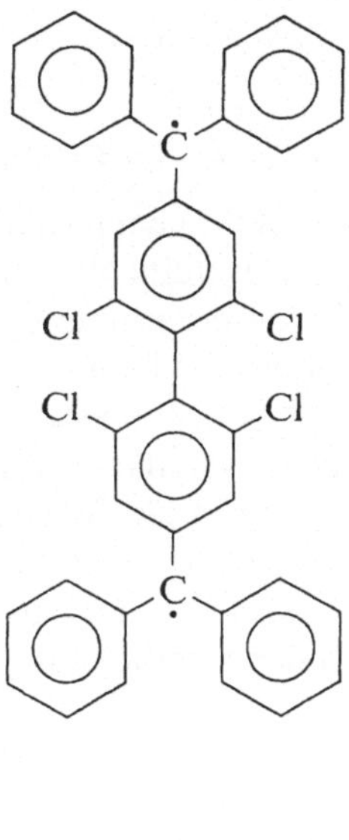

A biradical stable to oxygen is the blue heterocyclic compound "porphyridine" [19], which exists in triplet–singlet equilibrium with its diamagnetic form:

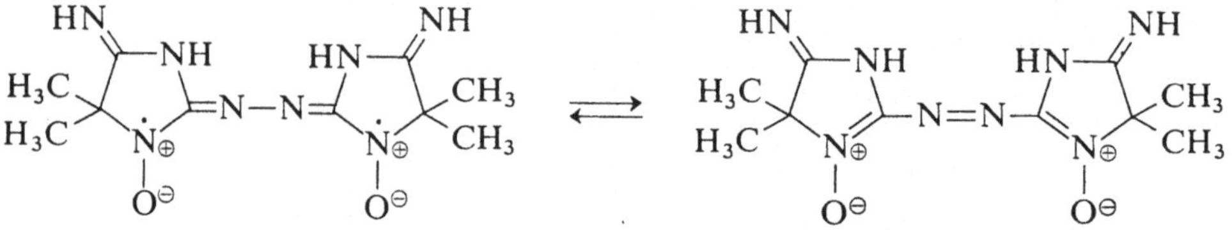

Examples of kinetically stable individual biradicals that are unaffected by air are bisaroxyl and bisnitroxyl radicals [20, 21]:

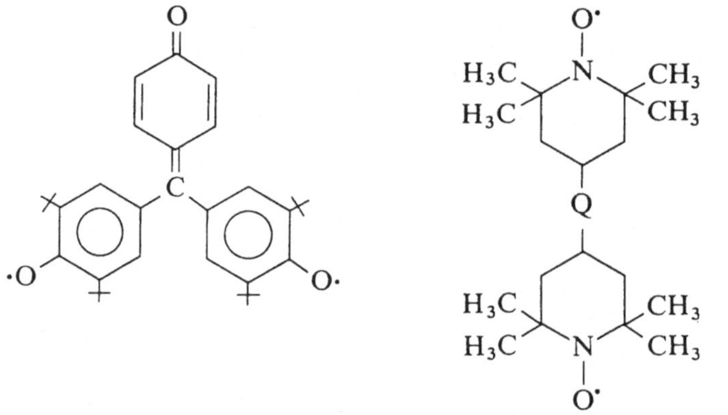

Finally, of individual compounds possessing more than two free valences (polyradicals), only tris- and tetrakisnitroxyl radicals—crystalline compounds stable to air—have been described in the literature up to 1966 [22].

2. Nomenclature and Symbols

An analysis of the chemical literature shows an avalanche of information on free radicals. It is quite obvious that in this situation questions connected with the regulation of terminology acquire great importance.

As is well known, the name of a radical is composed of the root of the mother molecule with the addition of a final "yl"—for example, methane–methyl, octane–octyl, alkane–alkyl. Derivatives of univalent oxygen are called oxyls—for example, hydroxyl, alkoxyl, aroxyl, and nitroxyl.

With the discovery of stable aromatic radicals, serious contradictions appeared with regard to nomenclature. The classification of radicals based on the formal valence indication of the "key" atom became widely used. The advocates of this classification distinguish radicals of trivalent oxygen, univalent oxygen, quatervalent nitrogen, and so on.

It was found that because of the delocalization of the unpaired electron over the aromatic nuclei, such different aromatic radicals as aroxyls, diarylpicrylhydrazyls, and diarylnitrogen oxides,

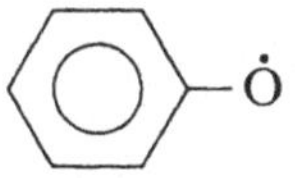

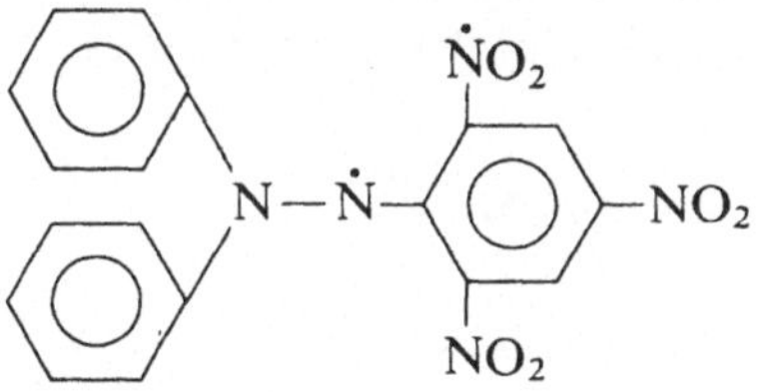

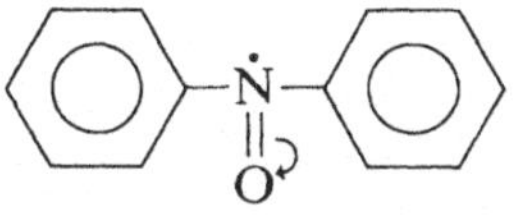

can quite well be regarded as compounds of a single class:

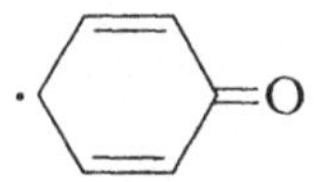

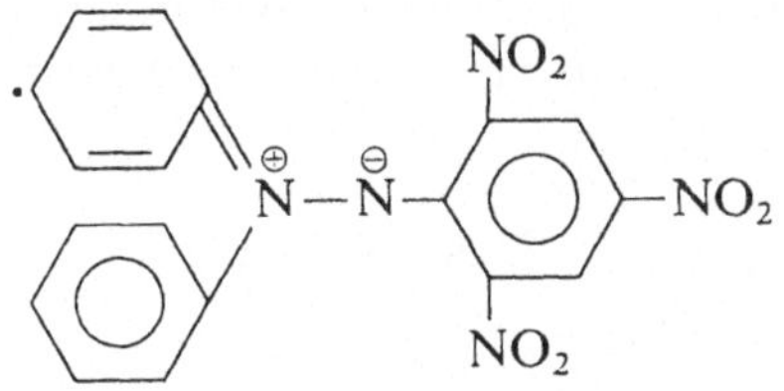

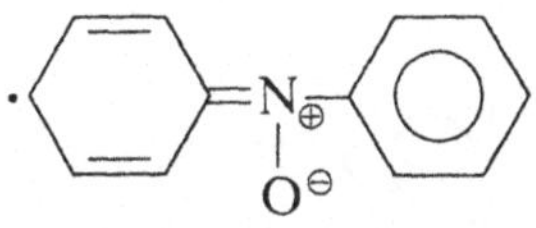

Up to the present time, a structural formula with "quatervalent nitrogen" has been widely used in the chemical literature to illustrate aromatic nitrogen oxide radicals [8, 23, 24]:

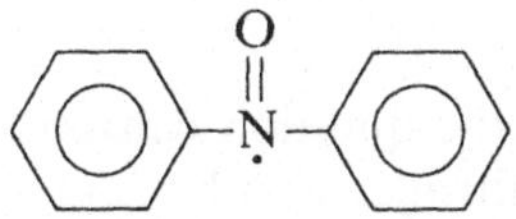

It is obvious that this classical method of representation has a particularly arbitrary nature, since nitrogen, which has five electrons in the external shell $1s^2 2s^2 2p^3$ cannot be quatervalent, but only tetracoordinate:

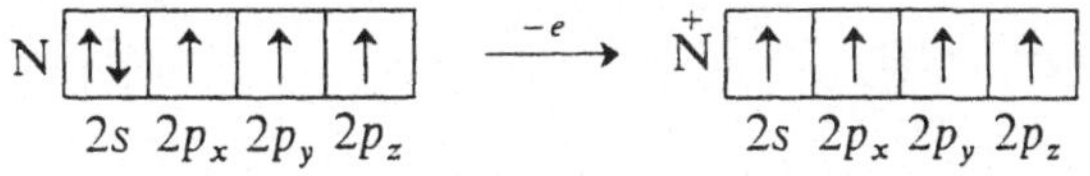

In order to "overcome" the contradiction, some authors continue to call these compounds radicals of quatervalent nitrogen using a formula with tervalent nitrogen [25]:

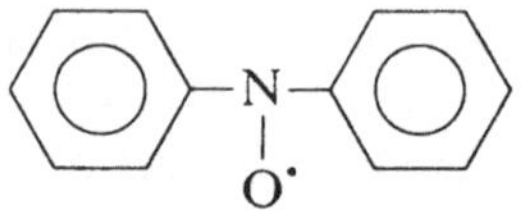

To represent nitroxide radicals, Linnett, on the basis of the double-quartet hypotheses that he has put forward [26], proposed a formula taking into account the mutual orientation of the electron spins:

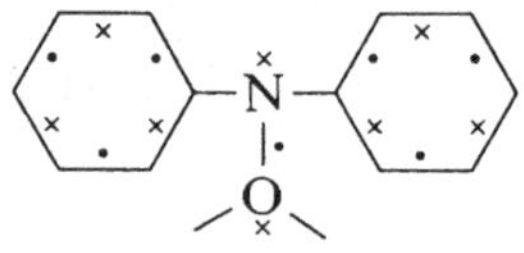

Dewar's formula, including a three-electron nitrogen–oxygen bond [27], is almost analogous to the traditional classical formula, and therefore this does not resolve the contradictions mentioned above either.

Since nitroxide radicals can be described by several valence structures,

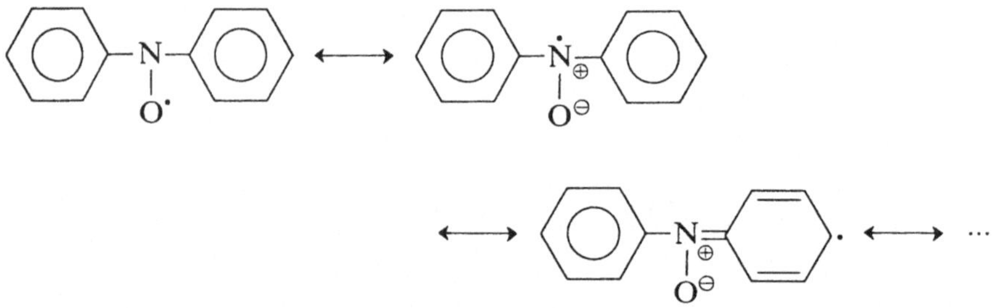

a compromise solution of the question can be achieved by using formulas with a semipolar oxygen–nitrogen bond.

For the new stable radicals in which the unpaired electron is almost completely localized on the nitrogen–oxygen bond,

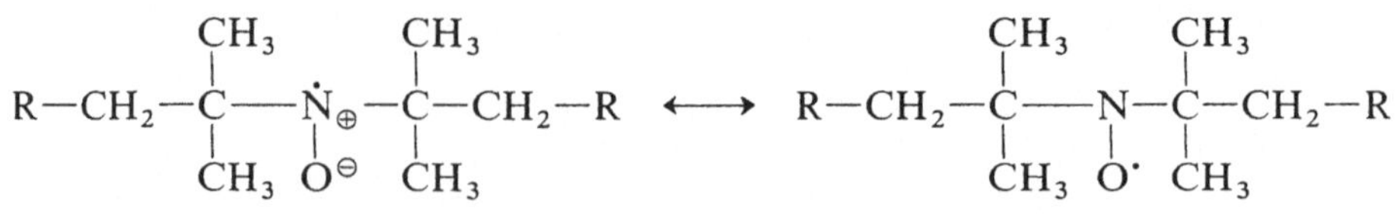

throughout the text the term "nitroxyl radicals" has been used. A special type of nitroxyl radical which has been uncovered recently is derived from oximes, and the name "iminoxyl" radical has been used for this species (see p. 41).

References

1. D. Ingram, Free Radicals As Studied by Electron Spin Resonance, Butterworths, London (1958).
2. J. Routh, 20th Century Chemistry, W. B. Saunders Co., Philadelphia (1963).
3. A. Inde, *Chem. Eng. News* **1966**, 90 (Oct. 3, 1966).
4. L. Pauling, The Nature of the Chemical Bond, Cornell University Press, New York, 3rd ed. (1960).
5. W. Waters, The Chemistry of Free Radicals, Clarendon Press, Oxford (1946).
6. V. N. Kondrat'ev, Free Radicals—An Active Form of Matter, Izd. Akad. Nauk SSSR, Moscow (1960).
7. E. Beckmann and T. Paul, *Ann. Chem.* **266**, 1 (1891).
8. L. V. Koshkin and Yu. S. Musebekov, Sketches on the History of Chemistry, Izd. Akad. Nauk SSSR, Moscow (1963), p. 190.
9. V. N. Kondrat'ev and É. G. Rozantsev, New Developments in Chemistry, a Collection of Papers, Nauka, Moscow (1964), p. 96.
10. G. Coppinger, *J. Am. Chem. Soc.* **79**, 501 (1957).
11. É. G. Rozantsev, *Usp. Khim.* **35**, 1549 (1966).
12. É. G. Rozantsev, *Izv. Akad. Nauk SSSR, Ser. Khim.* **1966**, 770.
13. L. A. Blyummenfel'd, V. V. Voevodskii, and A. G. Semenov, The Application of Electron Paramagnetic Resonance to Chemistry, Novosibirsk (1962).
14. M. S. Bluis *et al.* (ed.), Free Radicals in Biological Systems, Academic Press, New York (1961).
15. É. G. Rozantsev, L. A. Medzhidov, M. B. Neiman, and L. A. Skripko, Omagiu Raluca Ripan, [Homage to Raluca Ripan], Rumania (1966), p. 503.
16. A. E. Chichibabin, *Zh. Russ. Fiz.-Khim., Obshchestva* **39**, 925 (1907).
17. W. Schlenk and M. Brauns, *Ber.* **48**, 716 (1915).
18. E. Müller and E. Tietz, *Naturwiss.* **28**, 189 (1940).
19. O. Piloty and W. Vogel, *Ber.* **36**, 1283 (1903).
20. N. Yang and A. Kastro, *J. Am. Chem. Soc.* **82**, 6208 (1960).
21. É. G. Rozantsev, V. A. Golubev, and M. B. Neiman, *Izv. Akad. Nauk SSSR, Ser. Khim.* **1965**, 393.
22. É. G. Rozantsev, V. A. Golubev, and M. B. Neiman, USSR Authors' Certificate 175504, Oct. 24, 1964.
23. Yu. S. Musabekov, L. V. Koshkin, *Tr. Inst. Istorii, Estestvozn. i Tekh. Akad. Nauk SSSR* **35**, 246 (1961).
24. W. Hückel, Theoretische Grundlagen der organischen Chemie, Akademische Verlag, Leipzig, 5th ed. (1944).
25. A. L. Buchachenko, Stable Radicals, Izd. Akad. Nauk SSSR, Mscow (1963), p. 95.
26. J. Linnett and R. Rosenberg, *Tetrahedron* **20**, 53 (1964).
27. M. Dewar, The Electronic Theory of Organic Chemistry, Oxford University Press (1949).

Chapter I

NITROGEN OXIDES AND NITROXYLS

1. Paramagnetic Nitrogen Oxides

The simplest representative of this class of free radicals is nitric oxide. In spite of the presence of a free valence, nitric oxide is considerably less active chemically than atomic hydrogen or chlorine, and, under ordinary conditions, does not recombine to form N_2O_2 molecules, apart from those cases where it is present as a liquid (bp, $-151.7°C$) or solid (mp, $-163.6°C$) [1–5]. On the basis of spectroscopic investigations of dimeric nitrogen oxide, it may apparently be ascribed a Π-shaped configuration [6–8]. Because of this, liquid nitric oxide possesses a high specific heat capacity, with a marked dependence on temperature, and a high entropy of evaporation.

The ground state of the radical corresponds to the term $^2\Pi$. The energy of dissociation into atoms is 6.95 eV and the moment of inertia of the particle is 16.47×10^{-40} g-cm^2.

The valence state of nitric oxide can be described by the formula $NO[KK(N:2s)^2(O:2s)^2(2p\sigma)^2(2p\pi)^4(2p\pi*)]$, where the filled orbitals $(2p\sigma)^2$ and $(2p\pi)^4$ correspond completely to a triple bond, and $(2p\pi*)$ is the orbital with the unpaired electron. When the unpaired electron passes into the corresponding antibonding orbital, the excited NO particle behaves as if it possessed a triple bond [9]. Pauling explains the unusually low level of the ground state of the radical by the quantum-mechanical resonance of four valence structures [10]:

$$\overset{\oplus}{\underset{\ominus}{\overset{\displaystyle \cdot \text{N}}{\underset{\text{O}}{|}}}} \longleftrightarrow \underset{\text{O}}{\overset{\displaystyle \dot{\text{N}}}{||}} \longleftrightarrow \overset{\ominus}{\underset{\oplus}{\overset{\displaystyle \text{N}}{\underset{\dot{\text{O}}}{||}}}} \longleftrightarrow \underset{\dot{\text{O}}}{\overset{\displaystyle \text{N}}{|}}$$

Since the two partially ionic structures are represented in approximately equal amounts, the resulting dipole moment of the molecule must be close to zero. The dipole moment obtained experimentally [11] is actually very low (0.19 D).

Nitric oxide can be described by a single formula, using Pauling's idea of the three-electron bond: $N\!\equiv\!O$.

If the bond between the nitrogen and the oxygen were not more stable as the result of the superposition of a three-electron bond on the ordinary bond, the NO dimers would be stable not only in the liquid, but also in the gaseous state. However, in the gaseous state, nitric oxide exists exclusively in the form of monomeric paramagnetic particles [10].

A radical structure with a three-electron bond agrees well with the values of the interatomic distance determined experimentally, which proved to be intermediate between the values calculated for double and triple covalent bonds [12]: $N\!=\!O$ (1.18 Å); $N\!\equiv\!O$ (1.15 Å); and $N\!\equiv\!O$ (1.06 Å). The cation [13] and anion [14] of nitric oxide, accordingly, have the following electronic structures: $[\dot{N}\!\equiv\!\ddot{O}]^{\oplus}$ and $[:\dot{N}\!=\!\ddot{O}:]^{\ominus}$.

Nitric oxide is readily formed by the action of moderately concentrated nitric acid on copper:

$$8HNO_3 + 3Cu = 3Cu(NO_3)_2 + 4H_2O + 2NO$$

The well-known quantitative determination of nitric acid using the Lunge nitrometer is based on the measurement of the volume of nitric oxide obtained from the reaction [27]

$$6Hg + 2HNO_3 + 3H_2SO_4 = 3Hg_2SO_4 + 4H_2O + 2NO$$

With chlorine and bromine, nitric oxide gives nitrosyl halides, and with concentrated sulfuric acid in the presence of oxygen, it forms nitrosylsulfuric acid:

$$2H_2SO_4 + 2NO + \tfrac{1}{2}O_2 = 2NOHSO_4 + H_2O$$

Chromic, permanganic, and hypochlorous acids rapidly oxidize nitric oxide to nitric acid.

In the presence of water, sulfur dioxide readily reduces nitric oxide to nitrous oxide. Salts of bivalent chromium in neutral solution reduce it to ammonia and in an acid solution to hydroxylamine [16]. Hydrochloric acid solutions of stannous salts have a similar action.

Hydrogen chloride forms an interesting compound with nitric oxide. This dark red saltlike substance can be represented as nitrosonium chloride, $[HNO]^{\oplus}Cl^{\ominus}$.

Such elements as carbon, phosphorus, and magnesium burn vigorously in an atmosphere of NO, while burning sulfur is extinguished by it.

The gas-phase reactions of NO with chlorine [17, 18], bromine [19], and oxygen [20] take place mainly as triple collisions *via* an activated complex with an even number of electrons:

$$2NO + O_2 \longrightarrow [ONOONO] \longrightarrow 2NO_2 + 27\,kcal$$

With a very large excess of oxygen, dinitrogen trioxide is formed, since the bimolecular reaction

$$NO + NO_2 \longrightarrow N_2O_3 + 10\,kcal$$

takes place far more readily than the trimolecular reaction [21].

On the basis of Ogg's later investigations, it may be assumed that metastable intermediates arising in binary collisions take part in the formation of nitrogen dioxide [23]:

$$M + NO + O_2 \rightleftharpoons ON{-}O{-}O + M$$

$$ON{-}O{-}O + NO \rightleftharpoons 2NO_2$$

Similarly, the high-temperature gas-phase reaction of nitric oxide with hydrogen is basically trimolecular [24], but transition states with very short lives arising in binary collisions may be important [25].

Nitric oxide readily reacts with other free radicals [26–29], being, therefore, not only an indicator of chain reactions [30], but also an effective inhibitor of them [31, 32].

In the form of a neutral molecule, nitric oxide can combine with salts of many metals to form more or less stable complexes—for example nitrosyliron sulfate and sodium nitroprusside (sodium pentacyanonitrosylferrate) [33–36]: $Fe(NO)SO_4$ and $Na_2Fe(CN)_5NO$.

It is a fundamental feature that in almost all compounds of nitric acid, the additional covalent bond is formed predominantly with the nitrogen atom [37–39], which is in favor of the structure with the free valence on the nitrogen.

An analysis of the parameters of the hyperfine structure of the EPR spectrum of nitric oxide also leads to the conclusion that the greater part of the spin density (about 60%) is concentrated on the nitrogen atom [40–43]. The hyperfine structure (HFS) of the EPR spectrum is explained by the strong interaction between the moment of the unpaired electron and the rotation of the whole molecule [44].

The ground level of the molecule, $^2\Pi$, is split into a lower (diamagnetic) level, $^2\Pi_{1/2}$, and an upper (paramagnetic) level, $^2\Pi_{3/2}$. Because of interaction with the rotational motion of the molecule, the upper level is split into sublevels. The four components of the level with $S = \frac{3}{2}$ will give three absorption lines with the usual selection rule $m_S = \pm 1$. These lines have been observed experimentally, each line being additionally split into three components because of interaction with the spin of the nitrogen nucleus ($a_N = 14.2$ G).

Another stable free radical of this series is nitrogen dioxide. The gain in delocalization energy of the unpaired electron does not stabilize it sufficiently to prevent dimerization to dinitrogen tetroxide [45–49]:

$$2NO_2 \; \rightleftharpoons \; N_2O_4 + 147 \, kcal \quad (at \; 25°C)$$

Even in the vapor state (bp, 22.4°C), nitrogen dioxide is partially dimer-ized, and as a liquid and a solid (mp, − 10.2°C) it is dimeric. Only at 150°C does the gas dissociate completely into nitrogen dioxide, which, on further heating (about 600°C), decomposes into nitric oxide and oxygen [50–52].

When gaseous nitrogen dioxide is heated, its color becomes more intense. At a pressure of 1 atm, at 27, 50, 100, and 135°C, the gas contains 20, 40, 89, and 98.7%, respectively, of nitrogen dioxide.

The structure of NO_2 was established by Sutherland and Penney [53, 54]. From an analysis of the absorption bands, they concluded that the nitrogen dioxide molecule is not linear [55]. The molecular parameters

$$r(N-O) = 1.215 \, Å \quad and \quad \angle ONO = 132°$$

were determined by the electron diffraction method [56, 57]. Since the ONO angle is closer to 120° than to 90°, it may be considered, as an approximation, that the nitrogen atom is sp^2-hybridized, and the σ bonds are formed by using the electrons of molecular orbitals localized between the nitrogen and oxygen atoms. Besides ten electron pairs, there are three $2p$ electrons which can form a three-electron bond combining all three atoms [58–60]:

The overall bond order must therefore be more than 1 but less than 1.5, which corresponds to the bond length of 1.215 Å. This value lies between the figures for a normal N—O bond (1.43 Å) and a N=O double bond (1.16 Å) [61, 62].

The specific electrical conductivity of liquid nitrogen dioxide at 17°C is $1.3 \times 10^{-12} \, (\Omega\text{-cm})^{-1}$, the dielectric permeability at 18°C is 2.42, and the re-fractive index is n_D^{20} 1.420, the molecular refraction being 15.2 cm³. Therefore, it can be assumed that nitrogen dioxide is ionized to a considerable extent: $N_2O_4 \; \rightleftharpoons \; NO^\oplus + NO_3^\ominus$. Ionization can explain the reactions of dinitrogen tetroxide with, for example, diethylamine hydrochloride,

$$(C_2H_5)_2\overset{+}{N}H_2Cl^- + N_2O_4 \; \longrightarrow \; (C_2H_5)_2\overset{+}{N}H_2NO_3^- + NOCl$$

and with zinc

$$Zn + 2N_2O_4 \; \longrightarrow \; Zn(NO_3)_2 + 2NO^\oplus$$

$$Zn + 2NO^\oplus \; \longrightarrow \; Zn^{\oplus\oplus} + 2NO$$

With dinitrogen tetroxide, zinc nitrate forms a compound of the hydroxo acid type with the structure $(NO^{\oplus})_2[Zn(NO_3)_4]^{\ominus\ominus}$.

X-ray studies and entropy data are in favor of a symmetrical structure for dinitrogen tetroxide with the molecular parameters

$$r(N-O)=1.17\,\text{Å}; \quad r(N-N)=1.64\,\text{Å}; \quad \text{and} \quad \angle ONO=126°C.$$

Pauling, however, assumes that with such a structure, the neighboring nitrogen atoms have formal charges of the same sign, and therefore the asymmetrical structure

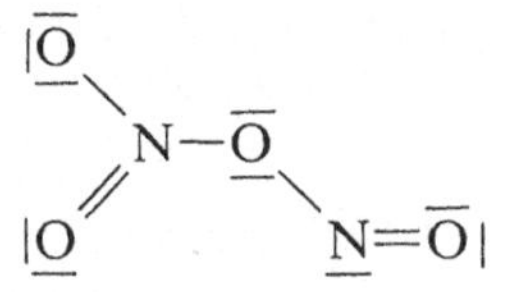

would be more stable.

The chemical properties of dinitrogen tetroxide agree better with the latter structure, because this compound can be used both for diazotization [63] and for nitration [64].

Thus, addition of dinitrogen tetroxide to carbon–carbon double bond yields nitro esters of nitrous acid exclusively [65, 66]:

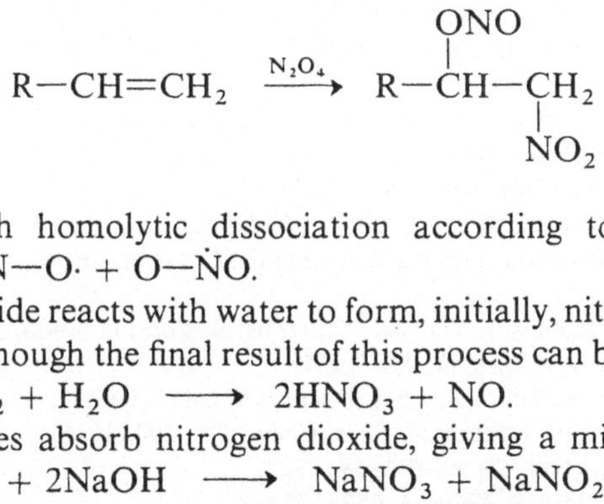

in agreement with homolytic dissociation according to the equation: $N_2O_4 \rightleftarrows ON-O\cdot + O-\dot{N}O$.

Nitrogen dioxide reacts with water to form, initially, nitric acid and dinitrogen trioxide, although the final result of this process can be represented by the equation $3NO_2 + H_2O \longrightarrow 2HNO_3 + NO$.

Caustic alkalies absorb nitrogen dioxide, giving a mixture of nitrites and nitrates $2NO_2 + 2NaOH \longrightarrow NaNO_3 + NaNO_2 + H_2O$.

Nitrogen dioxide is one of the most chemically reactive compounds, and carbon, sulfur, and phosphorus burn in an atmosphere of the gas. Its free-radical nature is clearly shown in recombination, addition, and initiation reactions [67–73].

Being a peculiar hybrid of valence-isomeric structures,

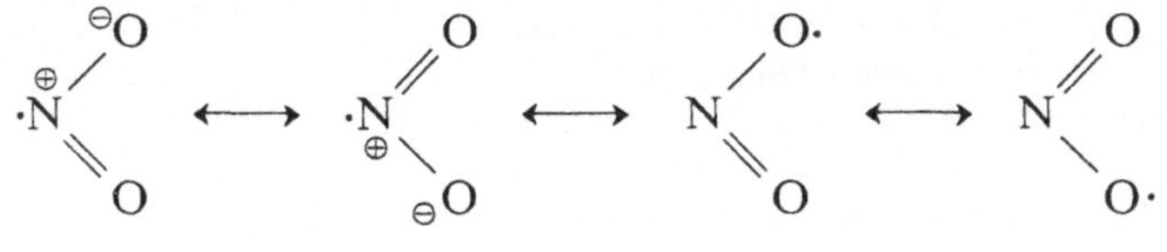

nitrogen dioxide almost always shows a well-defined dual reactivity in its reactions [74–75].

The EPR spectrum of nitrogen dioxide consists of a triplet with a distance of 47.1 gauss between the signals. Such a large value of the hyperfine splitting constant is apparently explained by a considerable admixture of the s-state in the ψ-function of the unpaired electron [76]: $\psi = a_1\psi_N(2s) + a_2\psi_N(2p_z) + B\psi_0(2p_z)$.

At the highest resolution of the spectrum, a large number of components of rotational-magnetic structure are observed in the HFS, due to the independent quantization in the magnetic field of the electron-spin, nuclear-spin, and magnetic moments of the particles [42, 77].

References

1. B. Field and C. Hardy, *Quart. Rev.* **18**, 361 (1964).
2. H. Urey, L. Dawsey, and F. Rice, *J. Am. Chem. Soc.* **51**, 3190 (1929).
3. H. Bizette and B. Tsay, *Compt. Rend.* **204**, 1638 (1937).
4. H. Johnston and W. Giauque, *J. Am. Chem. Soc.* **51**, 3194 (1929).
5. E. Lips, *Helv. Phys. Acta* **8**, 247 (1935).
6. L. D'Or and P. Tarte, *Bull. Soc. Roy. Sci., Liège*, **1953** (6–7), 276.
7. W. Fateley, H. Bent, and B. Crawford, *J. Chem. Phys.* **31**, 204 (1959).
8. W. Spindel and M. Stern, *J. Chem. Phys.* **32**, 1579 (1960).
9. R. Mulliken, *Rev. Mod. Phys.* **4**, 1 (1932).
10. L. Pauling, The Nature of the Chemical Bond, Cornell Univ. Press, New York, 3rd ed. (1960).
11. H. Watson, G. Rao, and K. Ramaswamy, *Proc. Roy. Soc.* **A143**, 558 (1934).
12. N. Jonathan, *J. Mol. Spectr.* **4**, 59 (1961).
13. L. Klinkenberg, *Rec. Trav. Chim.* **56**, 749 (1937).
14. J. Frazer and N. Long, *J. Chem. Phys.* **6**, 462 (1938).
15. H. Remy, Lehrbuch der anorganischen Chemie, Geest und Portig, Leipzig, 9th ed. (1959), 12th ed. (1963).
16. R. Heslop and P. Robinson, Inorganic Chemistry, London (1960).
17. M. Trautz, *Z. Anorg. Chem.* **88**, 285 (1964).
18. W. Krauss and M. Saracini, *Z. Phys. Chem.* **178**, 245 (1937).
19. M. Bodenstein and W. Krauss, *Z. Phys. Chem.* **175**, 294 (1936).
20. M. Bodenstein, *Z. Phys. Chem.* **100**, 68 (1922).
21. V. P'yankov, *Zh. Obshch. Khim.* **3**, 652 (1933).
22. R. Ogg, *J. Chem. Phys.* **21**, 2079 (1953).
23. S. Benson, Foundations of Chemical Kinetics, McGraw-Hill, New York (1960).
24. C. Hinshelwood and T. Green, *J. Chem. Soc.* **129**, 720 (1926).
25. J. Mitchell and C. Hinshelwood, *J. Chem. Soc.* **1936**, 378.
26. C. Cain and F. Wiselogle, *J. Am. Chem. Soc.* **62**, 1163 (1940).
27. W. Schlenk, L. Mair, and C. Bornhardt, *Ber.* **44**, 1170 (1911).
28. E. Arden and L. Phillips, *Proc. Chem. Soc.* **1962**, 354.
29. S. Brois, *Tetrahedron Letters* **1964**(7), 345.
30. L. Staveley and C. Hinshelwood, *Trans. Faraday Soc.* **35**, 845 (1939).
31. L. Phillips, *Proc. Chem. Soc.* **204** (1961).
32. A. Maschke, B. Shapiro, and F. Lampe, *J. Am. Chem. Soc.* **85**, 1876 (1963).

33. J. Lewis, R. Irwing, and G. Wilkinson, *J. Inorg. Nuclear Chem.* **7**, 32 (1958).
34. W. Griffith, J. Lewis, and G. Wilkinson, *J. Inorg. Nuclear Chem.* **7**, 38 (1958).
35. I. Bernal and S. Harrison, *J. Chem. Phys.* **34**, 102 (1961).
36. I. Bernal and E. Hockings, *Proc. Chem. Soc.* **1962**, 361.
37. B. Tyles, *Nature* **195**, 279 (1962).
38. J. Birchall, A. Bloom, R. Haszeldine, and C. Willis, *J. Chem. Soc.* **1962**, 3021.
39. L. Batt and B. Gowenlock, *Trans. Faraday Soc.* **56**, 682 (1960).
40. G. Dousmanis, *Phys. Rev.* **97**, 967 (1955).
41. R. Beringer and J. Castle, *Phys. Rev.* **78**, 581 (1950).
42. J. Castle and R. Beringer, *Phys. Rev.* **80**, 114 (1950).
43. C. Lin, K. Hijikata, and M. Sakamoto, *J. Chem. Phys.* **33**, 878 (1960).
44. R. Beringer, E. Rawson, and A. Henry, *Phys. Rev.* **94**, 343 (1954).
45. W. Giauque and J. Kamp, *J. Chem. Phys.* **6**, 40 (1938).
46. E. Leifer, *J. Chem. Phys.* **8**, 301 (1940).
47. D. Stronks and R. Wilkins, *Chem. Rev.* **57**, 743 (1957).
48. H. Siebert, *Z. Anorg. Allg. Chem.* **275**, 210 (1954).
49. F. Verhoek and F. Daniels, *J. Am. Chem. Soc.* **53**, 1250 (1931).
50. E. Archibald, The Preparation of Pure Inorganic Substances, John Wiley and Sons, New York (1932), p. 246.
51. L. Pauling, College Chemistry—An Introductory Textbook of General Chemistry, W. H. Freeman, San Francisco, 3rd ed. (1964).
52. B. Bodenstein and F. Lindner, *Z. Phys. Chem.* **100**, 82 (1922).
53. G. Sutherland and W. Penney, *Nature* **136**, 146 (1935).
54. G. Sutherland and W. Penney, *Proc. Roy. Soc.* **A156**, 678 (1936).
55. R. Gillespie and D. Millen, *Quart. Rev.* **2**, 277 (1948).
56. S. Claesson, J. Donohue, and V. Schomaker, *J. Chem. Phys.* **16**, 207 (1948).
57. L. Maxwell and V. Mosley, *J. Chem. Phys.* **8**, 738 (1940).
58. W. Orville-Thomas, *Chem. Rev.* **57**, 1179 (1957).
59. A. Walsh, *J. Chem. Soc.* **1953**, 2266.
60. J. Tanaka, *Nippon Kagaku Zasshi* **78**, 1643 (1957).
61. K. McEven, *J. Chem. Phys.* **32**, 1801 (1961).
62. H. Kato, T. Yonezawa, K. Motokuma, and K. Fukui, *Bull. Chem. Soc.* (*Japan*) **37**, 1710 (1964).
63. B. Houston and T. Johnson, *J. Am. Chem. Soc.* **47**, 3011 (1925).
64. P. P. Shorygin and A. V. Topchiev, *Zh. Obshch. Khim.* **5**, 549 (1935).
65. A. V. Fokin, V. A. Komarov, and I. N. Sorochkin, *Zh. Vses. Khim. Obschestva im. D. I. Mendeleeva* **10**, 354 (1965).
66. G. Hauptschein and R. Oesterling, *J. Org. Chem.* **28**, 1279 (1963).
67. R. Norrish and J. Wallace, *Proc. Roy. Soc.* **A145**, 307 (1934).
68. C. McDowell and J. Thomas, *Trans. Faraday Soc.* **46**, 1030 (1950).
69. F. Seel and H. Meier, *Z. Anorg. Allg. Chem.* **274**, 197 (1953).
70. J. Ogilvie, *Nature* **204**, 572 (1964).
71. J. Gray and D. Style, *Trans. Faraday Soc.* **49**, 52 (1953).
72. J. Thomas, *Trans. Faraday Soc.* **49**, 630 (1953).
73. C. McDowell and J. Thomas, *J. Chem. Soc.* **1950**, 1462.
74. A. Schaarschmidt and H. Hofmeier, *Ber.* **58B**, 1047 (1925).
75. B. Brooks and C. Boord, in: The Chemistry of Petroleum Hydrocarbons, Reinhold, New York, Vol. I (1954).
76. P. Atkins, N. Keen, and M. Symons, *J. Chem. Soc.* **1962**, 2873.
77. G. Bird, J. Baird, and R. Williams, *J. Chem. Phys.* **28**, 738 (1958).

2. Fremy's Salt

More than 100 years ago, Fremy obtained a yellow salt with the composition $K_2S_2O_7N$, which dissolved in water to give purple-violet solutions [1]. This salt is readily obtained by the action of potassium permanganate or lead dioxide on an alkaline solution of hydroxylaminedisulfonate [2, 3]:

$$(KO_3S)_2NOH \xrightarrow[KMnO_4]{-H\cdot} (KO_3S)_2NO$$

The precursor, potassium hydroxylaminedisulfonate, can be prepared by the reaction of potassium nitrite and bisulfite in aqueous solution [4].

In addition to the potassium salt, sodium [5], tetraphenylarsonium [6], and tetraphenylstibonium [7] Fremy's salts are known. By analogy with hyponitrous anhydride, Hantzsch ascribed to the solid salt the structure of a dimer which decomposed in solution into monomeric anions [8]:

$$(KO_3S)_4N_2O_2 \rightleftarrows 4K^\oplus + 2ON(SO_3^\ominus)_2$$

The astonishing analogy with nitrogen dioxide can well be seen from examples of the reactions of the salt with nitric oxide and the interaction of the resulting adduct with water [9]:

$$(KO_3S)_2NO + NO \longrightarrow (KO_3S)_2NONO$$

$$(KO_3S)_2NONO + H_2O \longrightarrow (KO_3S)_2NOH + HNO_2$$

The nitrogen–oxygen bond in Fremy's ion* is similar to a donor or semipolar bond, so that the unpaired electron is delocalized between the nitrogen and oxygen atoms of the NO group. The free ion-radical can be described by two mesomeric structures:

Formally, the structure of this radical can be represented by a single formula including a three-electron bond [10]. From this point of view, the stability† of Fremy's salt [12–18] in solutions can be explained by the low level of the ground state of a molecule possessing a three-electron bond [19–20].

The unique chemical properties of Fremy's salt have enabled it to be used as a selective one-electron oxidizing agent for aromatic amines and

---

*Fremy's anion can be found in the chemical literature under various names, such as peroxylaminodisulfonate, nitrosyldisulfonate, nitrosodisulfonate, or nitroxyl disulfonate.

†An aqueous solution of Fremy's salt is stable for several days under ordinary conditions [11].

phenols [21]. In these reactions, high yields have been obtained of many *ortho-* and *para*-quinone imines and quinones [22–42].

Isobutyraldehyde is also smoothly oxidized by Fremy's salt to α-hydroxybutyraldehyde, the reaction taking place *via* a one-electron oxidation of the corresponding enolate anion [43]:

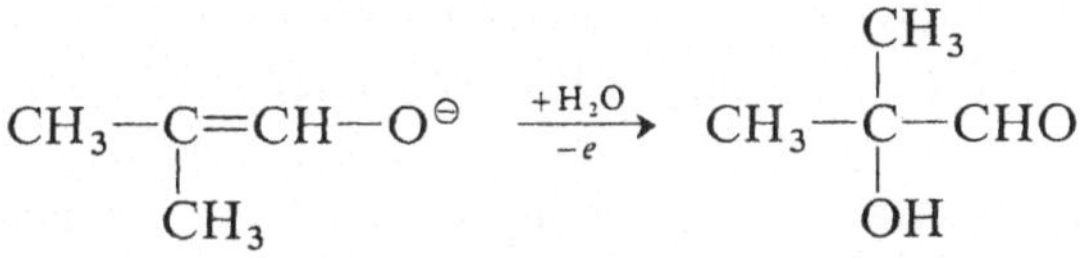

$$CH_3-\underset{\underset{CH_3}{|}}{C}=CH-O^{\ominus} \xrightarrow[-e]{+H_2O} CH_3-\underset{\underset{OH}{|}}{\overset{\overset{CH_3}{|}}{C}}-CHO$$

The facile reactions of Fremy's salt with hydrazine and hydroxylamine render it useful for the analytical determination of these reagents [44, 45]:

$$4(KO_3S)_2NO + N_2H_4 \longrightarrow 4(KO_3S)_2NOH + N_2$$

$$4(KO_3S)_2NO + 2NH_2OH \longrightarrow 4(KO_3S)_2NOH + N_2O + H_2O$$

The properties of this radical also enable it to be used in weakly alkaline solutions as a photographic developer [46]. The oxidation of 3,7-dimethoxyphenothiazine by Fremy's salt in an acid medium takes place in accordance with the following scheme [47]:

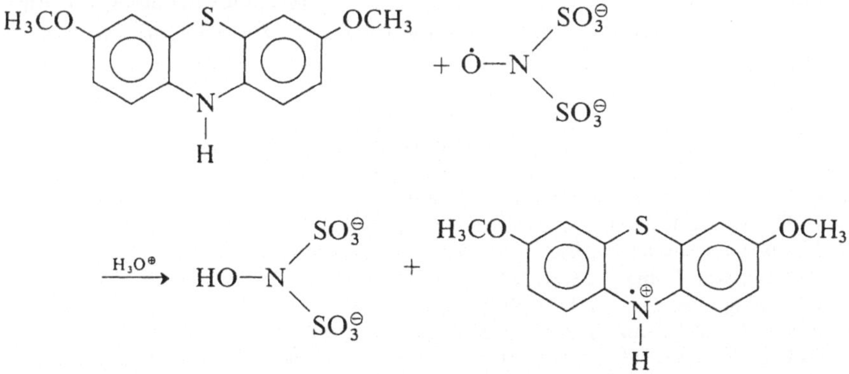

Other aromatic amines are oxidized similarly. It is possible to isolate the corresponding paramagnetic salts in the individual state and with high yields.

An analysis of the mechanism of the chemical reactions indicates that almost all radical reactions of Fremy's salt take place at the oxygen atom of the NO group, in agreement with the nitroxyl electronic structure of this radical,

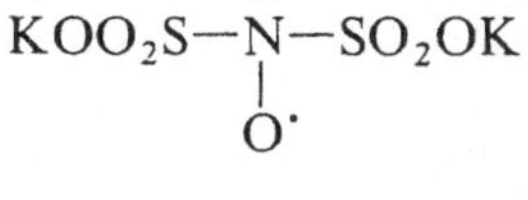

$$KOO_2S-\overset{-}{\underset{\underset{O^{\textbf{·}}}{|}}{N}}-SO_2OK$$

The EPR spectrum of Fremy's anion is a triplet with equidistant signals of similar intensities, the distance between which is about 13 G, while the width of each is less than 0.5 G [48–50]. The HFS of the EPR spectrum agrees well with the theoretically expected interaction of the unpaired electron with the nuclear spin of the nitrogen [51]. As the result of a detailed study of the fine features of the EPR spectrum of Fremy's salt [52–54], it has become possible to use it for calibrating radiospectrometers [7].

With the discovery of the dynamic nuclear polarization effect [55–57] (Overhauser–Abragam effect), Fremy's salt has unexpectedly acquired exceptional practical importance [58–60]. The Overhauser effect, upon which the method of double nuclear electron resonance is based, enables the HFS of EPR spectra of highly dilute solutions of free radicals to be studied in weak magnetic fields where the ordinary EPR method is insufficiently sensitive.

In recent years, solutions of stable radicals with a simple hyperfine structure of the EPR spectra, in particular, Fremy's salt, have come into use as working substances for high-sensitivity nuclear precession magnetometers [61–64] and for maser-type quantum generators [65].

By using Fremy's salt, it is possible by the spin echo method to study the microstructure of liquids and the mobility of particles in the liquid phase and also to resolve questions relating to energy-transfer mechanisms in condensed phases [66–68].

References

1. E. Fremy, *Ann. Chim. Phys.* **15**(3), 408 (1845).
2. G. Harvey and R. Hollingshead, *Chem. Ind.* (*London*) **1953**(11), 244.
3. F. Raschig, Schwefel und Stickstoffstudien, Leipzig and Berlin (1924), p. 148.
4. R. Asmussen, *Z. Anorg. Chem.* **212**, 317 (1933).
5. F. Raschig, *Ann. Chem.* **241**, 183 (1887).
6. B. Smaler and E. Yasaitis, *J. Chem. Phys.* **21**, 1905 (1953).
7. G. Pake, J. Townsend, and S. Weissman, *Phys. Rev.* **85**, 682 (1952).
8. A. Hantzsch and W. Semple, *Ber.* **28**, 2744 (1895).
9. H. Gehlen, *Ber.* **66B**, 292 (1933).
10. L. Pauling, The Nature of the Chemical Bond, Cornell University Press, New York, 3rd ed. (1960).
11. J. Candlin and R. Wilkins, *J. Chem. Soc.* **1961**, 3625.
12. E. Divers and T. Haga, *J. Chem. Soc.* **65**, 523 (1894).
13. T. Haga, *J. Chem. Soc.* **85**, 78 (1904).
14. J. Li and D. Ritter, *J. Am. Chem. Soc.* **75**, 3024 (1953).
15. J. Li and D. Ritter, *J. Am. Chem. Soc.* **75**, 5823 (1953).
16. H. Gehlen, *Ber.* **70B**, 1980 (1937).
17. S. Yanada and R. Tsuchida, *Bull. Chem. Soc. Japan* **32**, 721 (1959).
18. J. Burib and D. Ritter, *J. Am. Chem. Soc.* **74**, 3394 (1952).
19. A. Hoffman, *Chem. Eng. News* **1961**(51), 24.

20. J. Linnett, *J. Am. Chem. Soc.* **83**, 2643 (1961).
21. L. Fieser and M. Fieser, Advanced Organic Chemistry, Reinhold, New York (1961), p. 853.
22. H.-I. Teuber and G. Jellinek, *Ber.* **85**, 95 (1952).
23. R. Pumerer and F. Frankfurter, *Ber.* **47**, 1472 (1914).
24. H.-I. Teuber and W. Rau, *Ber.* **86**, 1036 (1953).
25. H.-I. Teuber and N. Götz, *Ber.* **87**, 1236 (1954).
26. L. Horner and K. Sturm, *Ber.* **88**, 329 (1955).
27. H.-I. Teuber and G. Staiger, *Ber.* **87**, 1251 (1954).
28. E. Müller, F. Günter, and A. Ricker, *Z. Naturforsch.* **18b**, 1002 (1963).
29. R. Magnusson, *Acta Chem. Scand.* **18**, 759 (1964).
30. H.-I. Teuber, *Angew. Chem.* **68**, 420 (1956).
31. H.-I. Teuber and G. Staiger, *Ber.* **89**, 489 (1956).
32. H.-I. Teuber, *Angew. Chem.* **69**, 308 (1957).
33. H.-I. Teuber, *Angew. Chem.* **68**, 420 (1956).
34. H.-I. Teuber, *Angew. Chem.* **68**, 628 (1956).
35. H.-I. Teuber and N. Götz, *Ber.* **89**, 2654 (1956).
36. H.-I. Teuber and G. Thaler, *Ber.* **91**, 2253 (1958).
37. H.-I. Teuber and G. Thaler, *Ber.* **92**, 667 (1959).
38. H.-I. Teuber and M. Hasselbach, *Ber.* **92**, 674 (1959).
39. H.-I. Teuber and W. Schmidtke, *Ber.* **93**, 1257 (1960).
40. H.-I. Teuber and H. Lindner, *Ber.* **92**, 921, 927 (1959).
41. H.-I. Teuber and H. Lindner, *Ber.* **92**, 932 (1959).
42. H.-I. Teuber and G. Staiger, *Ber.* **92**, 2385 (1959).
43. G. Allen and W. Waters, *J. Chem. Soc.* **1956**, 1132.
44. Von H. Gehlen, H. Elchlepp, and J. Cermak, *Z. Anorg. Allg. Chem.* **274**, 293 (1953).
45. H. Gehlen and G. Dase, *Z. Anorg. Allg. Chem.* **275**, 327 (1954).
46. G. Goldberg and J. LuValle, US Patent 2,938,793.
47. A. A. Medzhidov, É. G. Rozantsev, and M. B. Neiman, *Dokl. Akad. Nauk SSSR* **168**, 348 (1966).
48. T. Chu, G. Pake, D. Paul, I. Townsend, and S. Weissman, *J. Phys. Chem.* **57**, 504 (1953).
49. S. Weissman and D. Banfill, *J. Am. Chem. Soc.* **75**, 2534 (1953).
50. V. A. Sharpatyi, Yu. N. Molin, V. K. Ermolaev, T. S. Zhuravlev, and M. A. Proskurin, *Fiz. Problemy Spectroskopii* **2**, 100 (1963).
51. I. Townsend, S. Weissman, and G. Pake, *Phys. Rev.* **89**, 606 (1953).
52. I. Powles and M. Mosley, *Proc. Phys. Soc.* **77**, 729 (1961).
53. I. Burgess, *J. Phys. Rad.* **19**, 845 (1958).
54. I. Windle and A. Wiersema, *J. Chem. Phys.* **39**, 1139 (1963).
55. A. Abragam, J. Combrisson, and I. Solomon, *Compt. Rend.* **245**, 157 (1957).
56. A. Overhauser, *Phys. Rev.* **92**, 411 (1953).
57. A. Abragam, *Phys. Rev.* **98**, 1729 (1955).
58. V. M. Ryzhkov and A. P. Stepanov, *Geofiz. Priborostroenie* **12**, 35 (1962).
59. W. Müller-Warmuth, *Z. Naturforsch.* **15a**, 927 (1960).
60. W. Müller-Warmuth and P. Parikh, *Z. Naturforsch.* **15a**, 86 (1960).
61. A. P. Stepanov, USSR Authors' Certificate 148920 (1962).
62. A. P. Stepanov, *Pribory i Tekhn. Éksperim.* **1962**(3), 102.
63. A. I. Filatov, A. P. Stepanov, and V. M. Stotskii, *Pribory i Tekhn. Éksperim.* **1965**(1), 169.
64. J. Combrisson, *J. Phys. Rad.* **19**, 840 (1958).
65. E. Allais, *Compt. Rend.* **246**, 2123 (1958).
66. N. Bloembergen, E. Purcell, and B. Pound, *Phys. Rev.* **73**, 679 (1948).
67. E. Hahn, *Phys. Rev.* **80**, 580 (1950).
68. A. L. Buchachenko, Stable Radicals, *Izd. Akad. Nauk SSSR*, Moscow (1963).

3. Organic Nitroxyls with Delocalized Unpaired Electron

In the literature, paramagnetic compounds of this class are most frequently called radicals of quatervalent nitrogen. The first heterocyclic radical of quatervalent nitrogen was obtained in 1901 by Piloty and Graf Schwerin in the following way [1]:

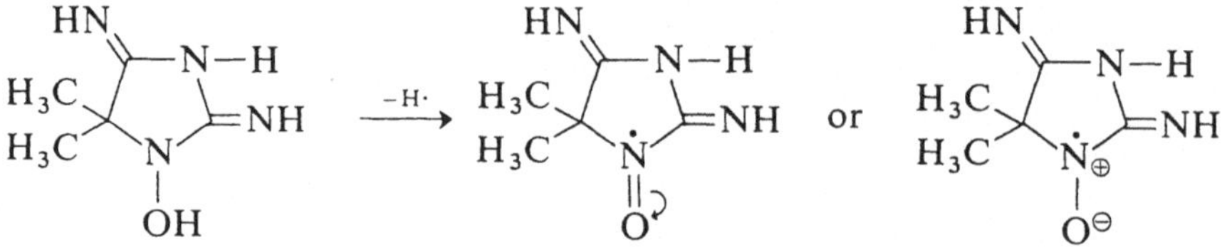

This brick-red substance, called porphyrexide (because of the similarity of its coloration to murexide, the ammonium salt of purpuric acid [2]) resembles nitrogen dioxide and Fremy's salt in its chemical properties. In contrast to the latter, porphyrexide readily extracts chlorine from hydrochloric acid, reacts with nitric acid and with alkalies, and, on reaction with chlorine, gives mono- and dichlorophorphyrexides [3]. Unfortunately, neither the structure nor the paramagnetism of porphyrexide derivatives has been studied. Subsequent work on its magnetic susceptibility [4], EPR spectra [5, 6], and oxidation potentials [7], has confirmed the free-radical nature of porphyrexide.

This stable organic free radical is a stronger oxidizing agent than nitric oxide. The oxidizing properties of porphyrexide ($E = 0.725$ V) are shown in reactions with hydriodic acid, catechol, hydroquinone, pyrogallol, ascorbic acid, cysteine, and alkaline solutions of formalin, glucose, and fructose. All these oxidative processes can be represented by the one-electron mechanism:

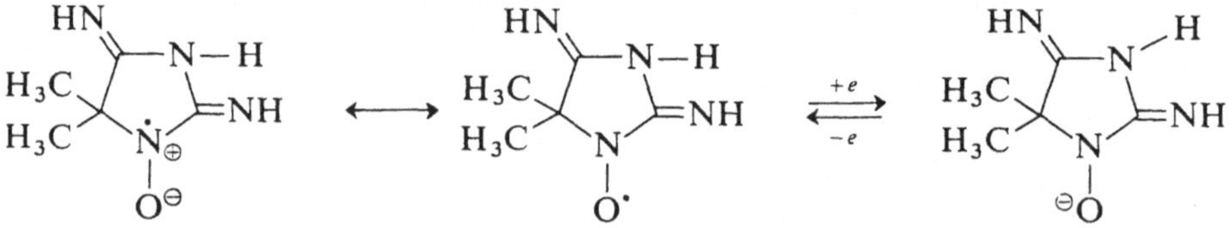

The EPR spectrum of porphyrexide is a triplet with a splitting between the components of about 10 G; however, the hyperfine structure of the components of the triplet does not permit an unambiguous interpretation. Hausser assumes that the complexity of the HFS of the spectrum can be explained by the existence in porphyrexide solutions of an equilibrium between several valence tautomeric forms [8],

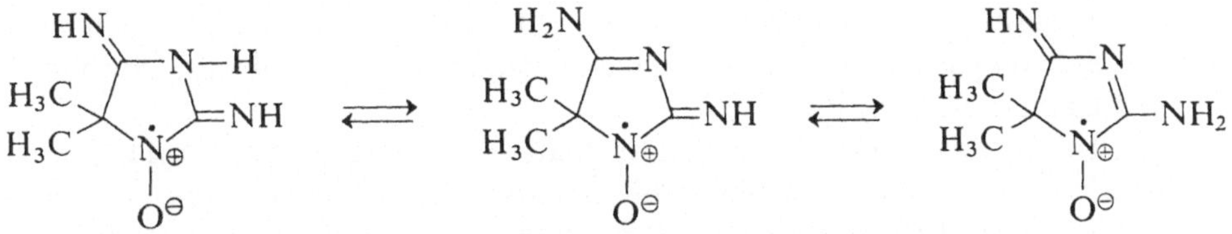

In all the tautomeric structures of porphyrexide, the bond between the nitrogen and the oxygen is a semipolar bond, which undoubtedly corresponds to a lower (more stable) energy level because of the possibility of the formation of a conjugated diene or triene system of bonds over which the unpaired electron is delocalized.

The individual diarylnitroxides can be obtained fairly reliably by making use of one of three methods: (1) the careful oxidation of diaryl-hydroxylamines [9],

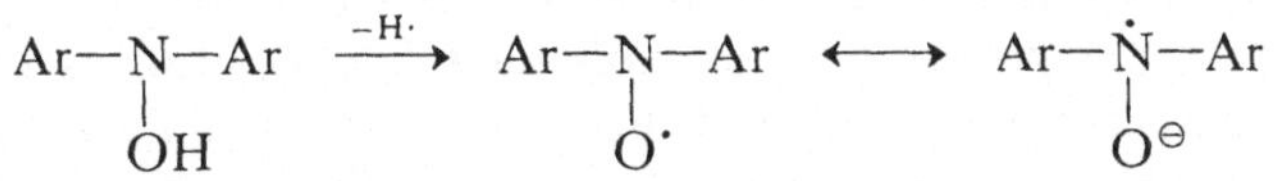

(2) the one-electron reduction of the perchlorates of the corresponding diarylimmonium oxides [10],

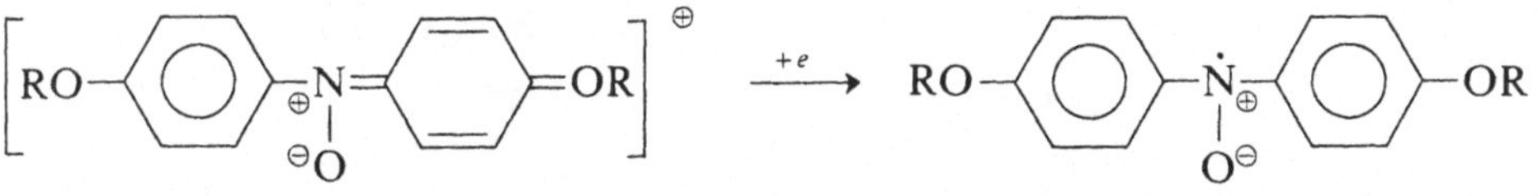

and (3) the action of organometallic compounds on nitro compounds [11],

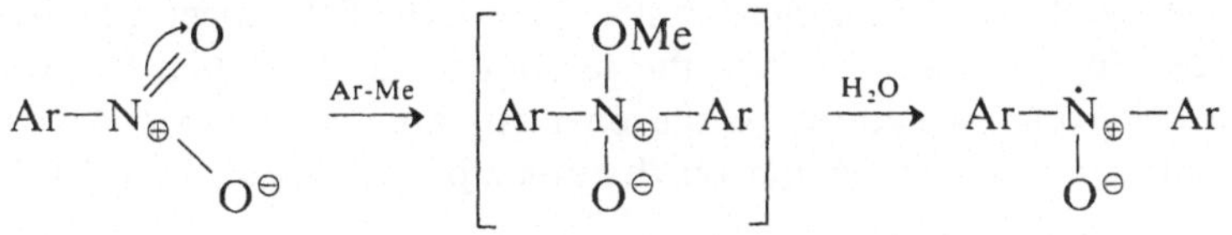

The radicals of this class are red crystalline compounds stable in air. An interesting exception is the blue ion-radical obtained in solution by Forrester and Thomson [12]:

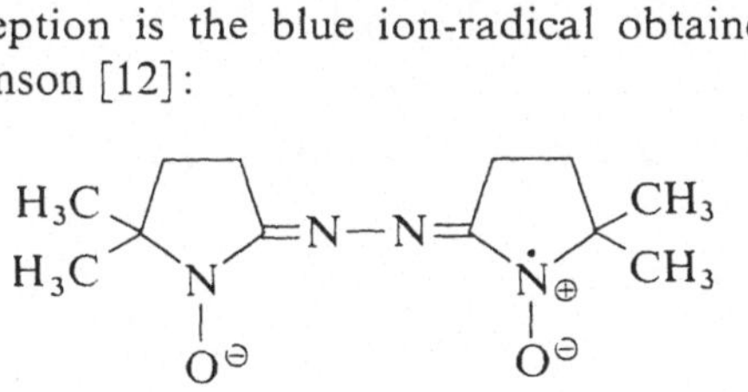

The high stability of aromatic nitroxides readily permits their detection in various types of oxidizing systems involving aromatic amines.

As early as 1956, Hoskins [13] observed the EPR spectrum of diphenyl-nitroxide on passing oxygen through a heated mixture of toluene, alcoholic alkali, and diphenylamine.

The oxidation of solutions of aromatic derivatives of naphthylamine, methylaniline, aniline, and phenothiazine also led to the appearance of the corresponding EPR signals [14]. It is quite obvious that stable nitroxides can be obtained from a large number of aromatic amines widely used as antioxidants and polymerization inhibitors. The oxidation of a variety of aromatic amines in the liquid phase has been studied by means of EPR [15, 16]. However, it must be borne in mind that the EPR spectrum in such cases does not always give unambiguous information on the chemical structure of the oxidation product. For example, the oxidation of N,N'-diphenyl-p-phenylenediamine, N,N'-di-β-naphthyl-p-phenylenediamine, and N-phenyl-N'-cyclohexyl-p-phenylenediamine gives not nitroxide radicals [17], but cation-radicals of the type of Wurster's salts [18].

As far as the preparation of individual free radicals is concerned, certain difficulties are encountered, because of which only a few diaryl-nitroxides have been obtained in the pure state in the last 30 years (Table 1).

Of the two possible structures of diphenylnitroxide,

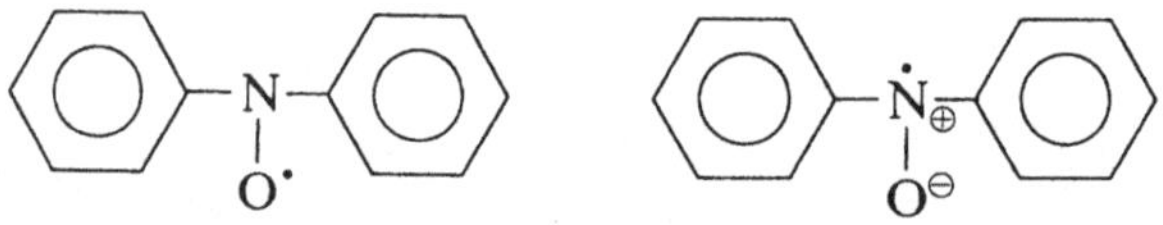

it is obvious that the latter represents the maximum gain in energy through conjugation of the nitrogen–oxygen bond with the aromatic system.

Investigations of the hyperfine structure of the EPR spectra of diphenyl-nitroxide [46] have shown that the latter can be described by means of several electronic structures of the covalent and ionic types, the relative contributions of which depend on the polar properties of the medium [47]

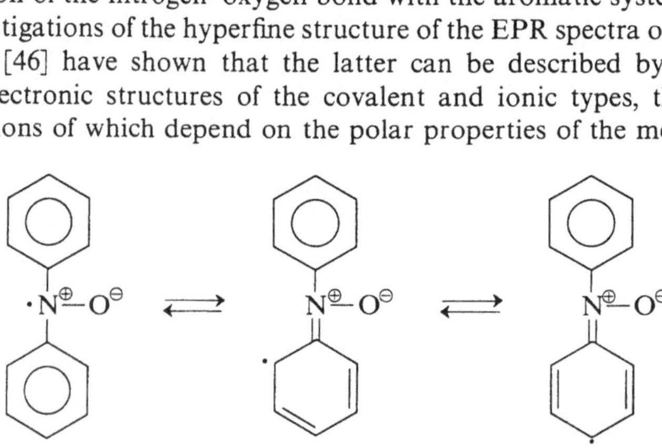

The HFS of the EPR spectrum of diphenylnitroxide (Fig. 1) consists of three groups of lines due to the splitting of the energy level of the unpaired

TABLE 1
Individual Diarylnitroxides

$$R_1-\!\!\!\bigcirc\!\!\!-\overset{\overset{\displaystyle\cdot}{\underset{|}{N}}\underset{\displaystyle O^{\ominus}}{}}{}\!\!-\!\!\!\bigcirc\!\!\!-R_2$$

(with substituents R_3, R_5 above and R_4, R_6 below)

R						Mp, °C	Stability	References
1	2	3	4	5	6			
H	H	H	H	H	H	64	Days	9, 19, 20, 28
CH_3	CH_3	H	H	H	H	59–60	Hours	19, 20
NO_2	NO_2	H	H	H	H	109	Months	19, 20
CH_3O	CH_3O	H	H	H	H	161	Years	10, 19, 20, 29, 44
CH_3O	CH_3O	CH_3O	H	CH_3O	H	121.5–122	Unstable	10, 30, 31, 45
CH_3O	CH_3O	CH_3	CH_3	CH_3	CH_3O	163	Weeks	10, 31
CH_3O	CH_3O	CH_3O	CH_3	CH_3O	CH_3	194	Years	31
H	H	CH_3O	CH_3O	CH_3O	CH_3O	199–200	—	11
CH_3O	H	H	H	CH_3O	CH_3O	114.5–115	—	11
C_2H_5O	C_2H_5O	H	H	H	H	153.5	—	10, 45

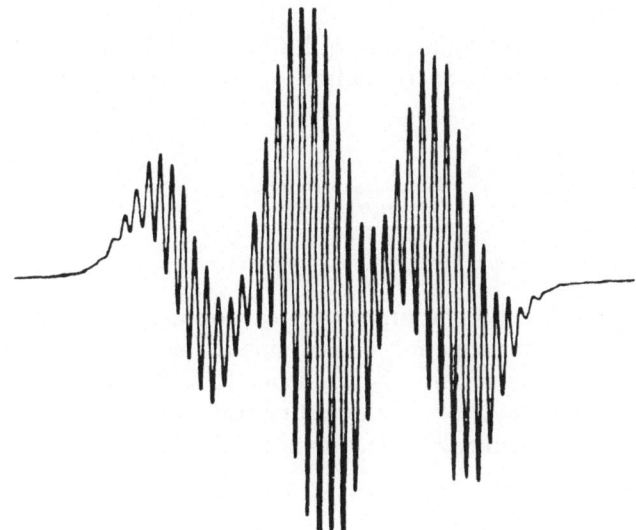

Fig. 1. EPR spectrum of diphenylnitroxide [49].

electron on the nitrogen nucleus. Each group is resolved into a large number of components because of the delocalization of the unpaired electron over the π-orbital system of aromatic nuclei and its interaction with the protons [48].

It follows from an analysis of the parameters of the EPR spectrum of diphenylnitroxide that about 40% of the spin density is concentrated on the nitrogen atom, 30% on the oxygen atom, and 30% in the aromatic nuclei [26, 46].

The delocalization of the unpaired electron involves a marked change in the optical and electrical properties of the molecule, which are shown, in particular, by the increase in the dipole moment on passing from diphenyl-hydroxylamine to diphenylnitroxide [50, 51].

The fundamental work on the synthesis and investigation of the reactivity of diarylnitroxides was carried out by Heinrich Wieland and his students (see Huisgen [52]).

Aromatic nitroxides readily recombine with nitric oxide, triphenyl-methyl, and nitrogen dioxide to form the corresponding derivatives at the nitrogen, oxygen, and carbon atoms [9],

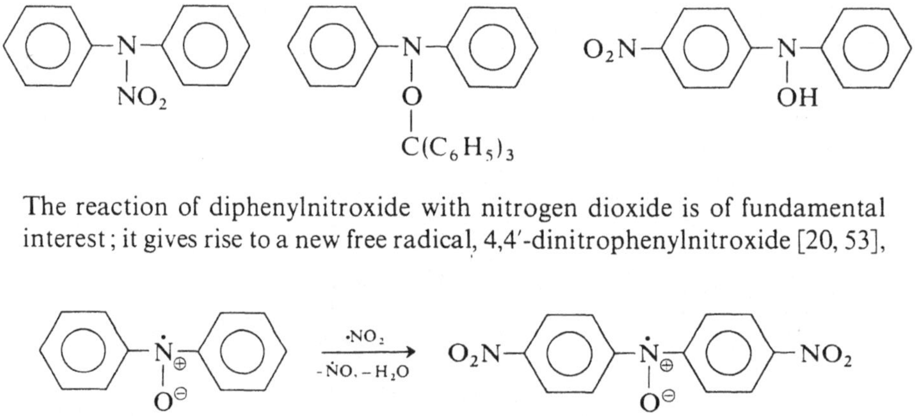

The reaction of diphenylnitroxide with nitrogen dioxide is of fundamental interest; it gives rise to a new free radical, 4,4'-dinitrophenylnitroxide [20, 53],

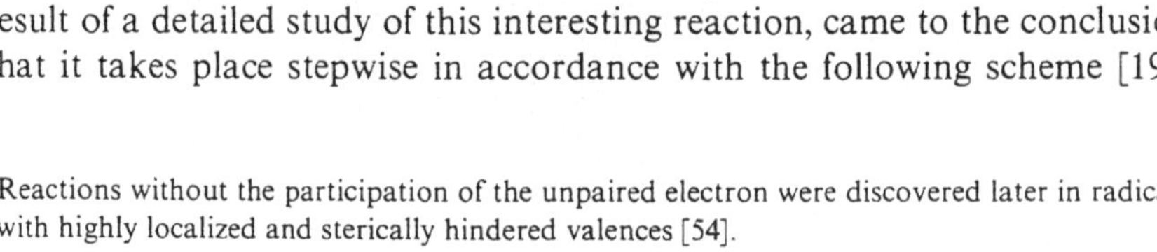

A superficial consideration of this reaction may indicate that it takes place without the participation of the unpaired electron* of the diphenyl-nitroxide, but this conclusion is erroneous [55]. Wieland and Roth, as a result of a detailed study of this interesting reaction, came to the conclusion that it takes place stepwise in accordance with the following scheme [19]:

---

* Reactions without the participation of the unpaired electron were discovered later in radicals with highly localized and sterically hindered valences [54].

A: $O-\overset{\ominus}{N}\overset{\oplus}{\cdot}$ + $\cdot NO_2$ ⟶ $\left[O-\overset{\ominus}{N}\overset{\oplus}{-}NO_2 \right]$ ⟶ $N-OH$ ... NO_2

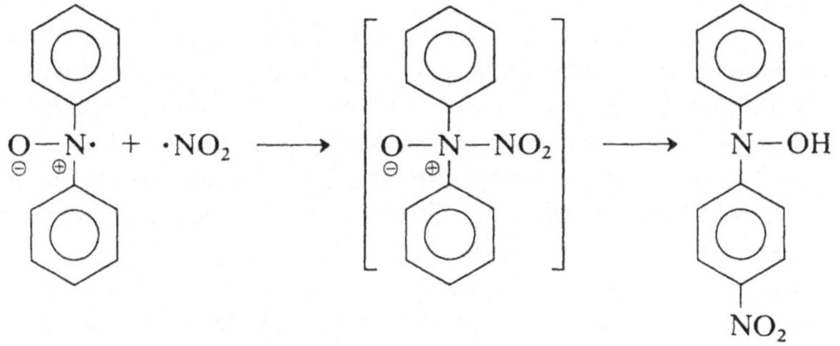

B: $2\ N-OH$ + $\cdot NO_2$ ⟶ $2\ \cdot\overset{\oplus}{N}-\overset{\ominus}{O}$ + H_2O + NO ... NO_2 ... NO_2

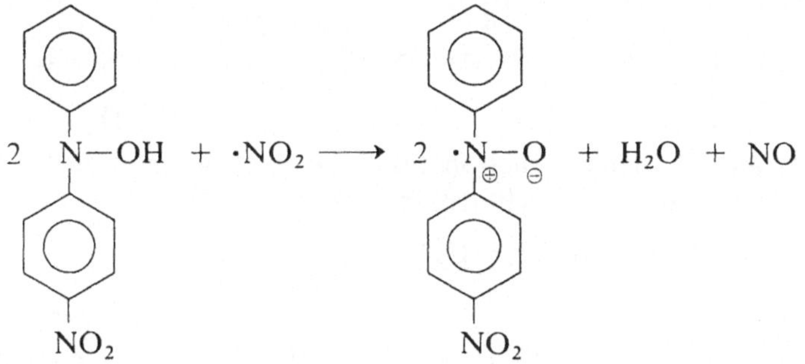

C: $O=\overset{\ominus}{N}\overset{\oplus}{\cdot}$ + $\cdot NO_2$ ⟶ $\left[O-\overset{\ominus}{N}\overset{\oplus}{-}NO_2 \right]$ ⟶ $N-OH$ ... NO_2

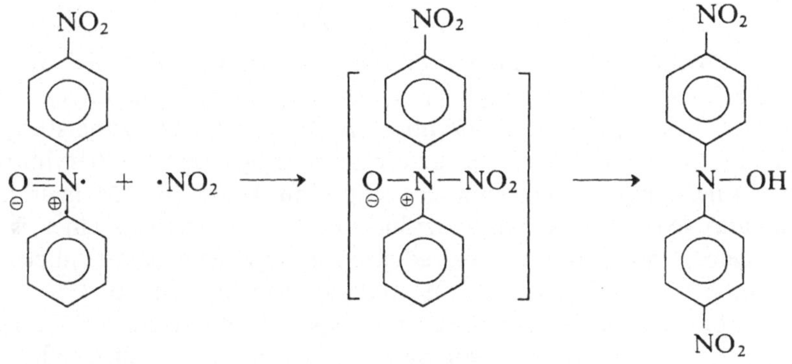

D: $2\ N-OH$ + $\cdot NO_2$ ⟶ $2\ \cdot\overset{\oplus}{N}-\overset{\ominus}{O}$ + H_2O + NO

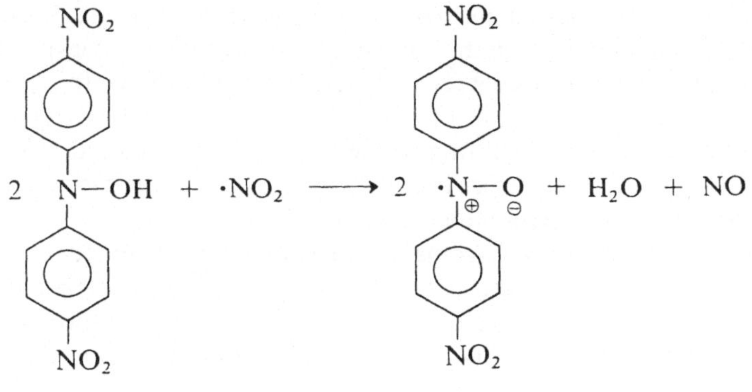

Thus, in actual fact, this reaction includes stages of recombination, rearrangement, and oxidation, and not one of them does not involve a free valence (unpaired valence electron). It is very likely that analogous reactions take place in the presence of chlorinated porphyrexide. The categorical statement that chlorporphyrexides are obtained as a result of the substitution of hydrogen by chlorine without affecting the unpaired electron [12] may be relegated to the area of chemical curiosities.

Nitroxides are readily reduced with hydrogen iodide, zinc dust, hydrazine, stannous chloride, sulfur dioxide, hydrogen sulfide, and hydrogen in the presence of hydrogenation catalysts:

$$\text{Ar}-\overset{\displaystyle\cdot}{\underset{\displaystyle O^{\ominus}}{N}}_{\oplus}-\text{Ar} \quad \overset{+H}{\underset{-H}{\rightleftarrows}} \quad \text{Ar}-\underset{\displaystyle OH}{N}-\text{Ar} \quad \overset{2H}{\underset{-H_2O}{\longrightarrow}} \quad \text{Ar}-\underset{\displaystyle H}{N}-\text{Ar}$$

A peculiar exchange reaction takes place between tetra(4-methoxyphenyl)hydrazine and diphenylnitroxide [20]:

$$2C_6H_5\overset{\displaystyle\cdot}{\underset{\displaystyle O^{\ominus}}{N}}_{\oplus}-C_6H_5 + N_2(C_6H_4-OCH_3-p)_4 \longrightarrow$$

$$2CH_3O-C_6H_4-\overset{\displaystyle\cdot}{\underset{\displaystyle O^{\ominus}}{N}}_{\oplus}-C_6H_4-OCH_3 + N_2(C_6H_5)_4$$

Subsequently, a similar exchange reaction was observed by Möbius and Schneider with diphenylpicrylhydrazyl and a peroxide radical [56].

In recent years, great attention has been devoted to the study of the parameters of the hyperfine structure of the EPR spectra of diarylnitroxides. As a rule, in liquid solutions at concentrations below $10^{-3}\,M$, in addition to the nitrogen triplet, a rich hyperfine structure is observed which is due to the delocalization of the unpaired electron over the aromatic nuclei and to its interaction with various *ortho*- and *para*-substituents [30].

For example, the EPR spectrum of 4,4'-dimethoxy-2,2',6,6'-tetramethyl-diphenylnitroxide (Fig. 2) consists of three groups of lines reflecting the interaction of the unpaired electron with the nitrogen nucleus. Each group includes HFS components with a binomial distribution of intensities due to the splitting of the electron levels at the 12 protons of the methyl groups of the aromatic rings [57].

The very fact of the splitting of the levels of the unpaired electron on the methyl groups of the aromatic nuclei is a direct experimental proof of the phenomenon of hyperconjugation [58].

A particular variety of aromatic nitroxides is represented by radicals in which the nitrogen–oxygen group is conjugated with only one aromatic nucleus. A long-known representative of this group of compounds is the Kenyon–Banfield radical, obtained by the oxidation of the product of the

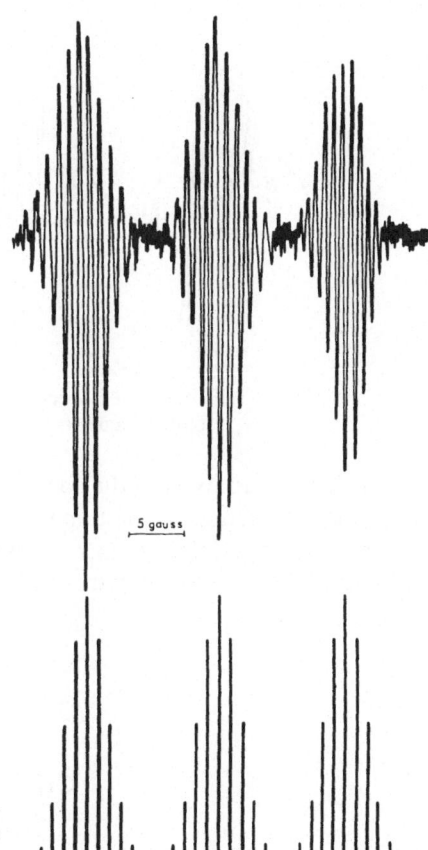

Fig. 2. HFS of the EPR spectrum of 4,4'-di-
methoxy-2,2',6,6'-tetramethyldiphenylnitroxide
[57].

condensation of acetone with phenylhydroxylamine [59, 60]. Several
individual compounds of this class have been described [61, 62]:

$$R\text{---}\underset{\underset{O^{\ominus}}{\overset{\bullet}{N}}}{}\overset{\underset{CH_3}{|}}{\underset{|}{\overset{CH_3}{C}}}\text{---}CH_2\text{---}\overset{\underset{|}{CH_3}}{C}\text{=}\underset{\overset{|}{O^{\ominus}}}{N}\text{---}R$$

where R = H, o-CH$_3$, m-CH$_3$, p-CH$_3$, m-F, p-F, m-Cl, p-Cl, m-Br, p-Br,
or m-NO$_2$.

A well-resolved EPR spectrum of the Kenyon–Banfield radical was
obtained by working with a deoxygenated tetrahydrofuran solution of the
substance. As can be seen from Fig. 3, each of the three quadruplets of the
HFS of the spectrum is split into six lines with a nonbinomial distribution of
intensities.

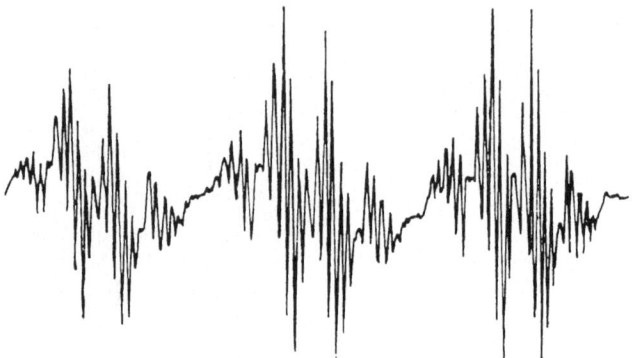

Fig. 3. HFS of the EPR spectrum of the Kenyon–Banfield radical in
solution in deoxygenated tetrahydrofuran [63].

A detailed analysis of the spectrum leads to the conclusion that there is
hyperfine interaction of the unpaired electron with a remote methyl group
which, conformationally, may closely approach the aromatic nucleus
containing the unpaired electron [63]:

Of other radicals of this series, phenyl-9-*trans*-decalylnitroxide [64, 67]
[(I); mp, 83°], 2,6-dimethoxyphenyl-tert-butylnitroxide [11] [(II); mp,
101–102.5°], and 2,6-dimethoxyphenyl-1′,1′-dimethylbenzylnitroxide [11]
[(III); mp, 77–78°] are known in the individual state:

All the substances mentioned were obtained by the reaction of organo-metallic compounds on the corresponding nitroso or nitro compounds.

Attempts to isolate phenyl-tert-butylnitroxide from solution in the pure state were unsuccessful [11, 68, 69], in spite of its obvious structural similarity to the radicals I–III.

A large number of aliphatic-aromatic nitroxides are now known in solution, but the assignment of a definite chemical structure to them is not always sufficiently well founded [70, 71].

Quite recently, the discovery of linear analogs of porphyrexide has been reported. Radicals of this class can be obtained by the oxidation of 1-arylamino-1,2-diarylnitrones with potassium ferricyanide [72]:

$$\text{Ar}'-\underset{\underset{H}{|}}{N}-\underset{\underset{O^\ominus}{|}}{\overset{\overset{\text{Ar}''}{|}}{C}}=N_\oplus-\text{Ar}''' \xrightarrow[-H\cdot]{K_3Fe(CN)_6} \text{Ar}'-N=\underset{}{\overset{\overset{\text{Ar}''}{|}}{C}}-\underset{\underset{O^\ominus}{|}}{\dot{N}_\oplus}-\text{Ar}'''$$

for the cases:

a. $\text{Ar}' = C_6H_5$, $\quad$ $\text{Ar}'' = C_6H_5$, $\quad$ $\text{Ar}''' = C_6H_5$.
b. $\text{Ar}' = p\text{-}CH_3C_6H_4$, $\text{Ar}'' = C_6H_5$, $\text{Ar}''' = C_6H_5$.
c. $\text{Ar}' = p\text{-}Cl\text{-}C_6H_4$, $\text{Ar}'' = C_6H_5$, $\text{Ar}''' = C_6H_5$.
d. $\text{Ar}' = C_6H_5$, $\quad$ $\text{Ar}'' = n\text{-}Cl\text{-}C_6H_4$, $\text{Ar}''' = C_6H_5$.
e. $\text{Ar}' = C_6H_5$, $\quad$ $\text{Ar}'' = C_6H_5$, $\quad$ $\text{Ar}''' = p\text{-}Cl\text{-}C_6H_4$.
f. $\text{Ar}' = C_6H_5$, $\quad$ $\text{Ar}'' = C_6H_5$, $\quad$ $\text{Ar}''' = p\text{-}(CH_3)_3CC_6H_4$.

The well-resolved EPR spectra of the new radicals are similar in general outline to the analogous spectra of the nitroxides already discussed.

Stable heterocyclic nitroxides of the hydrated quinoline series were first obtained in the form of solutions and were studied in detail by Razuvaev and Khidekel' [73, 74]. Only a few radicals of this class have been obtained subsequently in the pure state [75, 76].

References

1. O. Piloty and B. Graf Schwerin, *Ber.* **34**, 1870 (1901).
2. P. Karrer, Organische Chemie, Georg Thieme, Stuttgart, 12th ed. (1954), 14th ed. (1963) [English translation, Elsevier, Amsterdam, 4th ed. (1961)].
3. O. Piloty and B. Graf Schwerin, *Ber.* **34**, 2354 (1901).
4. E. Müller and I. Müller-Rodloff, *Ann. Chem.* **521**, 81 (1936).
5. N. Holden, W. Yager, and F. Merritt, *J. Chem. Phys.* **19**, 1319 (1951).
6. C. Hutchinson, R. Pastor, and A. Kovalsky, *J. Chem. Phys.* **20**, 534 (1952).
7. K. Kuhn and W. Franke, *Ber.* **68**, 1529 (1935).
8. K. Hausser, *Z. Naturforsch.* **14a**, 425 (1959).
9. H. Wieland and M. Offenbauher, *Ber.* **47**, 2111 (1914).

10. K. Meyer and H. Gottlieb-Billroth, *Ber.* **52**, 1476 (1919).
11. A. Hoffman, A. Feldman, and E. Gelblum, *J. Am. Chem. Soc.* **86**, 646 (1964).
12. A. P. Forrester, J. M. Hay, R. H. Thompson, *Organic Chemistry of Stable Free Radicals*, Academic Press, New York (1968).
13. R. Hoskins, *J. Chem. Phys.* **25**, 788 (1956).
14. J. Thomas, *J. Am. Chem. Soc.* **82**, 5955 (1960).
15. J. Baird and J. Thomas, *J. Chem. Phys.* **35**, 1507 (1961).
16. A. L. Buchachenko, *Opt. i Spektroskopiya* **13**, 795 (1962).
17. A. L. Buchachenko, Stable Radicals, Izd. Akad. Nauk SSSR (1963), p. 115.
18. É. G. Rozantsev, A. A. Medzhidov, M. B. Neiman, and L. A. Skripko, Omagiu Raluca Ripan, [Homage to Reluca Ripan], Rumania (1966), p. 503.
19. H. Wieland and K. Roth, *Ber.* **53**, 210 (1920).
20. H. Wieland and F. Kögl, *Ber.* **55**, 1798 (1922).
21. J. Osugi, M. Sato, and M. Sasaki, *Nippon Kagaku Zasshi* **85**, 307 (1964).
22. H. Lemaire, A. Rassat, and J. Ravet, *Tetrahedron Letters* **1964**, 3507.
23. K. Tokumaru, H. Sakuragi, and O. Simamura, *Tetrahedron Letters* **1964**, 3945.
24. O. A. Chaltykyan, Questions of Chemical Kinetics and Reactivity, A Collection of Papers, Moscow (1955), p. 354.
25. A. L. Buchachenko, Candidate's Thesis, Moscow (1961).
26. I. Deguchi, *Bull. Chem. Soc. Japan* **35**, 260 (1962).
27. O. Harle and J. Thomas, *J. Am. Chem. Soc.* **79**, 2973 (1957).
28. A. L. Buchachenko, E. N. Gur'yanova, L. A. Kalashnikova, and M. B. Neiman, *Dokl. Akad. Nauk SSSR* **148**, 95 (1963).
29. G. Schoffa, B. Wahler, and H. Thom, *Acta Biol. Med. Germ.* **1**, 114 (1958).
30. H. Thom, B. Wahler, and G. Schoffa, *Z. Naturforsch.* **13a**, 552 (1958).
31. K. Meyer and W. Reppe, *Ber.* **54**, 327 (1921).
32. H. Euler, H. Hasselquist, and O. Heidenberger, *Arkiv Kemi* **14**, 237 (1959).
33. E. Müller, I. Müller-Rodloff, and W. Bunge, *Ann.* **520**, 235 (1935).
34. N. Holden, C. Kittel, F. Merrit, and W. Yager, *Phys. Rev.* **77**(2), 147 (1950).
35. D. Bijl and A. Rose-Innes, *Phil. Mag.* **44**, 1187 (1953).
36. W. Heubner, B. Wahler, and C. Ziegler, *Z. Physiol. Chem.* **295**, 397 (1953).
37. F. Adrian, *J. Chem. Phys.* **28**, 608 (1958).
38. G. Bartlet, *Arch. Sci. Sonder* **10**(162), 98 (1957).
39. J. Burgess, *J. Phys. Rad.* **19**, 845 (1958).
40. W. Otting and H. Kainer, *Ber.* **87**, 1205 (1954).
41. A. Roggen, L. Roggen, and W. Gordy, *Phys. Rev.* **105**, 50 (1957).
42. F. Galavies, *Helv. Phys. Acta* **6**, 555 (1933).
43. É. G. Rozantsev, L. A. Kalashnikova, and M. B. Neiman, *Zh. Prikl. Khim.* **38**, 702 (1965).
44. A. Hanson, *Acta Cryst.* **6**, 32 (1953).
45. N. N. Volkotrub, É. G. Rozantsev, and L. A. Skripko, *Izv. Akad. Nauk SSSR, Ser. Khim.* **1967**, 179.
46. J. Deguchi, *Bull. Chem. Soc. Japan* **34**, 910 (1961).
47. S. Weissman, *J. Chem. Phys.* **25**, 890 (1956).
48. P. Fister and F. Neugebauer, *Z. Naturforsch.* **19a**, 1514 (1964).
49. A. Rassat and H. Lemaire, *J. Chem. Phys.* **1964**, 1576.
50. É. G. Rozantsev and E. N. Gur'yanova, *Izv. Akad. Nauk SSSR, Ser. Khim.* **1966**, 979.
51. L. Cambi, *Gazz. Chim. Ital.* **63**, 579 (1933).
52. R. Huisgen, *Proc. Chem. Soc.* **1958**, 210.
53. L. Cambi and L. Szegö, *Ber.* **64**, 2591 (1931).
54. É. G. Rozantsev and M. B. Neiman, *Zh. Organ. Khim.* **1**, 1337 (1965).
55. O. L. Lebedev and G. A. Razuvaev, *Zh. Organ. Khim.* **1**, 613 (1965).

56. K. Möbius and F. Schneider, *Z. Naturforsch.* **18a**, 428 (1963).

57. L. A. Kalashnikova, M. B. Neiman, É. G. Rozantsev, and L. A. Skripko, *Zh. Organ. Khim.* **2**, 1529 (1966).

58. É. G. Rozantsev, *Usp. Khim.* **35**, 1549 (1966).

59. F. Banfield and J. Kenyon, *J. Chem. Soc.* **1926**, 1612.

60. J. Kenyon and S. Sugden, *J. Chem. Soc.* **1932**, 170.

61. F. Tüdös, I. Heidt and I. Ero, *Magy. Kem. Foloirat* **70**, 329 (1964).

62. F. Tüdös, T. Berezsnich and M. Azori, *Acta Chem. Acad. Sci. Hung.* **24**, 91 (1960).

63. V. Griffiths and G. Parlett, *Nature* **204**, 69 (1964).

64. W. Hückel and W. Liegel, *Ber.* **71**, 1442 (1938).

65. V. Trkal, *Czech. J. Phys.* **6**, 770 (1957).

66. V. Trkal, *Czesk. Casopis. Fysiku* **7**, 748 (1957).

67. K. Ulbert and V. Trkal, *Collection Czech. Chem. Commun.* **25**, 1216 (1960).

68. G. Chapelet-Letourneux, H. Lemaire, and A. Rassat, *Bull. Soc. Chim. France* **1965**, 444.

69. A. Forrester and R. Thomson, *Nature* **203**, 74 (1964).

70. H. Lemaire, A. Rassat, and A.-M. Ravet, *Bull. Soc. Chim. France* **1963**, 1980.

71. H. Lemaire, Y. Marechal, R. Ramasseul, and A. Rassat, *Bull. Soc. Chim. France* **1965**, 372.

72. H. Aurich and F. Baer, *Tetrahedron Letters* **1965**, 2517.

73. G. A. Razuvaev, M. L. Khidekel', and V. B. Berlina, *Dokl. Akad. Nauk SSSR* **145**, 1071 (1962).

74. G. A. Razuvaev and M. L. Khidekel', *Korrelyats. Uravneniya v Organ. Khim. (Tartu)* **1**, 365 (1962).

75. A. B. Shapiro, É. G. Rozantsev, L. S. Povarov, and V. N. Grigos, *Izv. Akad. Nauk SSSR, Ser. Khim.* **1964**, 1725.

76. A. B. Shapiro, É. G. Rozantsev, L. S. Povarov, and V. N. Grigos, *Izv. Akad. Nauk SSSR, Ser. Khim.* **1965**, 1102.

4. Organic Nitroxyls with Localized Unpaired Electron

Free nitroxyl radicals are compounds that are formed as intermediates in the oxidation of secondary saturated amines [1] and hydroxylamines—for example,

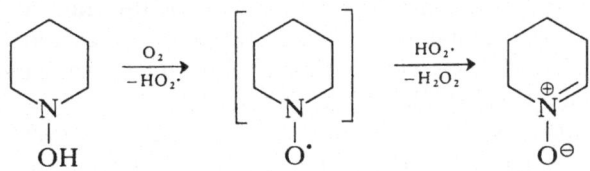

In contrast to nitroxides, the NO group of nitroxyls does not take part in the formation of a conjugated system of multiple bonds, and therefore the unpaired electron must be localized on the nitrogen and oxygen atoms to a considerably greater degree.

The existence of free nitroxyl radicals in solution was first postulated by Rogers *et al.* [2, 3]. They showed that free nitroxyls having no α-hydrogen atoms must be stable free radicals. For example, piperidine-1-oxyl is unstable, but the compound completely alkylated in positions 2 and 6 of the ring must be stable:

$$
\begin{array}{c}
H_3C \\
H_3C
\end{array}
\!\!\!\!
\underset{\underset{\displaystyle O^{\cdot}}{|}}{N}
\!\!\!\!
\begin{array}{c}
CH_3 \\
CH_3
\end{array}
$$

It is obvious that the substituents may be not only organic residues; what is important is that the nitroxyl is not capable of valence tautomeric transformations which are the condition of subsequent disproportionation in the following way [4–12]:

$$
R-\underset{\underset{\displaystyle O^{\cdot}}{|}}{N}-CHR_2 \longleftrightarrow R-\underset{\underset{\displaystyle O^{\ominus}}{|}}{\overset{\displaystyle \cdot}{N}}{}_{\oplus}-CHR_2
$$

$$
\downarrow
$$

$$
R-\underset{\underset{\displaystyle OH}{|}}{N}-CHR_2 \quad + \quad R-{}_{\oplus}\underset{\underset{\displaystyle O^{\ominus}}{|}}{N}=CR_2
$$

The first stable organic radical of this class was isolated in the individual state in the USSR in 1959. The radical was obtained by chance during a study of the action of the pertungstate ion on various amines with the object of following the individual stages of their oxidation [13]. The EPR spectrum of the radical dissolved in benzene showed a triplet with a hyperfine splitting constant of about 15.6 G [14, 15]. According to the results of other authors, the hyperfine splitting constant for solutions of this radical in benzene, chloroform, heptane, and tetrahydrofuran was also 15.6 G, and the spectrum itself consisted of a symmetrical triplet with equidistant components of equal intensities [16] (Fig. 4).

The published spectrum of the radical, which was first obtained by Lebedev and Kazarnovskii, in spite of more recent literature data, proved to be anisotropic (Fig. 5). In the opinion of the authors, the distortion of the symmetry of the spectrum was caused, on the one hand, by the fact that in weak fields, the HFS sublevels contain an admixture of various quantum states and, on the other hand, by the influence of the quadrupole moment of the nitrogen nucleus.

Samitov and Lebedev [17], rejecting the nitroxyl electronic structure as unlikely, assumed that the large value of the hyperfine splitting constant

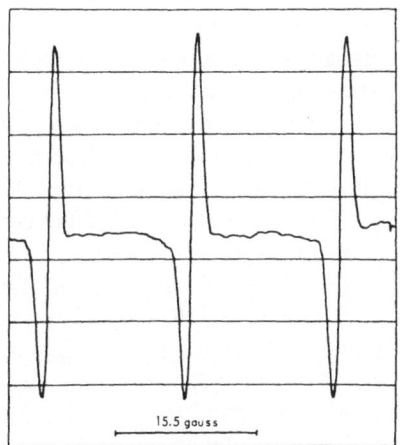

Fig. 4. HFS of the EPR spectrum of an ethereal solution of analytically pure 2,2,6,6-tetramethyl-piperidine-1-oxyl.

15.5 gauss

and the electronegativity* of the oxygen gave every reason for accepting a structure of this radical with the unpaired electron localized on the nitrogen atom.

As Fermi [18] has shown, the isotropic hyperfine splitting constant has the following form:

$$a = (16\pi/3)g_i\mu_N\beta\psi^2(0)\mathbf{I} \cdot \mathbf{S}$$

where g_i is the nuclear spectroscopic splitting factor, μ_N is the nuclear magneton, and $\psi^2(0)$ is the density of the unpaired electron at the atomic nucleus.

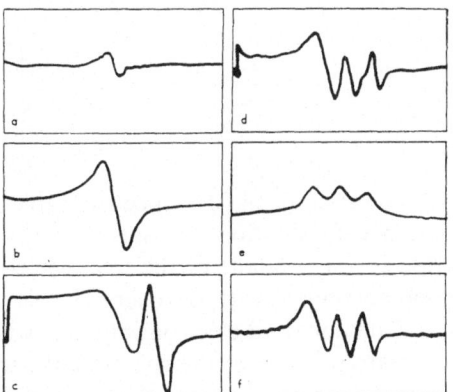

Fig. 5. Anisotropic EPR spectra of solutions of Lebedev's 2,2,6,6-tetramethylpiperidine-1-oxyl [17]: (a) 1,1-diphenylpicrylhydrazyl; (b) the nitroxyl (polycrystalline sample); (c) dilute ethereal solution; (d) aqueous ethanolic solution, $c = 0.18\%$ (123.9 MHz); (e) the same as (d) at 9468 MHz; (f) ethereal solution, $c = 0.09\%$.

*The electronegativities of oxygen and nitrogen on Pauling's scale are fairly similar, being 3.5 and 3.0, respectively.

It can be seen from this equation that the isotropic hyperfine splitting constant is actually proportional to the electron spin density at the nucleus of the corresponding atom of the molecule. But since the p orbitals and all the higher orbitals have a node at the point of the nucleus, only "pure" s electrons or electrons described by a hybridized wave function (with admixture of the s state) can interact with the nucleus.

Thus, a single value of the constant of the isotropic hyperfine splitting of the spectrum of a nitroxyl radical does not give exhaustive information on the degree of localization of the unpaired electron on the nitrogen atom.

Dousmanis [19] has calculated the density of the unpaired $2s$ electron on the nucleus of the nitrogen atom; the magnitude $\psi^2(0)$ proved to be 34×10^{24} cm^{-3}. The ratios of the spin densities $\psi^2(0)$ of the various nitrogen-containing radicals to the value 34×10^{24} cm^{-3} characterizes the degree of participation of the $2s$ state in the wave function of the corresponding unpaired electron.

It can readily be seen from Table 2 that, on the one hand, the low value of the Fermi constant for the free nitrogen radical corresponds to the complete localization of the unpaired electron on one atom and, on the other hand, the considerably greater value of the constant of isotropic hyperfine splitting in the spectra of nitrogen oxides and the Fremy ion is associated with the delocalization of the unpaired electron over several atoms.

Coppinger and Swallen [26] have demonstrated that oxidation of di- and trialkylamines with tertiary butyl peroxide yields the same nitroxyl radicals as the oxidation of the corresponding dialkylhydroxylamines:

$$
\begin{array}{ccc}
R{-}N{-}R & R{-}N{-}R & R{-}N{-}R \\
| & | & | \\
H & R & OH
\end{array}
$$

$$
\begin{array}{c}
R{-}N{-}R \\
| \\
O^{\bullet}
\end{array}
$$

An analysis of EPR spectra (Table 3) has shown that for all the radicals obtained in solution, the spectroscopic splitting factor (2.0038 ± 0.005) is close to the g-factor of the free electron, and the hyperfine structure is in good agreement with the energy levels expected theoretically.

The unpaired electron interacts with the nitrogen nucleus and with the α-carbon protons of the substituents, whereupon the ^{14}N nitrogen nucleus splits the spectral line into three components of equal intensity. In their turn, the four α-carbon protons split each of the three lines into five components with a binomial ratio of intensities ($1:4:6:4:1$).

TABLE 2

Parameters of the EPR Spectra of Some Nitrogen-Containing Free Radicals

Formula of the radical	a_N, G	$\psi^2(O) \times 10^{24}$	References
N·	3.7	0.23	20, 21
$H_2N·$	10.3	0.62	22
$^{\ominus}C_6H_5NO_2^{\cdot}$	11.1	0.66	23
$^{\ominus\ominus}(OSO_2)_2NO·$	13	0.78	24
ON·	14.2	0.85	19
$O_2N·$	47.1	—	25

Since all di-n-alkylnitroxyls possess similar Fermi constants, it may be considered that these radicals have similar electronic structures.

Diisopropyl- and di-sec-butylnitroxyls have only two α-carbon protons, and therefore each line of the nitrogen triplet is split into three components with a ratio of intensities of $1:2:1$ (Fig. 6). The lower value of proton splitting for these radicals apparently shows some change in the nature of the chemical bonds in the molecules, possibly through a partial change of hybridization.

Coppinger and Swallen, on comparing the Fermi constant of the interaction of the unpaired electron with the protons contained in various

TABLE 3

Parameters of the EPR Spectra of Free Radicals Formed in the Oxidation of Various Amines and Hydroxylamines with Tertiary Butyl Hydroperoxide

Initial compound	a_N, G	a_H, G	No. of lines
Diethylamine	15.5	10.6	15
Di-n-propylamine	15.5	10.6	15
Di-n-propylhydroxylamine	15.7	10.5	15
Di-n-butylamine	15.7	10.5	15
Diisopropylamine	16.0	4.7	9
Di-sec-butylamine	15.4	4.1	9
Dibenzylamine	15.6	9.5	15
Dibenzylhydroxylamine	15.6	9.1	15
Piperidine	17.4	11.7	15
Diallylamine	15.5	9.3	15
Diphenylamine	10.5	0.9	33
Tri-n-propylamine	15.7	10.5	15
Tri-n-butylamine	15.7	10.5	15

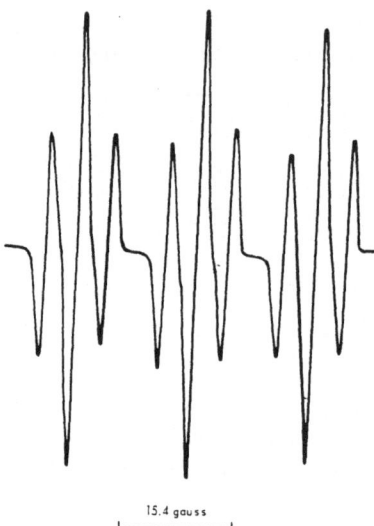

15.4 gauss

Fig. 6. EPR spectrum of a benzene solution of di-
sec-butylnitroxyl [27].

radicals (Table 4), came to the conclusion that the most probable structure
of the radicals obtained in the oxidation of aliphatic amines is the nitroxyl
structure:

$$\dot{R}-\ddot{N}-R$$
$$\underset{:\ddot{O}:}{|}$$

A consideration of the figures of Table 4 leads to the conclusion that
the proton splitting in the ethyl radical is greater for the methyl group than
for the methylene group (27.1 and 22.5 G), although the methyl protons are
further from the position of localization of the unpaired electron than the
methylene protons.

TABLE 4
Proton Parameters of the EPR Spectra of Some Free Radicals

Radical	a_N, G	Ref.
$H_2N\cdot$	23.9	22
$H_3C\cdot$	25.0	28–30
$H_3C-H_2C\cdot$	27.1, 22.5	31
$H_3C-\dot{C}H-CH_3$	25	26

This result is apparently connected with the fact that in the absence of orthogonal sp^2 hybridization of the methylene group, sp^3 hybridization of the methyl group provides for some admixture of the $2p\pi$ state.

It also follows from Table 4 that the amine radical has almost the same proton splitting constant as free methyl. It may be postulated that the α-carbon protons in the dialkylnitrogen (I) and in the isopropyl radical (II) must possess similar Fermi constants:

$$R-CH_2-\dot{N}-CH_2-R \qquad H_3C-\dot{C}-CH_3$$
$$\qquad\qquad\qquad\qquad\qquad\quad \underset{H}{|}$$

(I)   (II)

The values of the constants observed for Coppinger's radicals were, however, not 23–26 G but only 15–17 G (Table 3). This fact also casts doubt on the assumption that the unpaired electron in the free nitroxyls is completely localized on the nitrogen atom.

The hyperfine interaction of the unpaired electron with the nitrogen and hydrogen nuclei is apparently explained by hyperconjugation, in consequence of which the ground state of the radical can be represented by the participation of several valence isomeric structures [32, 33]:

$$\begin{array}{ccc}
\underset{\underset{R}{|}}{\overset{\overset{R}{|}}{H-C-N-C-H}} & \longleftrightarrow & \underset{\underset{R}{|}}{\overset{\overset{R}{|}}{H-C-\overset{\cdot}{N}_\oplus-C-H}} \longleftrightarrow \\
\underset{R}{|}\ \underset{O^\cdot}{}\ \underset{R}{|} & & \underset{R}{|}\ \underset{O^\ominus}{}\ \underset{R}{|}
\end{array}$$

$$\underset{\underset{R}{|}\ \underset{^\ominus O}{}\ \underset{R}{|}}{\overset{\overset{R}{|}\qquad\overset{R}{|}}{H\cdot C=N_\oplus-C-H}}$$

Closely connected with the dialkylnitroxyls is the class of labile alkylidene iminoxyls. The latter are obtained by the dehydrogenation of oximes:

$$\underset{R'}{\overset{R}{\diagdown}}C=N-OH \xrightarrow{-H\cdot} \underset{R'}{\overset{R}{\diagdown}}C=N-\dot{O}$$

The first radical of this type was detected in a monocrystal of dimethylglyoxime subjected to γ-irradiation [34–37]. Many alkylidene iminoxyls have been studied in liquid solutions by the flow method [38] using cerium ammonium nitrate [39] and lead tetraacetate [40] as oxidizing agents. Table 5 lists some radicals of this class studied in solutions by the EPR method. Alkylidene iminoxyls possess two interesting features. In the first place, the nitrogen hyperfine splitting constants are considerably larger and more characteristic than those of other organic nitroxyls.

In the second place, the EPR spectra of these radicals enable the *syn*- and *anti*-isomers of the alkylidene iminoxyls to be distinguished. Figure 7(a) shows a spectrum obtained in the oxidation of *syn*-benzaldehyde oxime. The group of 1,3-3,2,3-3,2,3-3,1 lines is explained by the superposition of the spectra of the *syn*- and *anti*-benzylidene iminoxyls present in approximately equal concentrations. The 1,2,2,1 group of lines relates to the *anti*-benzylidene iminoxyl, and the 3-3,3-3,3-3 group characterizes the *syn*-isomer. On the basis of Jerslev's paper [41], it may be assumed that the *syn*-isomer possesses a normal coplanar conjugated system of multiple bonds, and the EPR spectrum of the radical is due to the interaction of the unpaired electron with one imino carbon proton ($a_H = 6.2$ G) and two protons of the aromatic ring ($a_H = 1.4$ G). As far as the *anti*-isomer is concerned, here, according to Jerslev [41], the iminoxy group of the oxime is turned relative to the plane of the benzene ring by 19°, and this is the reason for the absence of resonance interaction of the unpaired electron with the protons of the aromatic nucleus.

Figure 7(b) shows the EPR spectrum of the radicals obtained by the oxidation of phenylglyoxaloxime. For one isomer, the hyperfine splitting constants are $a_N = 29.3$ and $a_H = 6.2$ G, and for the other (concentration about 2% with respect to the main isomer) they are, respectively, 31.3 and 26.8 G.

The iminoxyl obtained from *p*-nitrosophenol is characterized by a spectrum [Fig. 7(c)] in which three groups of lines are clearly shown, and each

TABLE 5

Hyperfine Splitting Constants of Some Alkylidene Iminoxyls in Methanol

Initial compound	Alkylidene iminoxyls		
	a_N, G	a_H, G	
		Imine carbon	Carbocycle
Acetone oxime	33	—	—
Acetaldehyde oxime	32.5	5.2	—
anti-Benzaldehyde oxime	29.2	26.9	—
syn-Benzaldehyde oxime	31.6	6.2	1.4 (2)
p-Benzoquinone oxime	33.0	—	1.2 (1)
Hydroxy-*p*-benzoquinone oxime	27.9	—	1.7 (1)
Benzophenone oxime	31.6	—	—
Acetophenone oxime	30.6	—	—

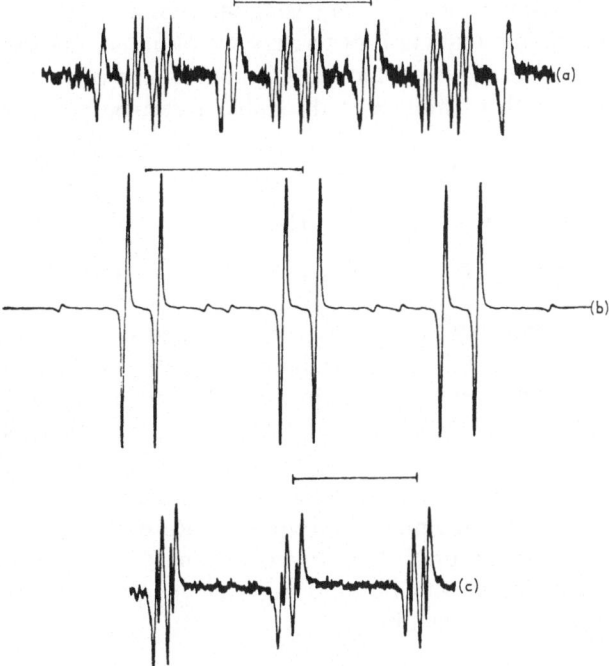

Fig. 7. EPR spectra of methanolic solutions of nitroxyl radicals obtained from (a) *syn*-benzaldoxime, (b) phenylglyoxaldoxime, and (c) *p*-nitrosophenol. Scales of the spectra: (a) 29.2; (b) 29.3; (c) 33.0 G [39].

group in turn is additionally split into two doublets. It may be assumed that the latter arise from the two ring protons, which are "nonequivalent" because of the proximity of the iminoxyl group to one of them:

The overwhelming majority of the alkylidene iminoxyls are extremely unstable, and therefore not one compound of this class has been obtained in the individual state.

In 1961, Hoffmann *et al.* [43, 44], using Maki's method [42], in the course of the polarographic reduction of tertiary nitrobutane, observed two EPR spectra (Fig. 8), one of which showed the appearance of an unstable anion radical while the other belonged to the stable di-tert-butylnitroxyl.

$$\underset{\displaystyle \overset{\displaystyle CH_3}{|}}{\underset{\displaystyle \underset{\displaystyle CH_3}{|}}{H_3C-C-NO_2^{\cdot -}}} \qquad \underset{\displaystyle \overset{\displaystyle CH_3}{|}\;\;\overset{\displaystyle CH_3}{|}}{\underset{\displaystyle \underset{\displaystyle CH_3\;O^{\cdot}}{|}\;\;\underset{\displaystyle CH_3}{|}}{H_3C-C-\;N-C-CH_3}}$$

Di-tert-butylnitroxyl, which has been obtained in the pure state by the reaction of tert-nitrobutane with sodium, is a red liquid stable in air up to 120°C [45, 46]. It is interesting that in the well-resolved hyperfine structure of the EPR spectrum of this radical it is possible to observe, in addition to the main lines, a system of low-intensity satellite doublets caused by the interaction of the unpaired electron with the nuclei of the natural carbon isotope ^{13}C (nuclear spin $\frac{1}{2}$) of the methyl groups [47]. Recently, a group of French scientists (at the Center for Nuclear Investigations, Grenoble) has developed a new method of obtaining this compound, which has important applications, *via* tert-butylmagnesium chloride [48].

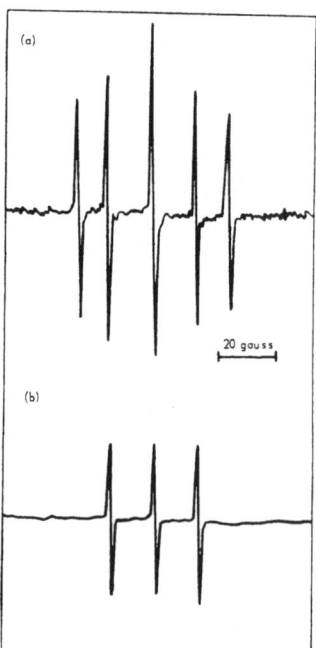

Fig. 8. EPR spectra of: (a) a mixture of an anion-radical and di-tert-butylnitroxyl; (b) pure di-tert-butylnitroxyl dissolved in ethylene glycol dimethyl ether [45].

In 1961, Lebedev *et al.* [49], by heating an aqueous alcoholic mixture of phorone, ammonium nitrate, and caustic soda at 100°C in a sealed tube, obtained a solution which was converted by the addition of Trilon B, hydrogen, peroxide, and sodium tungstate (and reheating to 50°C), into a yellow paramagnetic mass. For a few weeks, the complex mixture obtained gave an EPR signal in the form of three wide, equidistant lines of approximately equal intensity.

On the basis of this observation, the authors assumed that the reaction mixture contained a previously unknown stable radical with the structure of a nitrogen oxide of triacetoneamine. Unfortunately, these workers were unable to isolate the assumed radical in pure state.

As early as 1957, it was shown that the photochemical decomposition of trifluoronitrosomethane produces the intermediate formation of hexafluorodimethylnitroxyl radicals [50].

$$F_3C-NO \xrightarrow{h\nu} F_3C\cdot + NO$$

$$F_3C\cdot + F_3C-NO \longrightarrow F_3C-\overset{\cdot}{N}O-CF_3$$

$$F_3C-\overset{\cdot}{N}O-CF_3 + NO \longrightarrow (F_3C)_2NONO$$

Later, perfluorinated nitroxyls were observed by means of the EPR method in the polymerization of perfluoroethylene in the presence of trifluoronitrosomethane (Fig. 9). It is interesting that the copolymerization of tetrafluoroethylene with trifluoronitrosomethane begins in the absence of light at a temperature as low as -57°C in accordance with the following scheme [51]:

$$F_3C-NO + F_2C{=}CF_2 \longrightarrow F_3C-\overset{\cdot}{N}O-F_2C-\overset{\cdot}{C}F_2$$

$$F_3C-\overset{\cdot}{N}O-CF_2-\overset{\cdot}{C}F_2 + F_3C-NO \longrightarrow$$

$$F_3C-\overset{\cdot}{N}O-CF_2-CF_2-\overset{\cdot}{N}O-CF_3$$

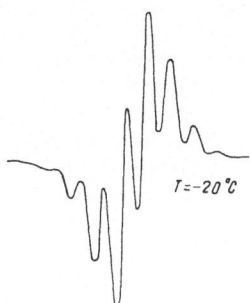

Fig. 9. EPR signal arising in the process of the copolymerization of tetrafluoroethylene with trifluoronitrosomethane in the absence of light at -20°C [51].

Attempts to obtain hexafluorodimethylnitroxyl in the pure state were successful only in 1964. The radical was obtained* by the action of silver oxide (or fluorine) on the corresponding hexafluorodimethylhydroxylamine [52, 53]:

$$F_3C-\underset{\underset{OH}{|}}{N}-CF_3 \xrightarrow[F_2]{-H\cdot} F_3C-\underset{\underset{O\cdot}{|}}{N}-CF_3$$

This radical was synthesized almost simultaneously by a group of Soviet scientists [54], who have studied its properties in detail [55]. Under the usual conditions, the new free radical exists in the form of a purple gas which, when the temperature is lowered, first condenses to a dark violet liquid and is then converted into a yellow crystalline mass containing about 1.5% of the free radical. In its electronic structure, the new radical apparently occupies an intermediate position between the nonfluorinated nitroxyls and Fremy's salt.

The hyperfine structure of the EPR spectrum of hexafluorodimethylnitroxyl consists of a nonuplet of lines as a consequence of the interaction of the unpaired electron with the nuclei of the fluorine atoms and the nucleus of the nitrogen atom:

$$F-\underset{\underset{F}{|}}{\overset{\overset{F}{|}}{C}}-\underset{\underset{O^{\ominus}}{|}}{\overset{\overset{\cdot\cdot}{}}{N_{\oplus}}}-\underset{\underset{F}{|}}{\overset{\overset{F}{|}}{C}}-F \longleftrightarrow F-\underset{\underset{F}{|}}{\overset{\overset{F}{|}}{C}}-\underset{\underset{O\cdot}{|}}{\overset{\overset{}{|}}{N}}-\underset{\underset{F}{|}}{\overset{\overset{F}{|}}{C}}-F \longleftrightarrow F-\underset{\underset{F}{|}}{\overset{\overset{F}{|}}{C}}-\underset{\underset{O^{\ominus}}{|}}{N_{\oplus}}=\underset{\underset{F}{|}}{\overset{\overset{F}{|}}{C}}\cdot F$$

References

1. H. La Mare, *J. Org. Chem.* **25**, 2114 (1960).
2. D. Johnson, M. Rogers, and G. Trappe, *J. Chem. Soc.* **1956**, 1093.
3. M. Rogers, *J. Chem. Soc.* **1956**, 2784.
4. R. Bonnett, R. Brown, V. Clark, J. Sutherland, and A. Todd, *J. Chem. Soc.* **1959**, 2094.
5. R. Bonnett, V. Clark, and A. Todd, *J. Chem. Soc.* **1959**, 2102.
6. C. Brown and M. Rogers, British Patent 850,418 (1960); *Chem. Abstr.* **55**, 6498 (1961).
7. R. Brown, V. Clark, and A. Todd, *J. Chem. Soc.* **1959**, 2105.
8. R. Brown, V. Clark, J. Sutherland, and A. Todd, *J. Chem. Soc.* **1959**, 2109.
9. P. Grammaticakis, *Compt. Rend.* **224**, 1066 (1947).
10. J. Thesing, A. Müller, and G. Michel, *Ber.* **88**, 1030 (1955).
11. J. Thesing and H. Mayer, *Ber.* **89**, 2159 (1956).
12. J. Thesing and H. Mayer, *Ann. Chem.* **609**, 46 (1957).
13. O. L. Lebedev and S. N. Kazarnovskii, Papers on Chemistry and Chemical Technology, Gor'kii (1959), p. 649.

*The work was carried out in the Rocket Fuels Laboratory of the Edwards Air Force Base of the US Air Force in California.

14. A. V. Il'yasov, *Zh. Strukt. Khim.* **3**, 95 (1962).
15. I. S. Garif'yanov, A. V. Il'yasov, and Yu. V. Yablokov, *Dokl. Akad. Nauk SSSR* **149**, 876 (1963).
16. A. L. Buchachenko and O. P. Sukhanova, *Zh. Strukt. Khim.* **6**, 32 (1965).
17. Yu. Yu. Samitov and O. L. Lebedev, Physical Problems of Spectroscopy, A Collection of Papers, Moscow (1963), Vol. 2, p. 103.
18. E. Fermi, *Z. Physik* **60**, 320 (1930).
19. G. Dousmanis, *Phys. Rev.* **97**, 967 (1955).
20. M. Heald and R. Beringer, *Phys. Rev.* **96**, 645 (1954).
21. W. Holloway and R. Novick, *Phys. Rev. Letters* **1**, 367 (1958).
22. S. Foner, E. Cochran, V. Bowers, and C. Jen, *Phys. Rev. Letters* **1**, 91 (1958).
23. R. Ward, *J. Chem. Phys.* **30**, 852 (1959).
24. G. Pake, J. Townsend, and S. Weissman, *Phys. Rev.* **85**, 682 (1952).
25. P. Atkins, N. Keen, and M. Symons, *J. Chem. Soc.* **1962**, 2873.
26. G. Coppinger and J. Swallen, *J. Am. Chem. Soc.* **83**, 4900 (1961).
27. J. Baird, *J. Chem. Phys.* **37**, 1879 (1962).
28. C. Luck and W. Gordy, *J. Am. Chem. Soc.* **78**, 3240 (1956).
29. W. Gordy and C. McCormick, *J. Am. Chem. Soc.* **78**, 3243 (1956).
30. B. Smaller and M. Matheson, *J. Chem. Phys.* **28**, 1169 (1958).
31. R. Fessenden and R. Schuller, *J. Chem. Phys.* **33**, 935 (1960).
32. A. McLachlan, *Mol. Phys.* **1**, 233 (1958).
33. E. Stone and A. Maki, *J. Chem. Phys.* **37**, 1326 (1962).
34. I. Miyagama and W. Gordy, *J. Chem. Phys.* **30**, 1590 (1959).
35. M. Symons, *J. Chem. Soc.* **1963**, 1189.
36. Y. Kurita, *J. Chem. Phys.* **41**, 3926 (1964).
37. Y. Kurita, *Nippon Kagaku Zasshi* **85**, 833 (1964).
38. T. Stone and W. Waters, *Proc. Chem. Soc.* **1962**, 253.
39. J. Thomas, *J. Am. Chem. Soc.* **86**, 1446 (1964).
40. M. Betoux, H. Lemaire, and A. Rassat, *Bull. Soc. Chim. France* **1964**,1985.
41. B. Jerslev, *Nature* **180**, 1410 (1958).
42. A. Maki and D. Geake, *J. Am. Chem. Soc.* **83**, 1853 (1961).
43. A. Hoffmann and A. Henderson, *J. Am. Chem. Soc.* **83**, 4671 (1961).
44. A. Hoffman and W. Hodgson, *J. Am. Chem. Soc.* **83**, 4675 (1961).
45. A. Hoffmann, W. Hodgson, D. Maricle, and W. Jura, *J. Am. Chem. Soc.* **86**, 631 (1964).
46. A. Hoffmann, A. Feldman, E. Gelblum, and W. Hodgson, *J. Am. Chem. Soc.* **86**, 639 (1964).
47. H. Lemaire, A. Rassat, P. Servoz-Gavin, and G. Berthier, *J. Chim. Phys.* **1962**, 1247.
48. R. Briers and A. Rassat, *Bull. Soc. Chim. France* **1965**, 378.
49. O. L. Lebedev, M. L. Khidekel', and G. A. Razuvaev, *Dokl. Akad. Nauk SSSR* **140**, 1327 (1961).
50. R. Hazeldine and B. Mattinson, *J. Chem. Soc.* **1957**, 1741.
51. L. Piette and G. Crawford, presented at 142nd National Meeting, American Chemical Society, Atlantic City, New Jersey, Sept. 1962.
52. W. Blackley and R. Reinhard, presented at 148th National Meeting, American Chemical Society, Chicago, Illinois, Sept. 1964.
53. W. Blackley and R. Reinhard, *J. Am. Chem. Soc.* **87**, 802 (1965).
54. S. P. Makarov, A. Ya. Yakubovich, S. S. Dubov, and A. N. Medvedev, *Dokl. Akad. Nauk SSSR* **160**, 1319 (1965).
55. S. P. Makarov, A. Ya. Yakubovich, S. S. Dubov, and A. N. Medvedev, *Zh. Vses. Khim. Obchestva im. D. I. Mendeleeva* **10**, 106 (1965).

5. Nitroxide Biradicals

In 1903, Piloty and Vogel [1], by treating porphyrexide with hydrazine and then with potassium ferricyanide, obtained a dark-blue crystalline compound which they called porphyrindine.

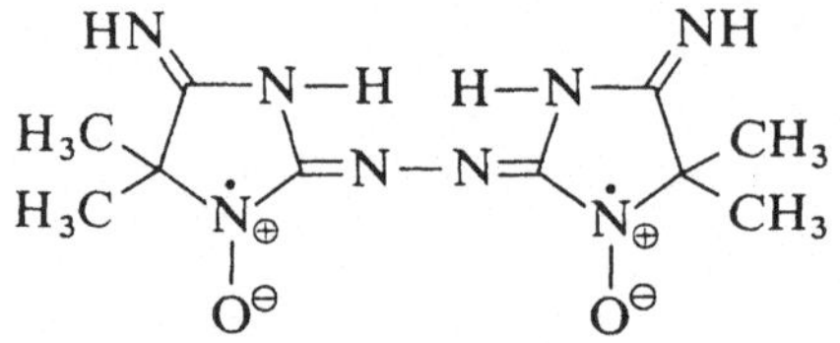

Porphyrindine, possessing two unsaturated electrons, is, on the one hand a biradical and, on the other hand, because of the nonobservance of Curie's law at low temperatures, an equilibrium mixture of paramagnetic and diamagnetic molecules. This mixture becomes enriched in the biradical form of porphyrindine as the temperature is raised.

The difference in the energies of the two forms is small and, as calculations based on measurements of magnetic susceptibility show, does not exceed 0.5 kcal. The increase in the magnetic susceptibility with a rise in the temperature leads to the conclusion that the ground state of porphyrindine is the singlet state [3, 4]:

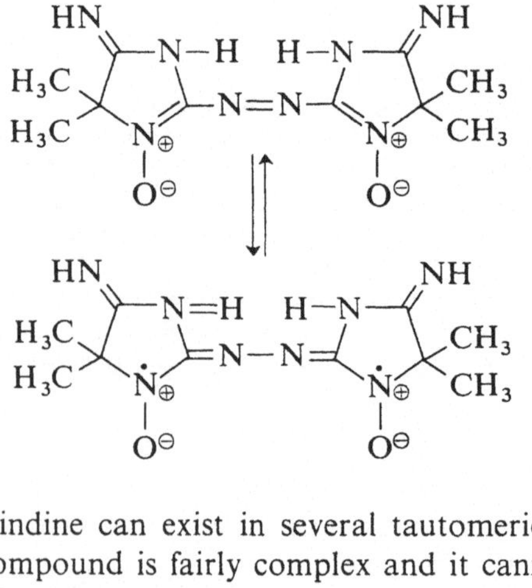

Since porphyrindine can exist in several tautomeric forms, the EPR spectrum of this compound is fairly complex and it cannot be interpreted accurately [5].

The HFS of the EPR spectrum consists of three groups of lines with a splitting of 11 G. Each group of nitrogen triplets consists of four lines, with an interline distance of 3 G (Fig. 10).

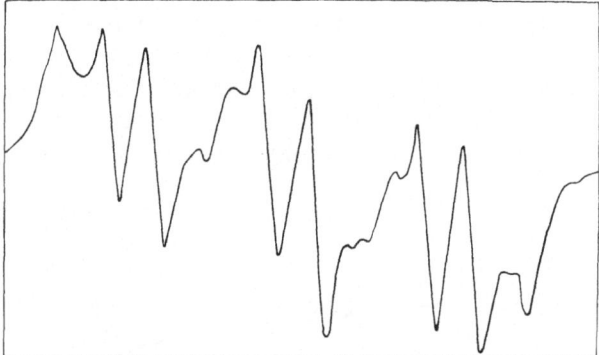

Fig. 10. EPR spectrum of porphyrindine with a constant of basic
splitting at the nitrogen nucleus of about 11 G [5].

An investigation of porphyrindine by the EPR method shows that the
biradical displays strong absorption at a g-factor of 2.0057. Such a large
deviation in the spectroscopic splitting factor from the g-factor of an unpaired
electron may apparently be considered the consequence of an extremely
weak bond between the unpaired electrons in porphyrindine [6].

The potentiometric titration of porphyrindine with hydroquinone leads
to a value of the index potential which corresponds to that expected theoretically for one-electron reduction [7].

Another biradical synthesized by Piloty and Vogel is the diacetyl derivative of porphyrindine. This substance has not been studied, and the positions
of the acetyl groups in it are unknown.

Niementowski's dye [8] may be included among the nitroxide biradicals:

As an investigation by Kuhn and Blau [9] has shown, this dark violet
dye can be obtained by the oxidation of 2,2'-bibenzimidazolyl with hydrogen
peroxide. The changes in the magnetic susceptibility lead to the conclusion
that this substance exists in the biradical form at room temperature, but when

the temperature is lowered to 77°K, it is converted completely into the singlet form.

In recent years, several publications have appeared on the preparation of a number of new nitroxide biradicals [10–12] whose ground state is apparently the singlet state, as in the case of porphyrindine:

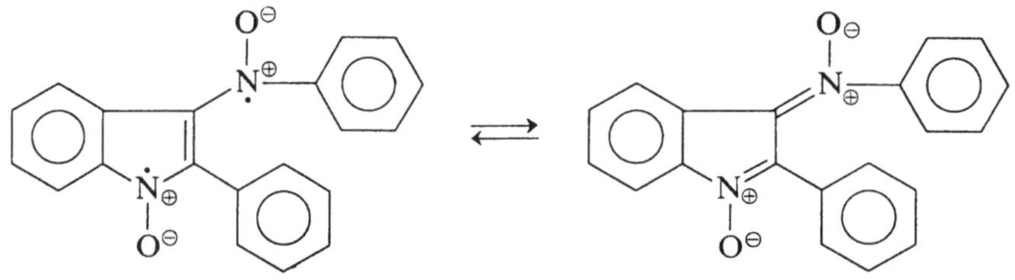

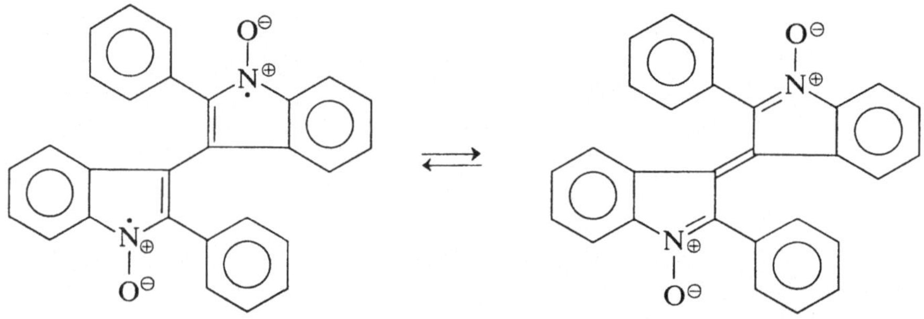

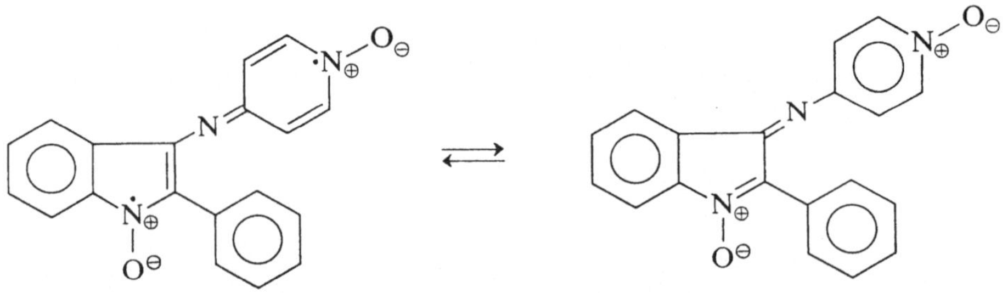

Unfortunately, the EPR spectra of these compounds have not been published, and the dependence of their magnetic susceptibility on the temperature has not been studied.

It may be expected that nitroxide biradicals of this type, like porphyrindine [13, 14], will find use for the study of proteins and the products of their hydrolysis.

References

1. O. Piloty and W. Vogel, *Ber.* **36**, 1283 (1903).
2. R. Kuhn, H. Katz, and W. Franck, *Naturwiss.* **22**, 808 (1934).
3. E. Müller and I. Müller-Rodloff, *Ber.* **69**, 665 (1936).
4. W. Hückel, Theoretische Grundlagen der organischen Chemie, Akademische Verlag, Leipzig, 9th ed. (1957)
5. K. Hausser, *Z. Naturforsch.* **14a**, 425 (1959).
6. C. Hutchison, R. Pastor, and A. Kowalsky, *J. Chem. Phys.* **20**, 534 (1952).
7. R. Kuhn and W. Franke, *Ber.* **68**, 1528 (1935).
8. St. Niementowski, *Ber.* **43**, 3012 (1910).
9. R. Kuhn and W. Blau, *Ann. Chem.* **615**, 99 (1958).
10. M. Colonna and P. Bruni, *Atti Accad. Naz. Lincei, Rend. Classe Sci. Fis., Mat. Nat.* **97**, 461 (1964).
11. M. Colonna and P. Bruni, *Gazz. Chim. Ital.* **95**, 1172 (1965).
12. M. Colonna and P. Bruni, *Atti Accad. Naz. Lincei, Rend. Classe Sci. Fis., Mat. Nat.* **40**, 872 (1966).
13. J. Greenstein and W. Jenrette, *J. Biol. Chem.* **142**, 175 (1942).
14. J. Greenstein, *J. Biol. Chem.* **125**, 501 (1938).

Chapter II

REACTIONS OF RADICALS NOT INVOLVING THE FREE VALENCES

1. Unusual Aspect of Radical Reactions

Radical reactions are generally considered to be those in which free radicals participate. But can there be nonradical chemical reactions in which free radicals take part? At first sight, the question appears paradoxical. Nevertheless, such reactions exist, although, until recently, almost no one had considered this possibility.

It follows from an analysis of the literature that all radical reactions can be divided into two groups: (1) reactions with saturation of the free valence of the radical (recombination, disproportionation, oxidation, reduction), for example:

$$CH_3O\cdot + CH_3CH_2\cdot \longrightarrow CH_3-O-CH_2-CH_3$$

and (2) reactions with retention of the free valence (decomposition, rearrangement, substitution, addition), for example

$$CH_3CH_2\cdot + CH_2{=}CH-C_6H_5 \longrightarrow CH_3CH_2-CH_2-\dot{C}H-C_6H_5$$

Both groups of reactions take place with the indispensable participation of the unshared electron, and this is what is meant by radical reaction. In other words, in any of the radical reactions mentioned, the free valence is always involved [1–6].

The classical idea of the first type of radical reaction was expressed with the utmost clarity in Hückel's fundamental monograph [7]: "If radicals do not react with their own kind," wrote Hückel, "reactions *always* take place as a result of which substances with an even number of electrons are formed."

It is not surprising that reactions of radicals without the involvement of the free valence were completely ignored until recently, and hypotheses on the realization of this type of reaction [8] were doubted [9, 10].

Nevertheless, it is natural to assume that when solid organic substances are irradiated, reactions of macroradicals, for example, may take place

53

without the involvement of the free valence [11]:

$$\cdots \sim CH_2-\dot{C}H-CH_2-CH_2 \sim \cdots \xrightarrow{-H\cdot}$$

$$CH_2-\dot{C}H-CH_2-\dot{C}H \sim \cdots + H\cdot$$

It is also not excluded that the formation of biradicals and polyradicals on irradiation may be connected with the general mechanism of the recombination of radicals in the solid phase.

The possibility of reactions of radicals by a nonradical mechanism also does not contradict the general theoretical situation. The possibility of the existence of such processes is determined by the ratio of the free energies of activation of the two competing reactions—the reaction with the participation of the unshared electron and the one without it—which, in turn, depends on the energy and entropy factors of the two routes. One of the conditions for the possibility of such reactions is a sufficiently reliable screening of the unshared electron not only through the steric effects of neighboring groups, but also through a system of bonds preventing the delocalization of the unshared electron. In practice, this means that a functional group participating in a reaction of this type must be isolated in the radical from the group bearing the unshared electron. It is natural that the greatest success here can be expected in the performance of reactions of a nonradical nature—for example, the condensation of a carbonyl group with amines [12].

Let us consider, for example, the probable reactions of a hypothetical ketone-radical with hydroxylamine:

$$
\begin{array}{c}
R-\underset{\underset{O}{\|}}{C}-CH_2-CH_3 + NH_2O\cdot \\
\overset{K_1}{\nearrow} \\
R-\underset{\underset{O}{\|}}{C}-CH_2-CH_2\cdot + NH_2OH \\
\overset{K_2}{\searrow} \\
R-\underset{\underset{N}{\|}}{C}-CH_2-CH_2\cdot + H_2O \\
\underset{OH}{\diagdown}
\end{array}
$$

It is easy to see that the rate of the reaction without the participation of the free valence of the radical will predominate if $\Delta F_2^{\ddagger} < \Delta F_1^{\ddagger}$, and since $\Delta F^{\ddagger} = \Delta E^{\ddagger} - T\Delta S^{\ddagger}$, K_2 may be greater than K_1 even if $\Delta E_2^{\ddagger}$ is greater than $\Delta E_1^{\ddagger}$, provided that the entropy of activation $\Delta S_1^{\ddagger}$ is considerably larger than $\Delta S_2^{\ddagger}$.

Many organic bimolecular reactions are known whose rates fall greatly with a rise in the steric factor [13, 14]. There is no doubt that with suitable steric screening of the center of localization of the unpaired electron, oximation without the involvement of the free valence can become the dominating reaction.

As a convenient model for the realization of such a reaction, we may consider, for example, the free radical 2,2,6,6-tetramethyl-4-oxopiperidine-1-oxyl:

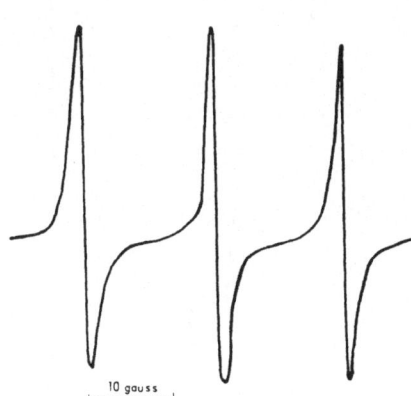

The first report of the existence of this radical in solution and its EPR spectrum was made by Lebedev et al. [15]. However, there has been no information in the chemical literature on the possibility of isolating this substance in the pure state [16, 17].

2. 2,2,6,6-Tetramethyl-4-oxopiperidine-1-oxyl, An Individual Free Radical

In accordance with the method of Lebedev et al. [15], a mixture of 1 ml of water, 0.2 mmole of ammonium nitrate, 0.02 ml of a 25% ethanolic solution of phorone, 0.4 ml of a 0.5 N solution of caustic soda, and 0.5 ml of ethanol was heated in a sealed tube at 100°C for 15 min. After the tube had been opened and its contents had been treated with 1 mg of Trilon B, 0.2 ml of 30% hydrogen peroxide and 0.1 ml of a unimolar solution of sodium tungstate was added, and the mixture was again heated, at 50°C, for 15 min.

For several days, the yellow reaction mixture obtained as a result of this procedure gave an EPR spectrum characteristic for stable nitroxyl radicals (Fig. 11).

Fig. 11. EPR spectrum of the reaction mixture obtained by Lebedev et al. [15].

10 gauss

An analysis of the reaction mixture by means of thin-layer chromatography on alumina led to the conclusion that the mixture under investigation contained not less than four organic compounds, of which only mesityl oxide and isophorone were reliably identified; it was not possible to accurately establish the structure of an amino acid also isolated from the reaction mixture. At the same time, two completely different paramagnetic substances "taken" from the paper chromatogram showed practically identical triplet EPR spectra.

The preparative chromatography on alumina of a chloroform extract obtained from the neutralized reaction mixture gave three pink zones.

The first paramagnetic fraction, a substance strongly adsorbed at the top of the chromatographic column, was a resin.

It was possible to elute the two following paramagnetic fractions separately. Evaporation of the chloroform from the first eluate gave orange needles melting at 36°C; the second eluate yielded a small amount of a red oily substance which was not further studied [18–21].

The elementary analysis and molecular weight of the substance with mp 36°C corresponded to pure, free 2,2,6,6-tetramethyl-4-oxopiperidine-1-oxyl [22, 23]. Found: C, 63.53; H, 9.60; N, 8.25%; 6.1×10^{23} spins/mole; mol. wt. 168.3, 175.1. $C_9H_{16}NO_2$. Calculated: C, 63.49; H, 9.47; N, 8.23%; 6.02×10^{23} spins/mole; mol. wt. 170.2.

As was to be expected, the solid substance possessed a singlet EPR spectrum (Fig. 12), which was transformed by dissolution of the sample in benzene into a triplet with a separation between the signals of 15.5 G (Fig. 13).

From a large number of experiments, it was possible to conclude that the formation of the complex range of byproducts interfering with the isolation of the 2,2,6,6-tetramethyl-4-oxopiperidine-1-oxyl is apparently connected with two main features of the process. The first feature consists in the use of

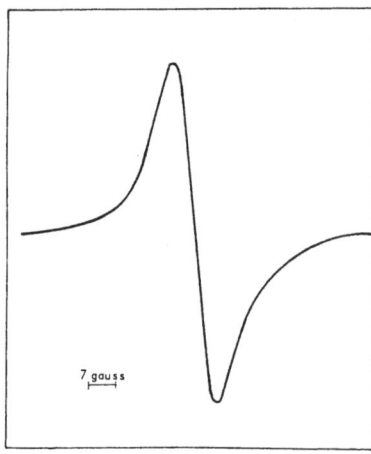

7 gauss

Fig. 12. EPR spectrum of a crystalline sample of 2,2,6,6-tetramethyl-4-oxopiperidine-1-oxyl [24].

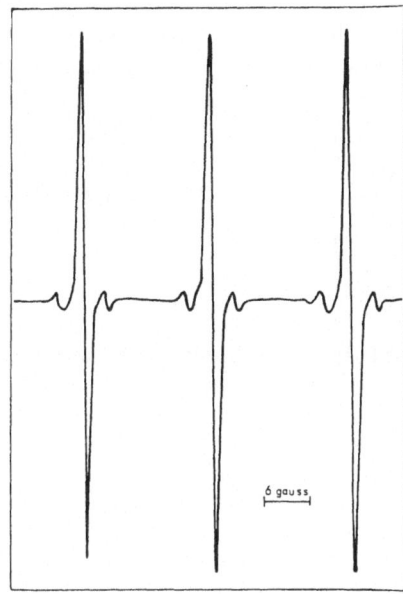

Fig. 13. EPR spectrum of pure 2,2,6,6-tetra-methyl-4-oxopiperidine-1-oxyl (benzene solution) [24].

phorone, which is difficult to purify and readily undergoes polymerization. The second feature consists in the excessively high temperature of the process, which undoubtedly favors the competing reaction of the oxidation at the carbonyl group of the ketone radical and the triacetoneamine formed as an intermediate [25, 26].

To obtain larger amounts of the new substance, Rozantsev [27, 28] subjected separately prepared pure triacetoneamine to the catalytic oxidation reaction. Hydrogen peroxide stabilized with Trilon B [29] was used as the oxidizing agent, the catalyst was sodium tungstate [30–34], and the oxidation itself was conducted in aqueous solution at room temperature.

The mechanism of this reaction falls within the framework of the general ideas advanced by Horner and Lunkermann [35] for peroxy compounds, and has been studied in detail by Kahr [36, 37], and by Lebedev and Kazarnovskii [38].

Under these conditions, the reaction takes place very selectively, and the ketone-radical obtained in 70% yield can be isolated in analytically pure state without the use of chromatography:

The starting triacetoneamine can be obtained by two different methods, (1) starting from phorone and ammonia in the following way [39–41]:

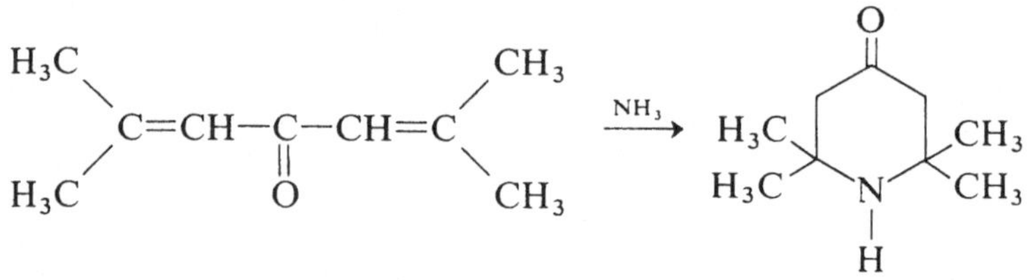

and (2) by the direct condensation of acetone with ammonia [42–46]:

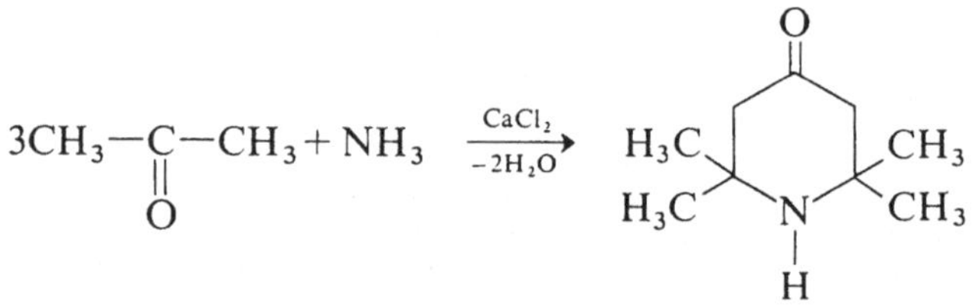

However, it must be noted that neither of the methods of condensation given by the present author and his colleagues can be regarded as satisfactory from the point of view of reproducibility and the constancy of the yields of the desired substance.

The cause of these phenomena is apparently the operation of saturating the mixture of calcium and acetone with gaseous ammonia. During this operation, calcium chloride ammoniate is formed, and this decomposes on heating with the liberation of half of the ammonia originally bound:

$$CaCl_2 + 8NH_3 \longrightarrow CaCl_2 \cdot 8NH_3 \underset{CaCl_2 \cdot 4NH_3}{\overset{4NH_3}{\diagup\diagdown}}$$

This undesirable process unavoidably accompanies the passage of ammonia into a mixture of acetone and calcium chloride, since heat is liberated, and the surface of the calcium chloride, the rate of passage of gaseous ammonia, and the intensity of stirring of the heterogeneous reaction mixture vary from experiment to experiment.

I became convinced that the defects mentioned in the methods of obtaining triacetoneamine can be eliminated to a considerable extent by using liquefied ammonia or independently prepared calcium chloride ammoniate.

Fig. 14. Structural model (Stuart–Briegleb) of 2,2,6,6-tetramethyl-4-oxopiperi-
dine-1-oxyl.

By making use of the methods mentioned above, the present author and his colleagues have prepared about 3 kg of the analytically pure free radical and have studied its properties.

2,2,6,6-Tetramethyl-4-oxopiperidine-1-oxyl readily crystallizes from hexane in the form of orange-yellow needles possessing a weak camphorlike odor. On dissolution in a number of polar solvents, the radical gives yellow solutions, while in ether, ethanol, benzene, and other organic liquids, it forms ruby-red solutions.

The crystalline radical possesses an exceptionally high stability, and can be stored for several months under ordinary conditions. Such a high stability can be explained by two factors: the low energy level of the ground state [47], and the pronounced steric screening of the center of localization of the unpaired electron by four methyl groups of the geminal type (Fig. 14).

During the investigation of the physical properties of the new radical, it was found that it is capable of forming so-called inclusion compounds with a whole series of normal paraffinic hydrocarbons [48], which is the name generally given to adducts of molecules bound to one another purely spatially with the systems of valence bonds of each component being un-changed [49].

An essential fact is that 2,2,6,6-tetramethyl-4-oxopiperidine-1-oxyl, unlike other free radicals of this class, possesses a peculiar twisted conforma-tion [50, 51]. This feature of the conformation of the radical apparently plays the decisive role in the building up of the skeletal lattice and the spatial formation of the bonds between the partners.

Adducts of the radical with unbranched paraffins are obtained in an extremely simple manner. It is sufficient to dissolve the 2,2,6,6-tetramethyl-4-oxopiperidine-1-oxyl (mp, 36°C) in the hot hydrocarbon and leave the solution to cool, whereupon crystals of the adduct begin to deposit immediately. The rate of formation and the stability of the compound increase with an increase in the molecular weight of the hydrocarbon. In contrast to the inclusion compounds of urea, the adducts of the radical possess a definite melting point (Table 6) which does not change when the crystals are washed with pentane or hexane. The least-stable compound of those listed in Table 6 is the adduct with n-heptane; it can be kept for only a few days at room temperature, while the compound with n-hexadecane undergoes no change in months.

When the adducts are melted in vacuum, the included substance is liberated and the residual 2,2,6,6-tetramethyl-4-oxopiperidine-1-oxyl melts normally at 36°C.

Since in the inclusion compounds, no bond exists between the included molecules and those of the skeleton, it is in no way surprising that when the adduct is dissolved it dissociates completely into its components.

The EPR spectra of the adducts scarcely differ from one another, forming curves identical (in the distances between the points of maximum slope) with the curve obtained for the initial radical.

In a study of the magnetic susceptibility [52] of the new radical in the range of temperatures from 1.4 to 300°K, it was found that there is a maximum in the width of the absorption line at $74 \pm 2°K$ (Fig. 15). The possibility is not excluded that there is a phase transition or a change in the crystal structure of the radical at the temperature of the maximum.

TABLE 6

Hydrocarbons Adducts of 2,2,6,6-Tetramethyl-4-oxopiperidine-1-oxyl

Hydrocarbon included	Mp of the adduct, °C	Color and crystal form	Found N, %
n-Heptane	46.5	Dark orange needles	8.19
n-Octane	54.0	Yellow-orange needles	7.87
n-Nonane	59.5	Light orange needles	7.84
n-Decane	57.0	Dark yellow needles	7.73
n-Dodecane	60.0	Yellow-orange needles	7.70
n-Hexadecane	63.0	Orange needles	6.73

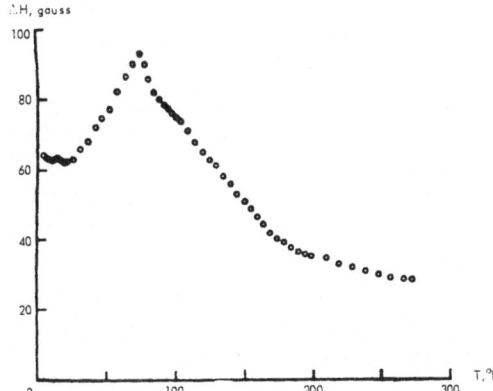

Fig. 15. Linewidth of the EPR signal of pure 2,2,6,6-tetramethyl-4-oxopiperidine-1-oxyl as a function of the temperature [52].

In the range of temperatures studied, the magnetic susceptibility of the radical obeys the Curie law, and the paramagnetic susceptibility reaches a maximum at a temperature below 1°K.

3. Reactions of 2,2,6,6-Tetramethyl-4-oxopiperidine-1-oxyl at the Carbonyl Group

The reaction of 2,2,6,6-tetramethyl-4-oxopiperidine-1-oxyl with hydroxylamine in aqueous solution gives a high yield of a new radical, an oxime, which has mp 180°C after recrystallization from methanol. Found: C, 58.45; H, 9.37; N, 15.25%; 6.0×10^{23} spins/mole. $C_9H_{17}N_2O_2$. Calculated: C, 58.51; H, 9.25; N, 15.12%; 6.0×10^{23} spins/mole. To prove the structure of the new radical, it was synthesized independently by the direct catalytic oxidation of triacetoneamine oxime [51]:

In special experiments, the oxime radical was obtained in an atmosphere of pure argon in order completely to exclude the possibility of the intermediate formation of the oxime 2,2,6,6-tetramethyl-1-hydroxy-4-oxopiperidine and its subsequent oxidation with air to the oxime radical.

For example, when the reaction was carried out in a Schlenk tube placed directly in the resonator of the radiospectrometer, the EPR signal did not disappear before the beginning and after the completion of the condensation reaction.

The EPR spectrum of the crystalline oxime radical is a singlet curve which, when the radical is dissolved in benzene, splits into three lines with a distance between them of 15.5 G and a spectroscopic splitting factor of 2.0064.

The conducted experiments show that in the synthesis of the oxime radical, the 2,2,6,6-tetramethyl-4-oxopiperidine-1-oxyl does in fact react with hydroxylamine without the involvement of the free valence.

The reactions of the ketone radical with semicarbazide and 2,4-dinitrophenylhydrazine also led to new free radicals:

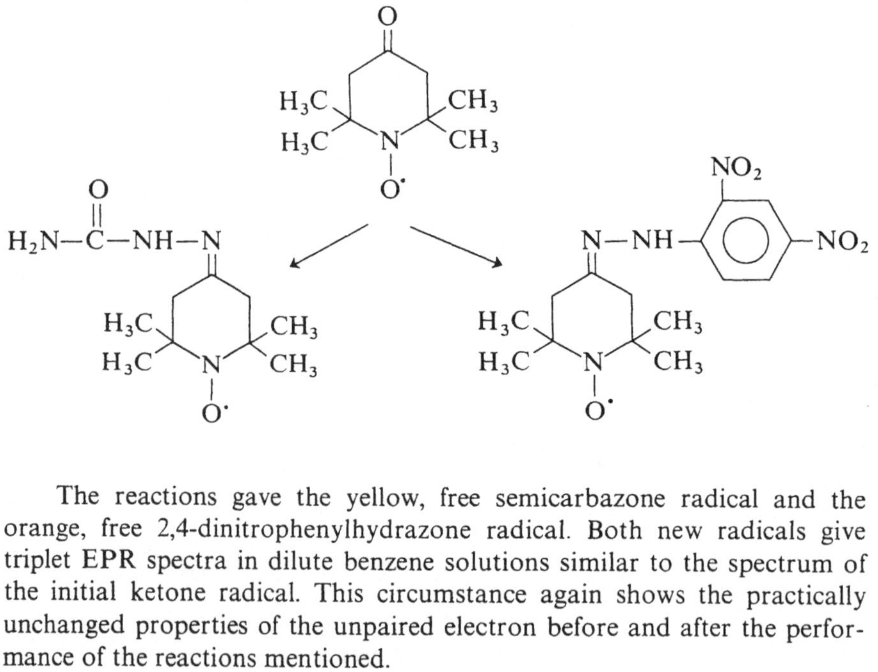

The reactions gave the yellow, free semicarbazone radical and the orange, free 2,4-dinitrophenylhydrazone radical. Both new radicals give triplet EPR spectra in dilute benzene solutions similar to the spectrum of the initial ketone radical. This circumstance again shows the practically unchanged properties of the unpaired electron before and after the performance of the reactions mentioned.

A paramagnetic substance was also obtained by the Wolff–Kishner reduction of the ketone radicals. The radical obtained in this way was identified by independent synthesis [21]:

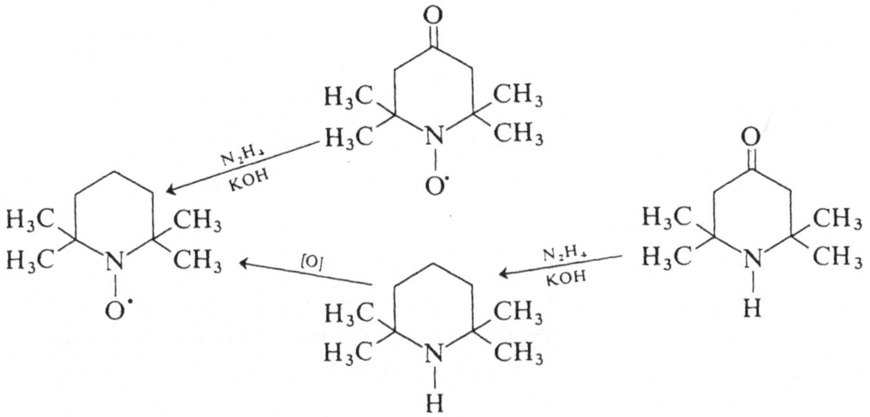

The reduction of the radical was carried out in a glass apparatus filled with argon and fitted with a mercury seal. From 100 g of diethyleneglycol, 10 g of 2,2,6,6-tetramethyl-4-oxopiperidine-1-oxyl, 9 g of 85% hydrazine hydrate, and 15 g of caustic potash was obtained 0.75 g (8%) of 2,2,6,6-tetramethylpiperidine-1-oxyl with mp 37–38°C. A mixture with the substance obtained by the catalytic oxidation of 2,2,6,6-tetramethylpiperidine melted without depression.

A paper-chromatographic analysis of the two radicals obtained by the different methods gave the same retention time. The ultraviolet and infrared spectra of the radical obtained by the catalytic oxidation of 2,2,6,6-tetramethylpiperidine were identical with the corresponding spectra of the substance obtained by reducing the ketone radical without the involvement of the free radical.

Since the Wolff–Kishner reduction is assumed to take place ionically [53], the very small yield of the free radical apparently shows the dominating role in this case of trivial reactions with the participation of the unshared electron—for example,

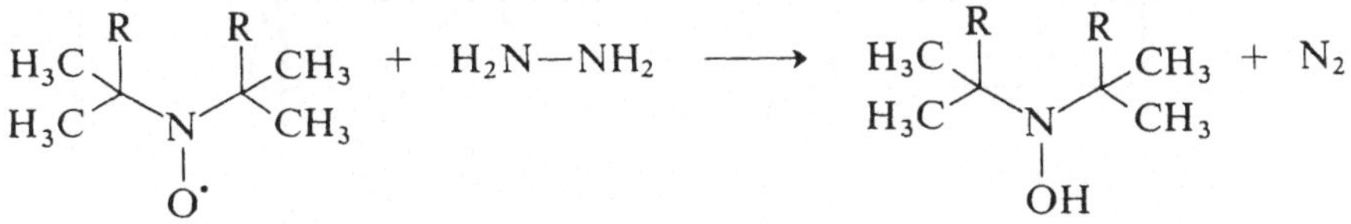

In conclusion, let us consider the reaction of 2,2,6,6-tetramethyl-4-oxopiperidine-1-oxyl with ethylmagnesium bromide [20]:

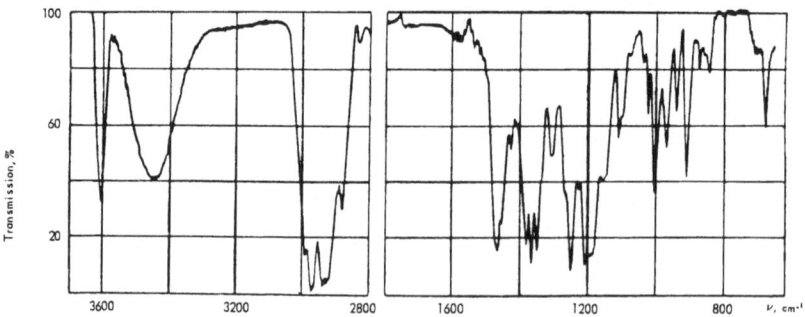

Under the usual conditions of the Grignard reaction [54], from 24 g of magnesium, 109 g of ethyl bromide, 220 ml of absolute ethanol, and 56.7 g of 2,2,6,6-tetramethyl-4-oxopiperidine-1-oxyl, the present author and his colleagues obtained 5.6 g of a bright red oil readily soluble in organic solvents.

Chromatographic analysis showed that the reaction product was a mixture of two radicals, one of which was 2,2,6,6-tetramethyl-4-oxopiperidine-1-oxyl. The separation of the reaction products on a column of alumina enabled us to isolate 5 g (7.4%) of a paramagnetic substance with mp 67–69°C (from a mixture of ether and n-hexane). Found: $\%H_{act}$, 0.480; mol. wt., 199.4; 5.0×10^{23} spins/mole. Calculated: $\%H_{act}$, 0.503; mol. wt., 200.3; 6.0×10^{23} spins/mole.

The new radical is an orange-yellow substance soluble in organic solvents and stable without appreciable decomposition for several months.

The absorption spectra of the radical in the ultraviolet and visible regions of the spectrum resembled the spectra of the ketone radical and some of its derivatives (λ_{max} = 227, 285, and 460 nm). The infrared spectrum lacked the band at 1720 cm^{-1} characteristic of a carbonyl group, and two new bands characteristic for a hydroxy group appeared: a narrow band in the 3590-cm^{-1} region and a broad band close to 3460 cm^{-1} (Fig. 16).

Fig. 16. IR spectrum of 4-ethyl-4-hydroxy-2,2,6,6-tetramethylpiperidine-1-oxyl in 10 N carbon tetrachloride solution [55].

The free radical 4-ethyl-4-hydroxy-2,2,6,6-tetramethylpiperidine-1-oxyl obtained by independent synthesis [55]:

was shown by spectroscopic studies and the results of a chromatographic analysis to be identical with the substance obtained by the reaction of the ketone radical with ethylmagnesium bromide.

From an analysis of the results obtained, it can be stated with confidence that reactions of free radicals without the participation of the unshared electron have been demonstrated with 2,2,6,6-tetramethyl-4-oxopiperidine-1-oxyl as an example, and are easy to realize.

References

1. C. Walling, Free Radicals in Solution, John Wiley and Sons, New York (1957).
2. A. M. Bass and H. P. Broida (eds.), The Formation and Stabilization of Free Radicals, A Collection of Papers, Academic Press, New York (1960).
3. M. S. Blois *et al.* (ed.), Free Radicals in Biological Systems, A Collection of Papers, Academic Press, New York (1961).
4. V. N. Kondrat'ev, Free Radicals—An Active Form of Matter, Izd. Akad. Nauk SSSR, Moscow (1960).
5. N. N. Semenov, Some Problems of Chemical Kinetics and Reactivity, Pergamon Press, Oxford (1958–1959).
6. E. Steacie, Atomic and Free Radical Reactions, Reinhold, New York, Vol. 1 (1954).
7. W. Hückel, Theoretische Grundlagen der organischen Chemie, Akademische Verlag, Leipzig, 9th ed. (1957).
8. E. Schneider, *Discussions Faraday Soc.* **19**, 158 (1965).
9. D. Ingram, M. Symons, and G. Townsend, *Trans. Faraday Soc.* **54**, 409 (1958).
10. R. Abraham, H. Melville, D. Ovenall, and D. Whiffen, *Trans. Faraday Soc.* **54**, 1133 (1958).
11. M. Fujimoto and D. Ingram, *Trans. Faraday Soc.* **54**, 1304 (1958).
12. A. L. Buchachenko, Stable Radicals, Izd. Akad. Nauk SSSR, Moscow (1963), p. 132.
13. W. Vaughan, *J. Am. Chem. Soc.* **54**, 3863 (1962).
14. J. Harkness, G. Kistiakowsky, and W. Mears, *J. Chem. Phys.* **5**, 682 (1937).
15. O. L. Lebedev, M. L. Khidekel', and G. A. Razuvaev, *Dokl. Akad. Nauk SSSR* **140**, 1327 (1961).
16. É. G. Rozantsev, "The structure and reactivity of stable iminoxyl mono- and polyradicals" (*in Russian*), presented at the International Symposium on the Structure and Reactivity of Organic Compounds, Sofia, 1966.
17. É. G. Rozantsev and M. B. Neiman, *Zh. Orgab. Khim.* **1**, 1337 (1965).
18. M. B. Neiman, Yu. G. Mamedova, and É. G. Rozantsev, *Azerb. Khim. Zh.* **1962**, 37.

19. M. B. Neiman, É. G. Rozantsev, and Yu. G. Mamedova, *Nature* **196**, 472 (1962).

20. É. G. Rozantsev, M. B. Neiman, and Yu. G. Mamedova, *Izv. Akad. Nauk SSSR, Ser. Khim.*, **1963**, 1509.

21. É. G. Rozantsev, Yu. G. Mamedova, and M. B. Neiman, *Izv. Akad. Nauk SSSR, Otd. Khim. Nauk* **1962**, 2250.

22. M. B. Neiman, É. G. Rozantsev, and Yu. G. Mamedova, *Nature* **200**, 256 (1963).

23. É. G. Rozantsev and M. B. Neiman, *Angew. Chem.* **76**, 618 (1964).

24. Yu. G. Mamedova, Thesis, Moscow (1965).

25. V. N. Belov, L. A. Kheifits, and S. I. Virezub, Reaktsii i Metody Issledovan. Org. Soedinenii, Coll. 10, Moscow (1961).

26. C. H. Hassall, Organic Reactions, John Wiley and Sons, New York, Vol. 9 (1957), p. 73.

27. É. G. Rozantsev, USSR Authors' Certificate 166032, August 27, 1962.

28. É. G. Rozantsev, *Izv. Akad. Nauk SSSR, Ser. Khim.* **1964**, 2218.

29. W. Schumb, C. Satterfield, and R. Wentworth, Hydrogen Peroxide, Reinhold, New York (1955).

30. Swedish Patent 146,038 (1954).

31. US Patent 2,718,528 (1955).

32. German 953,069 (1956).

33. French Patent 1,113,022 (1956).

34. German 951,933 (1956).

35. L. Horner and H. Lunkermann, *Ann. Chem.* **591**, 53 (1955).

36. K. Kahr and C. Berther, *Ber.* **93**, 132 (1960).

37. K. Kahr, *Angew. Chem.* **72**, 135 (1960).

38. O. L. Lebedev and S. N. Kazarnovskii, *Tr. po Khim. i Khim. Tekhnol.* **1959**(3), 649.

39. J. Guareschi, *Atti Acc. Sci. Torino: Classe Sci. Fis., Mat. Nat.* **29**, 680 (1894).

40. J. Guareschi, *Ber.* **28**, 160 (1895).

41. C. Sandris and G. Quarisson, *Bull. Soc. Chim. France* **1958**, 345.

42. F. Francis, *J. Chem. Soc.* **1927**, 2897.

43. H. Hall, *J. Am. Chem. Soc.* **79**, 5444 (1957).

44. E. Mailey and A. Day, *J. Org. Chem.* **22**, 1061 (1957).

45. W. Lutz, S. Lazarus, and R. Meltzer, *J. Org. Chem.* **27**, 1695 (1962).

46. I. B. Simon and V. P. Vvedenskii, *Med. Prom. SSSR*, **1963**(5), 9.

47. Yu. A. Lebedev, É. G. Rozantsev, M. B. Neiman, and A. Ya. Apin; *Zh. Fiz. Khim* **1956**, 2340.

48. É. G. Rozantsev, *Teor. i Éksperim. Khim.* **1966**(2), 286.

49. F. Cramer, Einschlussverbindungen, Springer Verlag, Berlin (1954).

50. É. G. Rozantsev and E. N. Gur'yanova, *Izv. Akad. Nauk SSSR, Ser. Khim.*, **1966**, 979.

51. É. G. Rozantsev and M. B. Neiman, *Tetrahedron* **20**, 121 (1964).

52. Yu. S. Karimov, É. G. Rozantsev, *Fiz. Tverd. Tela* **8**, 2787 (1966).

53. H. Szmant, H. Harnsberger, T. Bulber, and W. Barie, *J. Am. Chem. Soc.* **74**, 2724 (1952).

54. M. Kharasch and O. Reinmuth, Grignard Reactions of Nonmetallic Substances, Prentice-Hall, New York (1954).

55. É. G. Rozantsev, *Izv. Akad. Nauk SSSR, Ser. Khim.* **1964**, 2187.

Chapter III

THE SYNTHESIS OF INDIVIDUAL RADICALS

1. The Oxidation of Secondary Amines and Hydroxylamines

In spite of the fact that free radicals always arise as intermediates in the oxidation of secondary amines, only a few oxidizing agents have acquired preparative importance for the synthesis of these radicals.

As already mentioned, stabilized hydrogen peroxide in the presence of ions of tungstic, molybdic, or vanadic acid, which give the corresponding peracids under these conditions [1–3],

$$3H_2O_2 + O_2W \begin{array}{c} O^{\ominus} \\ \diagup \\ \diagdown \\ ONa \end{array} \rightleftarrows \begin{array}{c} HOO \\ \diagdown \\ \diagup \\ NaOO \end{array} WO_2 + HO_2^{\ominus} + 2H_2O$$

form the mildest and most selective oxidizing agents for certain amines.

In all cases, the application of this type of oxidizing system to amines gives substances possessing a nitrogen–oxygen bond—oximes, hydroxylamines, and amine oxides [4–9].

On the basis of numerous investigations on the oxidation of amines with the pertungstate ion, the mechanism of this reaction for the case of strongly hindered secondary amines can be illustrated by the following scheme:

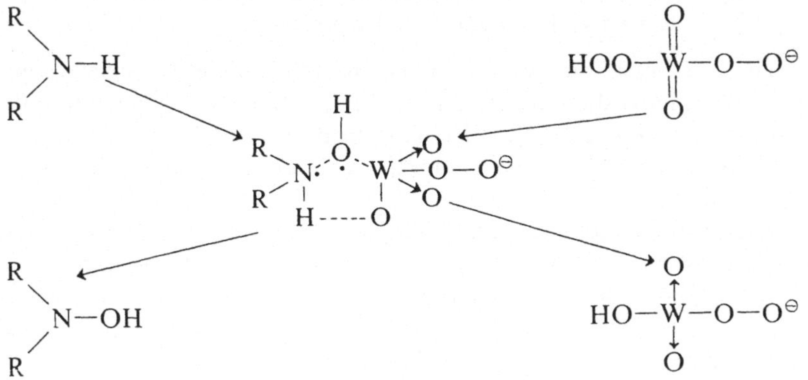

The hindered hydroxylamine formed undergoes further oxidation to an unstable hydroxylamine nitrogen oxide, which then decomposes into a nitroxyl and hydroxyl [10]:

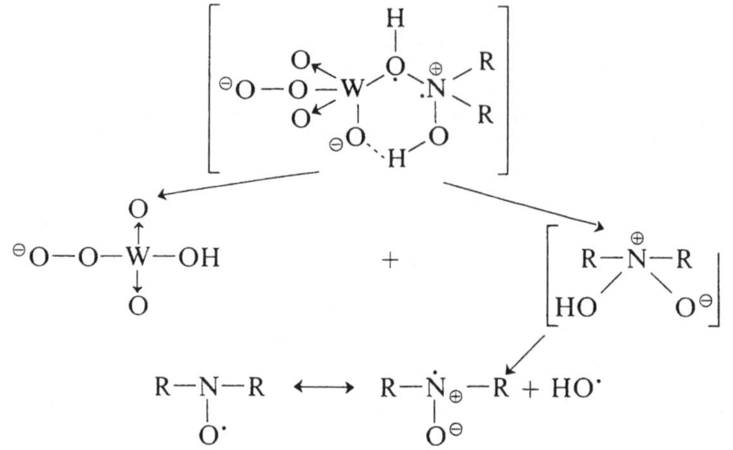

$$
\begin{bmatrix}
{}^{\ominus}O{-}O{-}W \cdots N \\
\end{bmatrix}
$$

The method of oxidizing amines with the pertungstate ion was used by Lebedev and Kazarnovskii [11] to obtain the first individual nitroxyl radical, 2,2,6,6-tetramethylpiperidine-1-oxyl.

We have used the same method for the synthesis of many nitroxyl radicals. Thus, the oxidation of hindered heterochain amines gave a series of previously unknown acyclic nitroxyls [12],

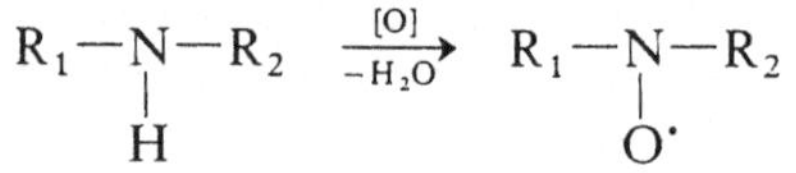

$$
R_1{-}\underset{H}{N}{-}R_2 \xrightarrow[-H_2O]{[O]} R_1{-}\underset{O^\cdot}{N}{-}R_2
$$

for the cases:

    a. $R_1 = R_2 =$ tert-amyl, 3,3-dimethylpropynyl;
    b. $R_1 =$ tert-butyl, $R_2 =$ tert-amyl, 3,3-dimethylpropynyl;
    c. $R_1 =$ tert-amyl, $R_2 =$ 3,3-dimethylpropynyl.

The direct catalytic oxidation of heterocyclic amines has given preparative yields of individual nitroxyls of the heterocyclic series, including those of dihydropyrroles [13, 14] and tetrahydropyrroles [15, 16], dihydro- and tetrahydropyridines [17, 18], homopiperazinone, and carboline, and also tetrahydroquinoline [19–22].

A large number of individual radicals have been obtained with the use of phosphotungstic acid by the French organic chemist Rassat and his co-workers [23, 24]. The structure and individuality of the nitroxyl radicals and nitrogen oxides obtained in this way are confirmed by elementary analysis, their chemical properties, and their UV, Raman, and EPR spectra.

The presence of a substituent in position 4 of the piperidine heterocycle apparently has no fundamental influence on the rate of oxidation of the secondary amino group, and with a suitable selection of the experimental conditions, the reaction can easily be stopped at the radical state. Confirmation of this fact are the very high yields of nitroxyl radicals formed from secondary amines containing groups inert to oxidation.

However, the presence of groupings active with peroxides leads to a substantial lowering in the yield of the radical.

By a suitable selection of the experimental conditions, it is sometimes possible to exclude undesirable reactions. Thus, by lowering the temperature and strongly diluting the reagents, the oxidation of the carbonyl group of triacetoneamine takes place to only a very small extent, and the ketone-radical is the main product.

Analogously, in the low-temperature oxidation of 2,2,6,6-tetramethyl-1,2,5,6-tetrahydropyridine, the double bond is only slightly affected and the yield of the radical is fairly high [25].

In the case of more labile substituents—for example, the readily oxidizable cyano, amino, or mercapto groups—side reactions cannot be excluded, which places serious limitations on the use of the oxidation method for the synthesis of the nitroxyl radicals.

In agreement with this, attempts to obtain a nitroxyl radical by oxidizing 2,2,6,6-tetramethyl-4-iodopiperidine,

were unsuccessful, not, apparently, because of the low stability of the iodine-containing radical, but because of the side reactions involving an iodoalkyl group. Confirmation of this hypothesis is the marked decrease in radical yield (from 82% to 36%) on passing from chlorotramethylpiperidine to its bromine-containing analog, although the amines corresponding to these radicals have almost identical stabilities.

The method of oxidation of amines can be used not only for the synthesis of monoradicals, but also for the production of individual biradicals. This method was used for the preparation of 2,2,2',2',6,6,6',6'-octamethyl-4,4'-bipiperidyl-1,1'-dioxyl, the first individual representative of this series of nitroxyl biradicals in which the paramagnetic centers are separated by a system of single carbon–carbon bonds [26]:

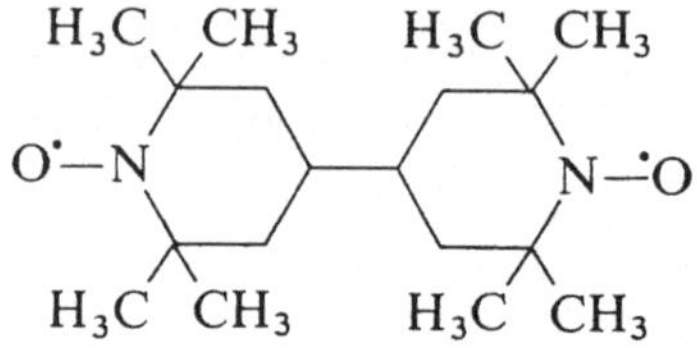

$$
\begin{array}{c}
\text{H}_3\text{C}\quad\text{CH}_3\qquad\text{H}_3\text{C}\quad\text{CH}_3 \\
\\
\text{O}^\cdot-\text{N}\qquad\qquad\qquad\text{N}-{}^\cdot\text{O} \\
\\
\text{H}_3\text{C}\quad\text{CH}_3\qquad\text{H}_3\text{C}\quad\text{CH}_3
\end{array}
$$

The oxidation of the corresponding octamethylbipiperidyl containing two isolated amino groups screened from one another takes place in stages and is accompanied by the formation of a monoradical. However, when an excess of hydrogen peroxide is used, the monoradical is easily converted into the biradical, which makes it possible to oxidize the diamine to the biradical almost quantitatively in one stage.

The following individual substances have been obtained in a similar way [27, 28]: sym-di(2,2,6,6-tetramethyl-1-oxyl-4-piperidyl)urea, and 1,4-bis(4-hydroxy-2,2,6,6-tetramethyl-1-oxyl-4-piperidyl)butadiyne and 1,4-bis(4-hydroxy-2,2,6,6-tetramethyl-1-oxyl-4-piperidyl)butane, given by

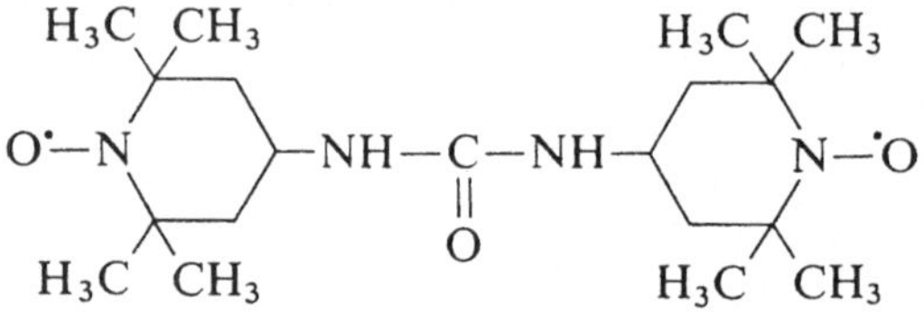

$$
\begin{array}{c}
\text{H}_3\text{C}\quad\text{CH}_3\qquad\qquad\qquad\qquad\text{H}_3\text{C}\quad\text{CH}_3 \\
\\
\text{O}^\cdot-\text{N}\qquad-\text{NH}-\underset{\underset{\text{O}}{\|}}{\text{C}}-\text{NH}-\qquad\text{N}-{}^\cdot\text{O} \\
\\
\text{H}_3\text{C}\quad\text{CH}_3\qquad\qquad\qquad\qquad\text{H}_3\text{C}\quad\text{CH}_3
\end{array}
$$

and

$$
\begin{array}{c}
\text{H}_3\text{C}\quad\text{CH}_3\qquad\text{H}_3\text{C}\quad\text{CH}_3 \\
\qquad\qquad\text{HO} \\
\text{O}^\cdot-\text{N}\qquad-\text{Q}-\qquad\text{N}-{}^\cdot\text{O} \\
\qquad\qquad\text{OH} \\
\text{H}_3\text{C}\quad\text{CH}_3\qquad\text{H}_3\text{C}\quad\text{CH}_3
\end{array}
$$

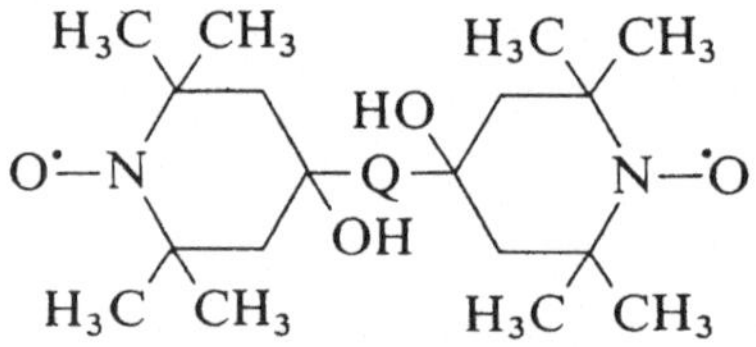

with $Q = -C\equiv C-C\equiv C-$ and $-CH_2-CH_2-CH_2-CH_2-$, respectively.

Besides amines, the corresponding substituted hydroxylamines, for example,

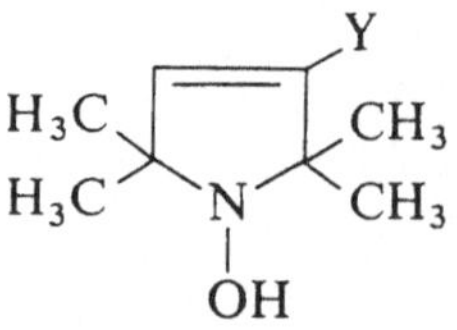

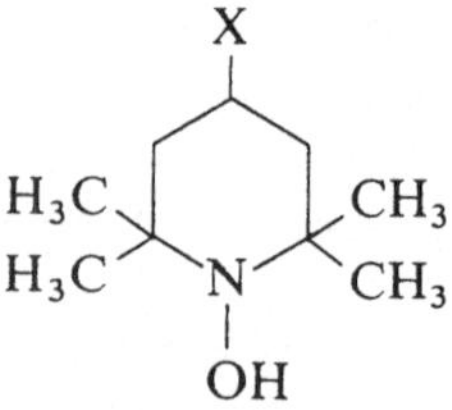

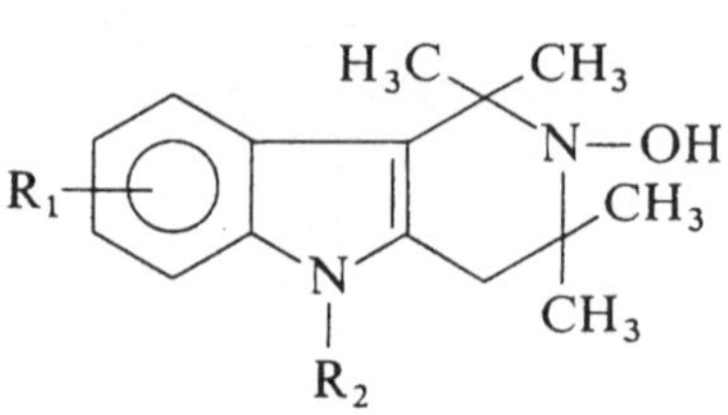

$$
\begin{array}{ccc}
& & \text{X} \\
& \text{Y} & \\
\text{H}_3\text{C} & \text{CH}_3 & \text{H}_3\text{C}\quad\text{CH}_3 \\
\text{H}_3\text{C} & \text{CH}_3 & \text{H}_3\text{C}\quad\text{CH}_3 \\
& \text{N} & \text{N} \\
& | & | \\
& \text{OH} & \text{OH}
\end{array}
$$

have been used to obtain free radicals.

These compounds, which were difficult to obtain until recently, can easily be converted into radicals by a large number of oxidizing agents. The most suitable for this purpose are the oxides of lead, mercury, and silver, with which the reaction takes place even in the cold, while it may be accelerated by vigorous stirring and heating.

The use of this method to obtain, for example, 2,2,6,6-tetramethyl-4-oxopiperidine-1-oxyl and 4-hydroxy-2,2,6,6-tetramethylpiperidine-1-oxyl enables the time necessary for the oxidation of the amines to be shortened in some cases, and leads, in practice, to quantitative yields of the corresponding radicals [29, 30].

In spite of the fact that the nitroxyl radicals formed may undergo further oxidation in the presence of the metal oxides, when the process is stopped at the appropriate time, the yields of radicals undergo almost no reduction. The optimum time of oxidation, which must be known for each given case, can be determined by monitoring the reaction by thin-layer chromatography.

An alkaline solution of potassium ferricyanide may be used for preparative oxidation. However, an excess of this oxidizing agent is harmful because of further oxidation with formation of a complex mixture of byproducts.

Atmospheric oxygen may also be used to obtain nitroxyl radicals, since all hindered hydroxylamines of this type are gradually converted into radicals in air [29].

The ready oxidizability of hindered hydroxylamines makes it possible to choose selective oxidizing agents which, by not affecting other reaction centers, enable the required radicals to be obtained in good yields. The main difficulty of obtaining nitroxyl radicals by this method is that up to the present time there have been no simple methods of obtaining sterically hindered hydroxylamines. The only known route for obtaining derivatives of 1-hydroxy-2,2,6,6-tetramethylpiperidine consists in the condensation of phorone with hydroxylamine in the presence of sodium methoxide; however, the condensation product described by Harries and Lehman [31, 32] is apparently not 1-hydroxy-2,2,6,6-tetramethyl-4-oxopiperidine, but another compound the structure of which has so far remained obscure.

In some cases, stabilized hydrogen peroxide in the presence of sodium tungstate has proved to be a convenient oxidizing agent for obtaining unstable aromatic radicals including mercurated radicals [33–35]:

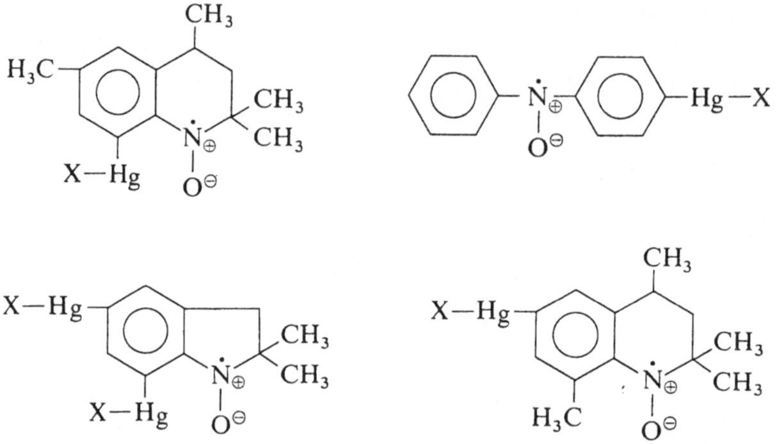

Since aromatic amines are not, as a rule, oxidized by pertungstate, the authors obtained a number of aromatic nitrogen oxides in solution by heating the amines with a benzene solution of tert-butyl hydroperoxide in the presence of catalytic amounts of cobalt stearate.

The formation of free radicals in solution under these conditions apparently takes place in accordance with the following scheme:

$$Ar_2NH + RO_2\cdot \longrightarrow Ar_2N\cdot + RO_2H$$

$$Ar_2N\cdot + Ar_2H\cdot \rightleftarrows Ar_2N-NAr_2$$

$$Ar_2N\cdot + RO_2\cdot \longrightarrow Ar_2\dot{N}O + RO\cdot$$

By this method paramagnetic nitrogen oxides of the acridane series have been obtained [36]:

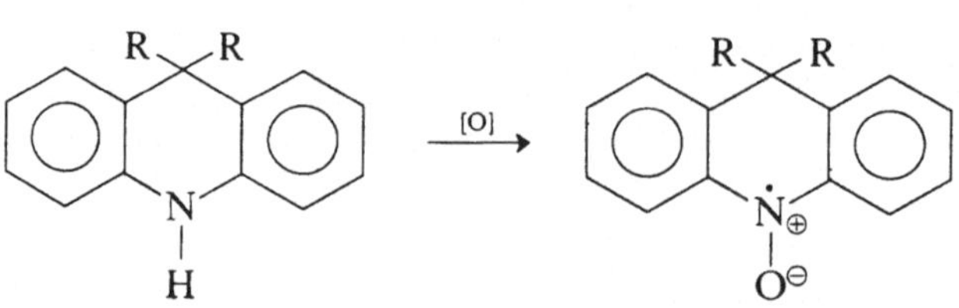

where $R = CH_3, C_6H_5, p\text{-}CH_3C_6H_4, p\text{-}CH_3OC_6H_4$, and $p\text{-}C_2H_5OC_6H_4$.

However, it must be noted that this method cannot be used to obtain free radicals possessing easily oxidized functional groups, since the oxidation may be accompanied by a number of side reactions which sometimes change the chemical structure and elementary composition of the radical to a considerable extent.

Similar remarks also apply in full measure to the method of oxidation using a pertungstate. To illustrate these remarks, reference may be made, for example, to the oxidation of 4-amino-2,2,6,6-tetramethylpiperidine, which is accompanied by the transformation of the primary amino group [37],

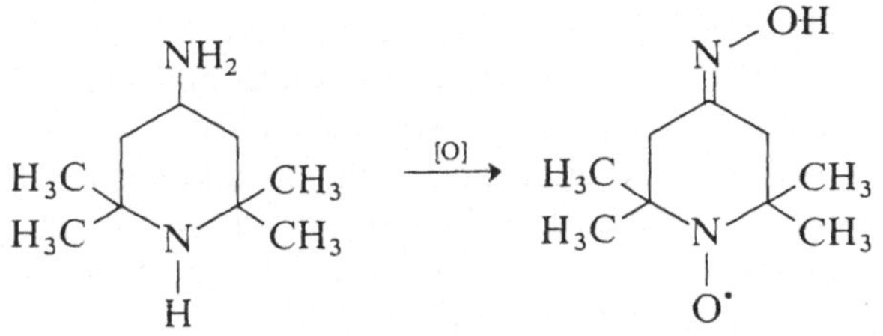

and the oxidation of 3-cyano-2,2,5,5-pyrroline, which results in the simultaneous transformation of the nitrile group into an amide group [13]:

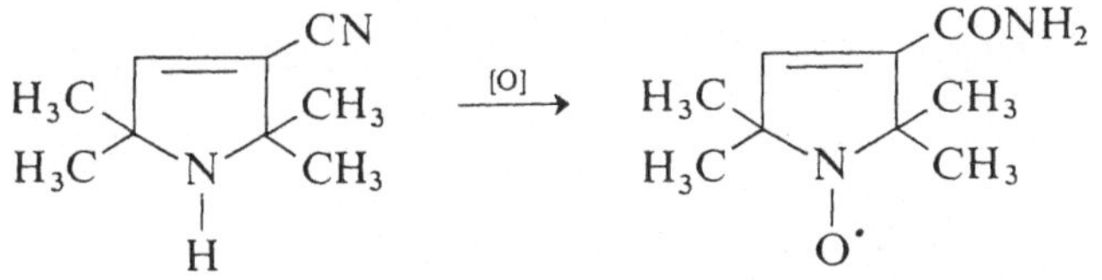

2. Selective Reactions of Radicals without the Involvement of the Free Valence

In the subsequent study of nitroxyl radicals, it has been possible to considerably broaden the range of their reactions without the involvement of the free valence and to establish general laws and limits for the use of such reactions.

As already mentioned, 2,2,6,6-tetramethyl-4-oxopiperidine-1-oxyl gives derivatives at the carbonyl group with a whole series of strong bases (2,4-dinitrophenylhydrazine, semicarbazide, hydroxylamine) [38]. In all cases, the nitroxyl group remains unaffected, but the low yields of the main products show that in some cases, the process is complicated by side reactions.

The particular feature of the interaction of this ketone with nucleophilic reagents is that the latter, possessing unshared electron pairs, can use them not only to form a new bond with the carbon atom of the carbonyl group, but also to reduce the nitroxyl group. When certain conditions are observed, the reduction reaction may become the dominating one, and therefore in choosing nucleophilic reagents one must take their reducing properties into account.

It is natural to expect that the most reactive compounds containing an active methylene group will react with the ketone-radical without saturating its free valence. Of the large number of compounds possessing high CH acidity, succinic acid esters, malonic acid derivatives, and bromoacetic ester have been studied. These compounds were of interest because it could be expected that they would lead to radicals with other reactive substituents.

The reaction of succinic esters with ketones and aldehydes, which is known as the Stobbe condensation, takes place only in the presence of very strong bases—for example, sodium amide or metal alkoxides or hydrides [39]. Of the condensing agents mentioned, the most suitable for the present case seem to be the alkoxides, since at temperatures close to room temperature, they do not react with the nitroxyl group:

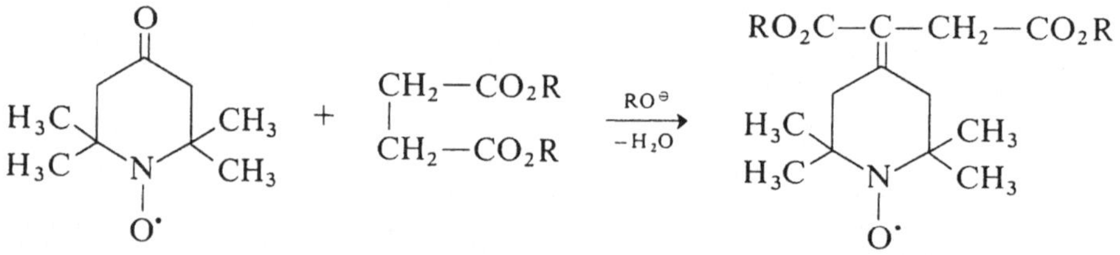

where $R = CH_3, C_2H_5$.

The best results were obtained by using potassium tert-butoxide [17]. The lower yields of esters of 2,2,6,6-tetramethyl-1-oxylpiperidylidenesuccinic acid when sodium ethoxide was used are apparently explained by the partial reduction of the ketone group to a hydroxy group, as is confirmed by the isolation of 1-hydroxy-2,2,6,6-tetramethylpiperidine-1-oxyl from the reaction mixture.

The position of the double bonds in the compounds synthesized was determined from an analysis of their electronic spectra: the absorption maxima in the ultraviolet spectra of the monoesters are located at 211 nm, which shows the presence of an exocyclic double bond in the α-position to the carboxyl group [40].

The monoesters obtained were subjected to further reactions, which also took place without the involvement of the free valence. Hydrolysis of the monoesters in barium hydroxide solution formed the salt of the corresponding acid

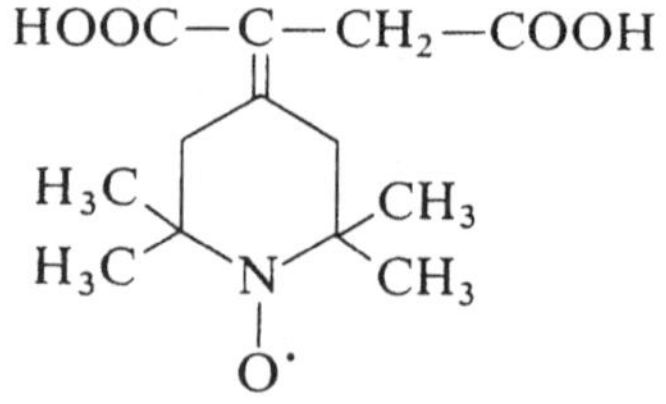

The absence of side reactions in the hydrolysis is not surprising, since hydroxyl ions, which are weaker bases than alkoxide ions, also have feebler reducing properties. The absorption maximum at 211 nm in the UV spectrum of the acid shows that hydrolysis is not accompanied by a shift of the double bond.

It is considerably more difficult to effect the esterification of the carboxy group than the hydrolysis of the ester group of the radical. The reason for this is that the acid catalysts used for esterification react with the nitroxyl group, while in the absence of acids the reaction takes place very slowly.

Practically the only method for the preparative production of esters from carboxylic acids containing a nitroxyl group is the reaction of the latter with diazo compounds. Thus, esterification of the paramagnetic monoesters and the dicarboxylic acid with diazomethane give the corresponding diester-radicals in quantitative yields

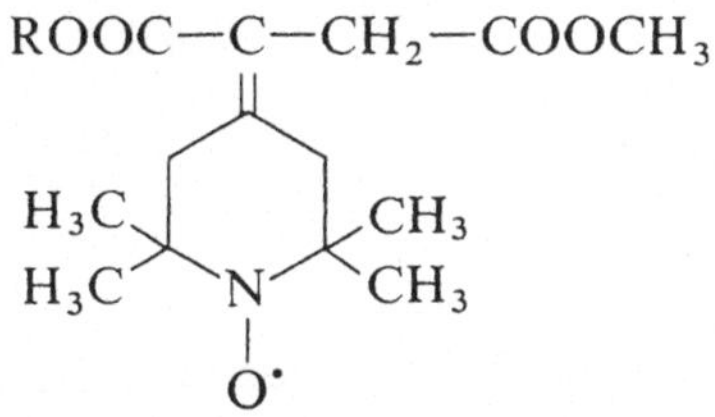

$$ROOC-C-CH_2-COOCH_3$$

(where $R = CH_3, C_2H_5$), which is in good agreement with a nonradical mechanism of the reaction of diazo compounds with carboxylic acids [41, 42].

The Darzens reaction, which is widely used for the synthesis of glycide esters from carbonyl compounds, has much in common with the Stobbe condensation [43]. A study of the condensation of chloroacetic ester with 2,2,6,6-tetramethyl-4-oxopiperidine-1-oxyl in the presence of alkali-metal alkoxides has shown that the carbonyl group of the ketone-radical is fairly active and readily takes part in the reaction [44]:

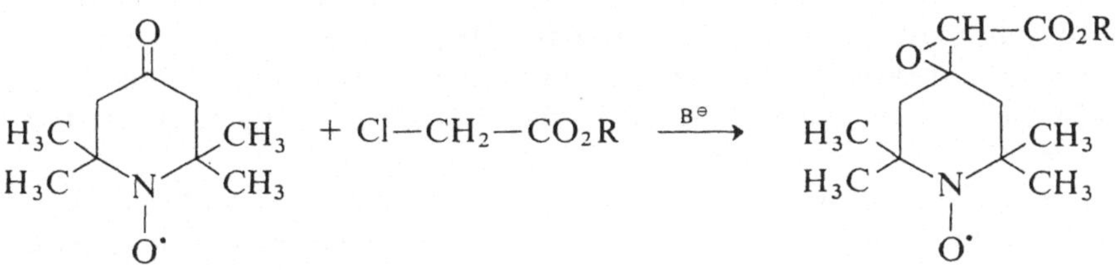

where $R = H, C_2H_5$, tert-C_4H_9.

The yield of paramagnetic glycide ester formed by this condensation depends largely on the nature of the condensing agent, and is a maximum when potassium tert-butoxide is used.

In contrast to the majority of glycide esters, which are readily converted into α-keto esters on distillation [45], the radicals obtained undergo no change under similar conditions, their chromatographic constants being the same before and after distillation.

The successive treatment of a glycide ester with equivalent amounts of sodium ethoxide and water gives a quantitative yield of the paramagnetic sodium salt of the acid from which, on careful acidification, the acid itself can be isolated.

The increase in the stability of the acid as compared with its nonradical analogs [46] gives grounds for assuming that its structure is that of a β-unsaturated α-hydroxycarboxylic acid:

$$HO-CH-COOH$$

However, the absence of absorption in the 1550–1650-cm^{-1} region that is characteristic for a double bond excludes this possibility. The similar structures of the glycide acid and its ethyl ester is confirmed by the similarity of their IR spectra. The small shift in the absorption band of the carboxy group in the spectrum of the acid (1734 cm^{-1}) as compared with the band of the glycide ester (1743 cm^{-1}) is characteristic for all carboxy compounds, and the appearance of a band at 935 cm^{-1}, lacking in the ester, is apparently explained by deformation vibrations of the OH group [47]. The IR spectrum of the sodium salt of the acid differs considerably from the spectrum of the acid itself and its ester, but this is apparently explained not by isomerization, but only by the change in the polar nature of the carboxy group. In the spectrum, the bands previously present in the 1730–1740-cm^{-1} and 1185–1205-cm^{-1} regions disappear and new bands arise at 1620 cm^{-1} and 1405 cm^{-1} which can be referred to the stretching vibrations of the CO_2 group [47].

The UV spectra of the compounds obtained also confirm the structure put forward. Thus, the spectrum of the glycide acid taken in methanolic solution has three maxima, at 428 nm, 240 nm, and 209 nm. The first two are connected with the absorption of the N—Ȯ group and correspond to $\pi \rightarrow \Pi$ and $\Pi \rightarrow \Pi^*$ electronic transitions [48]. The maximum at 209 nm is close to the absorption maximum of the analogous α-unsaturated carboxy compounds [33] and is obviously due to the carbonyl group, since the epoxide group causes approximately the same bathochromic displacement as a double bond [49, 50].

The Darzens condensation with the free ketone-radical is not only of preparative interest, but also gives additional information on the very mechanism of this reaction [51]. If a biradical were formed as an intermediate, it would be impossible to carry out the reaction quantitatively without involving the free valence of the nitroxyl radical.

The result of this condensation therefore provides further proof of the fact that this reaction does not occur by a radical mechanism [52].

However, 2,2,6,6-tetramethyl-4-oxopiperidine-1-oxyl does not react so smoothly with all CH-acid compounds. With some of them, the reaction does not take place at all or is complicated by side reactions involving the free valence.

Attempts to effect the condensation of the ketone radical with malonic ester proved unsuccessful. All the most effective catalytic systems used for the performance of reactions of this type—for example, ammonium acetate, β-alanine, piperidine, and pyridine [53]—were ineffective in this case.

Besides 2,2,6,6-tetramethyl-4-oxopiperidine-1-oxyl, another characteristic example of a free radical with a dual reactivity is the free 4-hydroxy-2,2,6,6-tetramethylpiperidine-1-oxyl, obtained in the following way [54]:

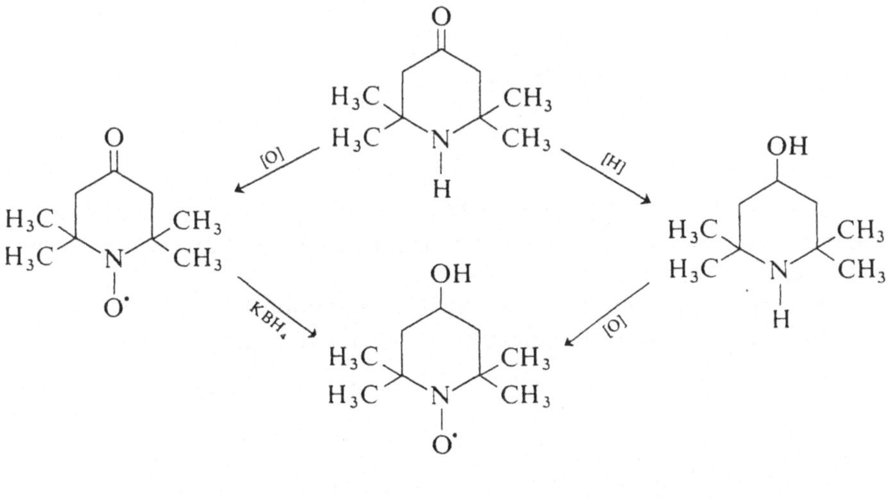

The presence of a hydroxy group in this compound has only a slight effect on its stability, since the oxidation potential of the hydroxy group here is not sufficiently high for the nitroxyl group to be reduced at the expense of the hydroxy group at ordinary temperatures. The inductive influence of the two groups on one another is also very small, and therefore both the hydroxyl and the nitroxyl groups should take part in reactions practically independently of one another.

As is well known, hydroxyl-containing compounds are capable of reacting with electrophilic reagents, and therefore it was of interest to study the reaction of 4-hydroxy-2,2,6,6-tetramethylpiperidine-1-oxyl with carbonyl compounds.

The first reactions of the new radical without the involvement of the free valence were its condensation with certain acid chlorides and phenyl isocyanate, which gave paramagnetic esters of 4-hydroxy-2,2,6,6-tetramethylpiperidine-1-oxyl [55]:

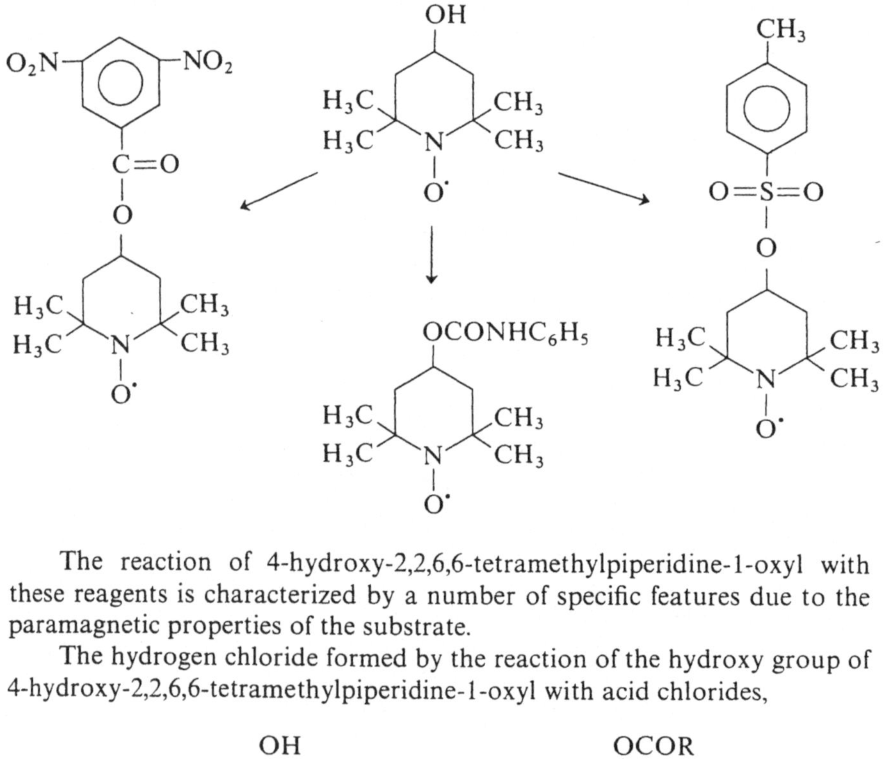

The reaction of 4-hydroxy-2,2,6,6-tetramethylpiperidine-1-oxyl with these reagents is characterized by a number of specific features due to the paramagnetic properties of the substrate.

The hydrogen chloride formed by the reaction of the hydroxy group of 4-hydroxy-2,2,6,6-tetramethylpiperidine-1-oxyl with acid chlorides,

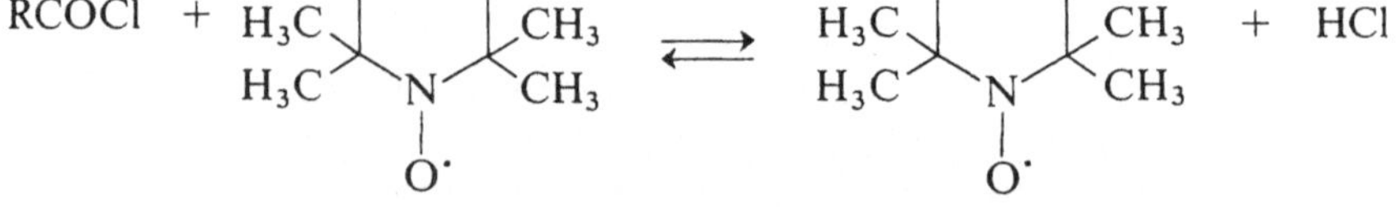

not only displaces the equilibrium in the direction of the starting materials, but also reacts vigorously with the nitroxyl group [56]:

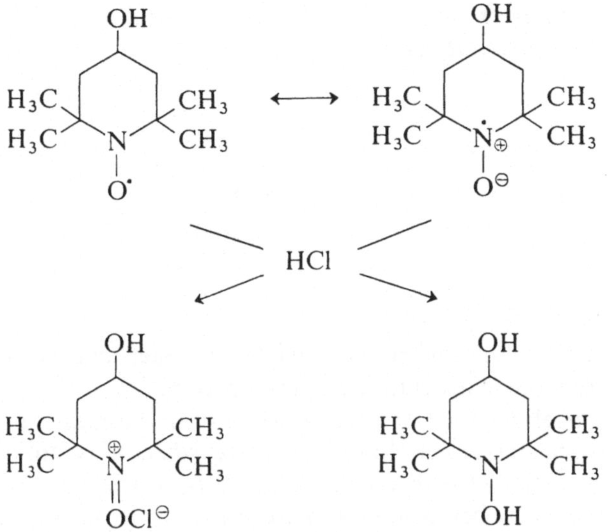

To remove the hydrogen chloride from the sphere of the reaction in the esterification of ordinary alcohols, alkali or tertiary amines are used. The use of the latter is preferable, since, through the formation of a complex, they considerably raise the reactivity of the acid chlorides. However, the presence of the nitroxyl group in the compound together with the hydroxy group imposes serious limitations to the selection of the amine. The reason for this is that some amines can react comparatively readily with nitroxyl radicals [57–61]:

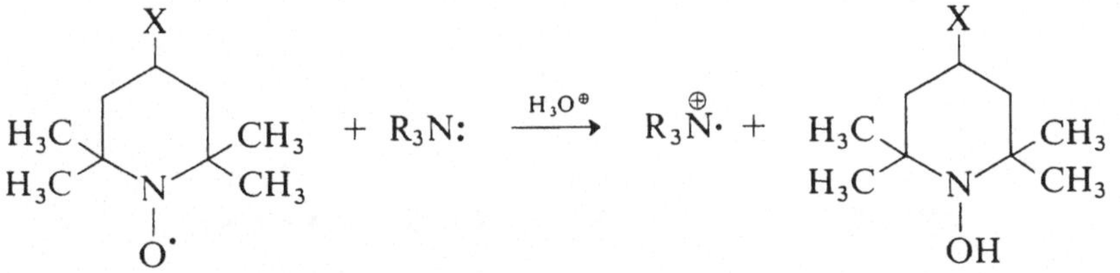

Since the transfer of an electron from the amine to the radical takes place particularly readily in an acid medium, the hydrogen chloride formed catalyzes this reaction. Only amines with the highest oxidation potentials can be used as proton acceptors. Those suitable for this purpose have proved to be pyridine and trialkylamines, which do not react with nitroxyl radicals and, being strong bases, bind the hydrogen chloride almost completely.

The resulting quaternary salts of these amines, in contrast to some alkyl-pyridinium salts [62, 63], possess low oxidation potentials, and interaction between pyridinium chloride and the nitroxyl radical in the manner shown below does not apparently take place:

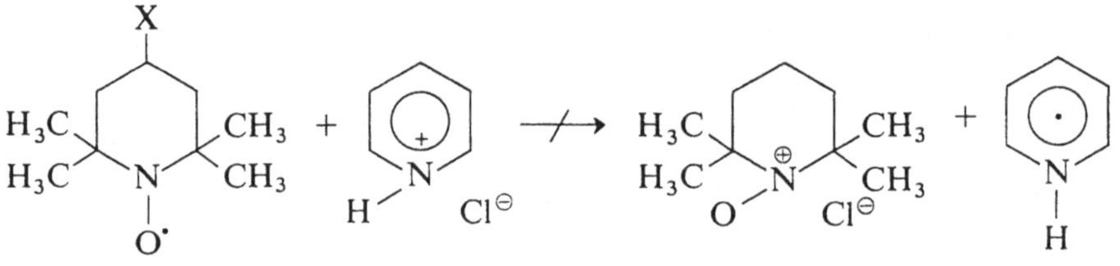

This is all the more the case if one considers the fact that when anhydrous amines are used, the hydrochloride forms a precipitate.

The careful elimination of traces of water from the amines used is necessary for other reasons as well. In the first place, when moist amines are used, the water will partially hydrolyze the amine salts, which leads to an increase in the equilibrium concentration of hydrogen chloride. In the second place, the water will react with the acid chloride, converting it into the inactive carboxylic acid.

At low temperatures, the acid chlorides themselves are either completely inert with respect to nitroxyl radicals, or react with them to a very small extent in the following way:

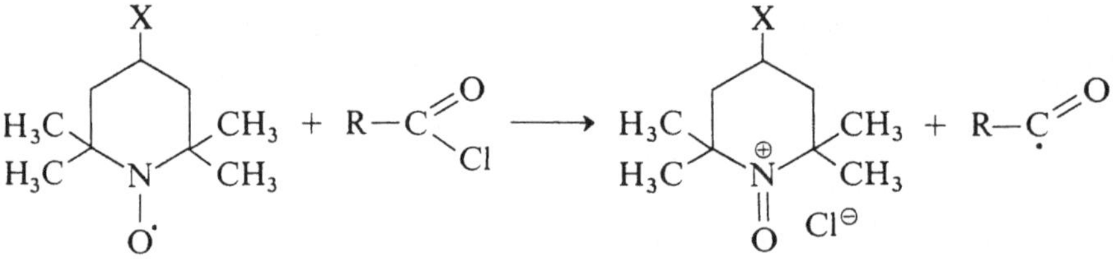

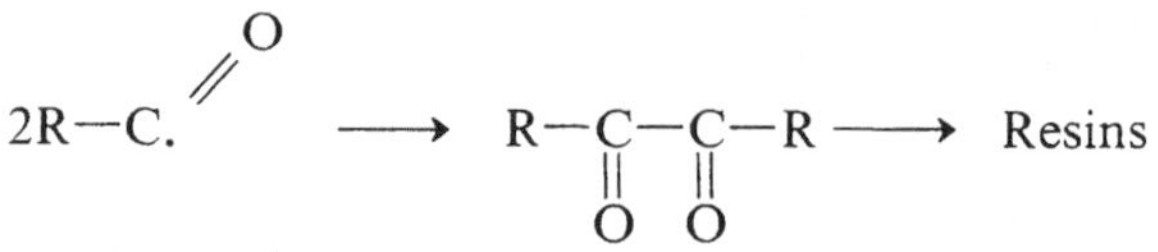

At high temperatures, these reactions apparently play a substantial part, since the process is always accompanied by pronounced resinification.

On the basis of what has been said above, it may be concluded that the optimum conditions for the esterification of 4-hydroxy-2,2,6,6-tetramethyl-piperidine-1-oxyl are a low temperature and an excess of anhydrous triethylamine or pyridine. In the case of chlorides of carboxylic acids, the most suitable proton-acceptor is pyridine.

When this base is used, the isolation of the reaction product is greatly simplified, since the initial compounds, unlike the final esters, are readily soluble in water.

The use of mild conditions considerably increases the duration of the reaction. The average time necessary for practically complete reaction is 10–12 hr, depending on the chemical activity of the chloride and the concentration of the reactants, as has been shown on the basis of the synthesis of the radicals 4-benzoyloxy-2,2,6,6-tetramethylpiperidine-1-oxyl and 4-(3,5-dinitrobenzoyloxy)-2,2,6,6-tetramethylpiperidine-1-oxyl [16]. Where necessary, the end of the reaction can be determined qualitatively by comparing the intensities of the spots on chromatograms in thin layers on alumina.

Nitroxyl radicals practically incapable of preparation by other methods can be synthesized by means of reactions of radicals not involving the free valence. For example, attempts to obtain 3-hydroxy-2,2,5,5-tetramethyl-pyrrolidine-1-oxyl by the direct catalytic oxidation of 3-hydroxy-2,2,5,5-tetramethylpyrrolidine were unsuccessful, but the reduction of the carbonyl group of the corresponding ketone-radical with aluminum isopropoxide gives the alcohol-radical in good yield [64]:

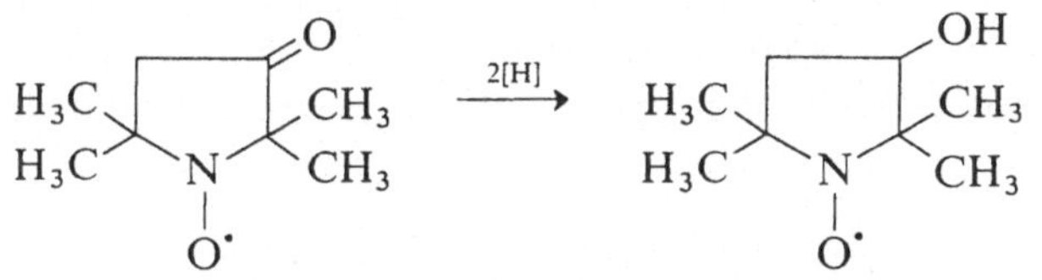

The dehydration of 3-carbamoyl-2,2,5,5-tetramethylpyrroline-1-oxyl and 3-carbamoyl-2,2,5,5-tetramethylpyrrolidine-1-oxyl with *p*-toluenesulfonyl chloride in pyridine gives the corresponding nitrile-radicals

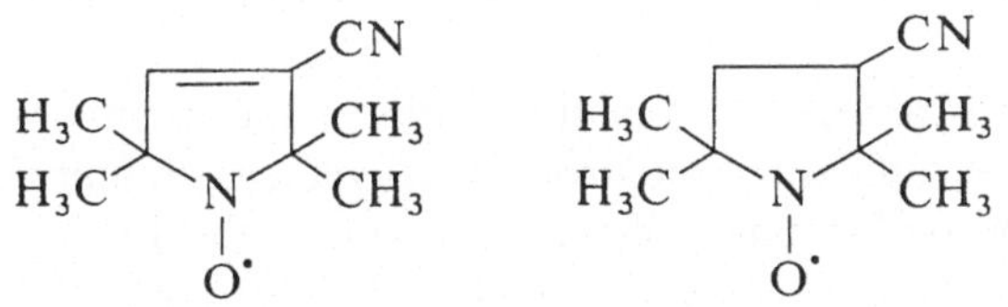

The synthesis of the free radicals given above by the oxidation of 3-cyano-2,2,5,5-tetramethylpyrroline and of 3-cyano-2,2,5,5-tetramethylpyrrolidine with hydrogen peroxide is impossible, since, under these conditions, the nitrile group undergoes hydrolysis. Thus, the oxidation of 3-cyano-2,2,5,5-tetramethylpyrroline with hydrogen peroxide in the presence of sodium tungstate gives the corresponding amide radical.

3-Oxy-2,2,5,5-tetramethylpyrroline-1-oxyl and 3-amino-2,2,5,5-tetra-methylpyrrolidine-1-oxyl have been obtained by the reactions of 3-carba-moyl-2,2,5,5-tetramethylpyrroline-1-oxyl and 3-carbamoyl-2,2,5,5-tetra-methylpyrrolidine-1-oxyl with sodium hypobromite in accordance with the following reactions:

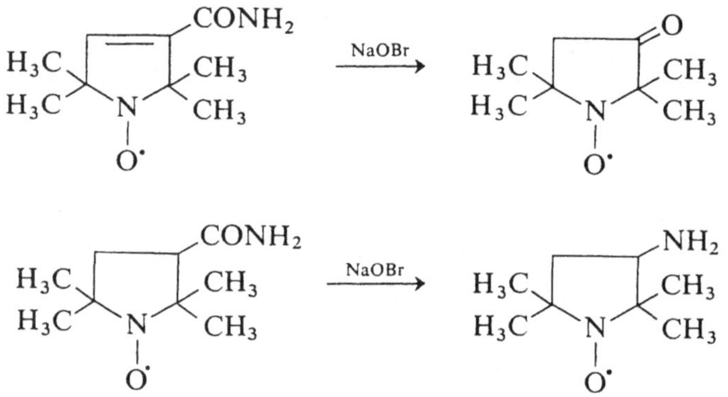

In a study of the reaction of the free organomercury radical 8-acetoxy-mercuri-2,2,4,6-tetramethyl-1,2,3,4-tetrahydroquinoline nitrogen oxide with thallium triisobutoxide, the replacement of mercury by thallium was ob-served with the formation of an organothallium free radical:

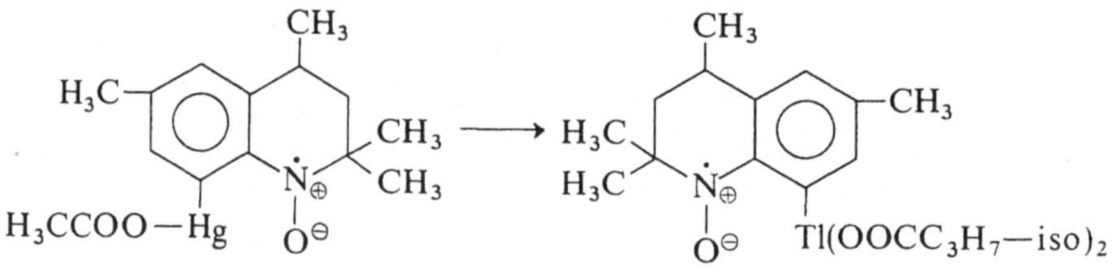

This is an example of a reaction without the involvement of the free valence of a free radical with a strongly delocalized unpaired electron.

During the performance of this reaction, in the EPR spectrum of the reaction mixture (Fig. 17) a new doublet ($a_1 = 240$ G) of triplets ($a_2 = 1.2$ G) arises with a simultaneous decrease in the intensity of the spectral lines of the mercury doublet ($a_{Hg} = 60$ G, $\mu = 0.55$, $S = \frac{1}{2}$) by a factor of approximately two as compared with the initial spectrum.

The HFS of the secondary EPR spectrum (Fig. 18) agrees well with the appearance of a new radical in which there is Fermi interaction of the unpaired electron with the ^{14}N nitrogen nucleus and with the two magnetic isotopes of thallium, ^{203}Tl ($\mu = 1.61$, $S = \frac{1}{2}$) and ^{205}Tl ($\mu = 1.62$, $S = \frac{1}{2}$) [34].

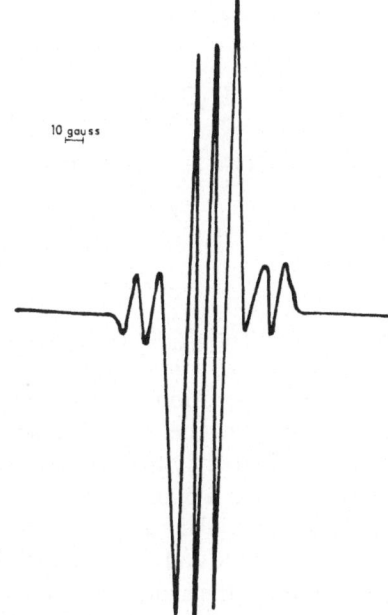

Fig. 17. HFS of the EPR spectrum of the nitroxide of 8-acetoxymercuri-2,2,4,6-tetramethyl-1,2,3,4-tetra-hydroquinoline.

The results obtained in the synthesis and study of monoradicals have been used subsequently to obtain nitroxyl biradicals possessing a series of new and interesting properties.

The reaction of 4-hydroxy-2,2,6,6-tetramethylpiperidine-1-oxyl with dicarboxylic acid chlorides takes place smoothly and is accompanied by the formation of paramagnetic esters containing two unshared electrons in a

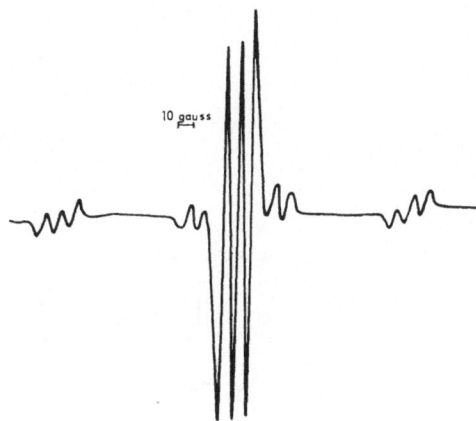

Fig. 18. Appearance in the EPR spectrum of a reaction mixture of an additional doublet of triplets due to the presence of an organothallium radical in the solution.

single molecule [28]:

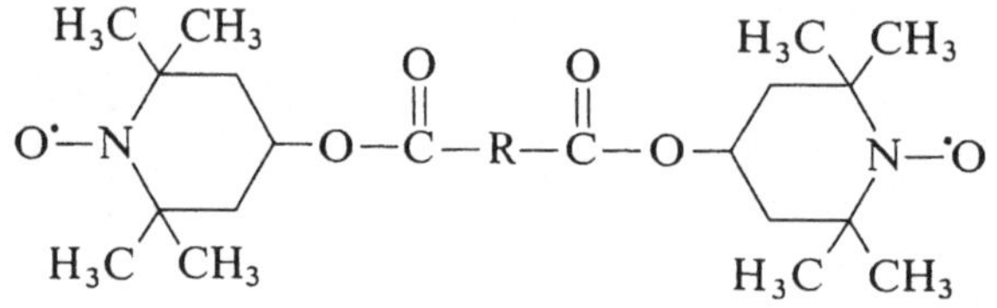

where $R = (CH_2)_2$, $(CH_2)_4$, $(CH_2)_5$, $(CH_2)_8$, $p\text{-}C_6H_4$, $o\text{-}C_6H_4$, $m\text{-}C_6H_4$, $o\text{-}C_6H_4\text{-}C_6H_4\text{-}o$, $o\text{-}C_6H_4\text{-}C_6H_4\text{-}m$, etc.

The preparation of such biradicals is more conveniently carried out in anhydrous pyridine. In this case, the slight excess of monoradical used is easily eliminated by treating the reaction product with water, and the mono-ester formed as a byproduct can be eliminated by recrystallizing the reaction product from polar solvents or by chromatography on alumina.

This simple and reliable method of obtaining nitroxyl biradicals has also been used for the synthesis of individual polyradicals [65].

By using polycarboxylic acid chlorides to "cross-link" piperidine-oxyl radicals into a single molecule, the author and his colleagues have obtained in the pure state compounds containing more than two unpaired electrons in a single molecule. The triradical [66]

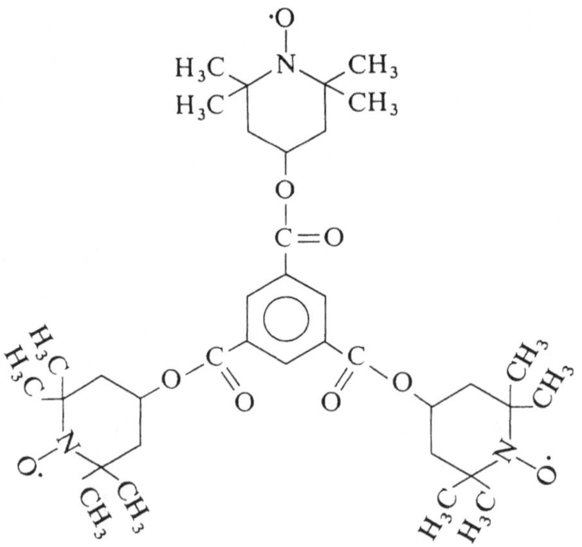

and the analogous type of tetraradical [67, 68],

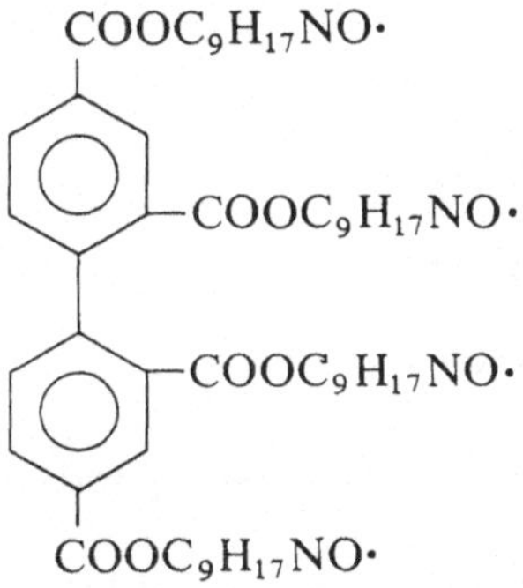

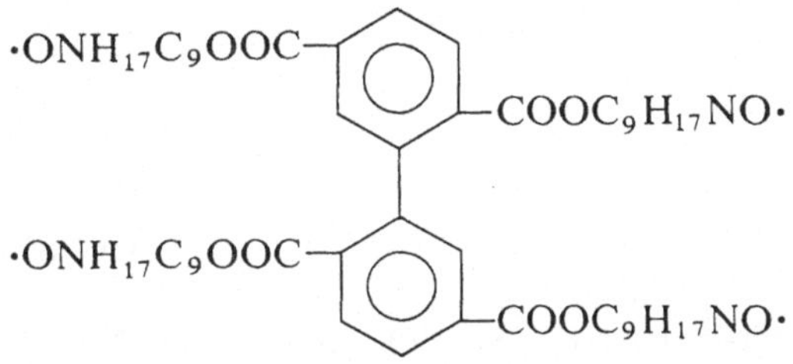

$COOC_9H_{17}NO\cdot$

where $COOC_9H_{17}NO\cdot$ represents the nitroxyl fragment, were synthesized in this way.

By using as starting materials 4-hydroxy-2,2,6,6-tetramethylpiperidine-1-oxyl and polyacrylic acid chloride, the first multivalent multiradical was obtained with, apparently, the structure [69]:

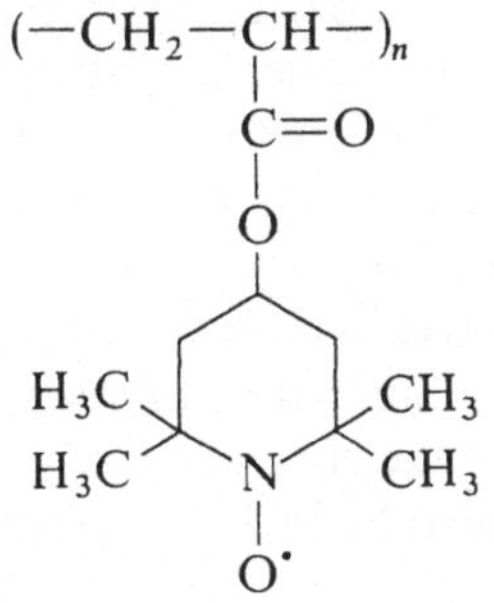

The light red transparent paramagnetic resin formed in the reaction was insoluble in the usual solvents and apparently had a cross-linked structure, which complicates its study.

Polyradicals can be obtained not only by using carboxylic acid chlorides, but also with chlorides of phosphorus and silicon. The biradicals, triradicals, and tetraradicals obtained in this way are completely stable compounds:

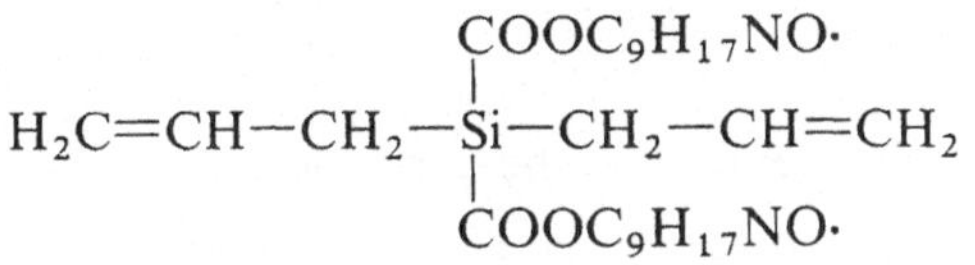

$$COOC_9H_{17}NO\cdot$$
$$H_2C{=}CH{-}CH_2{-}\underset{\underset{COOC_9H_{17}NO\cdot}{|}}{\overset{\overset{|}{}}{Si}}{-}CH_2{-}CH{=}CH_2$$

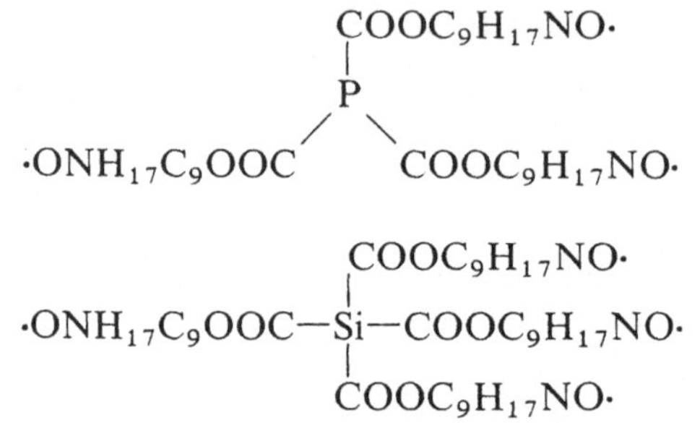

$$\cdot ONH_{17}C_9OOC \underset{\nearrow \quad \searrow}{\overset{\underset{|}{COOC_9H_{17}NO\cdot}}{P}} COOC_9H_{17}NO\cdot$$

$$\cdot ONH_{17}C_9OOC - \underset{\underset{|}{COOC_9H_{17}NO\cdot}}{\overset{\overset{|}{COOC_9H_{17}NO\cdot}}{Si}} - COOC_9H_{17}NO\cdot$$

where $COOC_9H_{17}NO\cdot$ denotes the 2,2,6,6-tetramethyl-1-oxyl-4-piperidyl-oxy group.

The reaction of sulfur dichloride with diallyldi(2,2,6,6-tetramethyl-4-piperidyloxy)silane gave a ruby-red glasslike polymer containing 2.0×10^{21} unpaired electrons per gram.* The new paramagnetic polymer dissolved in benzene, chloroform, and methylene chloride and was stable in air [70].

In the preparation of paramagnetic esters of phosphorous or orthosilicic acid, it is more convenient to use triethylamine as the hydrogen-chloride acceptor, since the use of pyridine and subsequent treatment of the reaction product with water leads to the partial hydrolysis of the ester formed.

The yield of nitroxyl polyradicals synthesized by this method depends to a considerable extent on the purity of the starting acid chlorides. Consequently, all the acid chlorides used in the syntheses were carefully purified by redistillation or recrystallization.

The chemically less active carboxylic anhydrides have also been used to obtain paramagnetic esters. Since the full esters of carboxylic acids are more conveniently obtained from the chlorides, only the synthesis of monoesters of decarboxylic acids from the anhydrides corresponding to them is of preparative interest.

In view of the fact that acid anhydrides are considerably less reactive than the chlorides, their inertness with respect to the nitroxyl group is not a matter of doubt, and the reaction gives compounds containing a free carboxyl group [71]:

where $COOC_9H_{17}NO\cdot$ represents the corresponding nitroxyl moiety.

*It is worth noting that 1 g of copper contains 3.2×10^{21} unpaired electrons.

The possibilities of modifying 4-hydroxy-2,2,6,6-tetramethylpiperidine-1-oxyl are not limited to its reaction with acid anhydrides and chlorides. It is obvious that this substance will undergo changes not affecting the unpaired electron on interaction with all compounds possessing a sufficiently high electrophilicity and a low activity with respect to oxidation–reduction reactions.

Among the substances belonging to this group, isocyanates have also been used, because they are of interest for the synthesis of biradicals and polyradicals. As has been shown previously [55], isocyanates are fairly inert with respect to the nitroxyl group even at high temperatures. The paramagnetic derivatives of carbamic acids formed as a result of the reaction are stable compounds. The same can be said about the corresponding biradicals [28]:

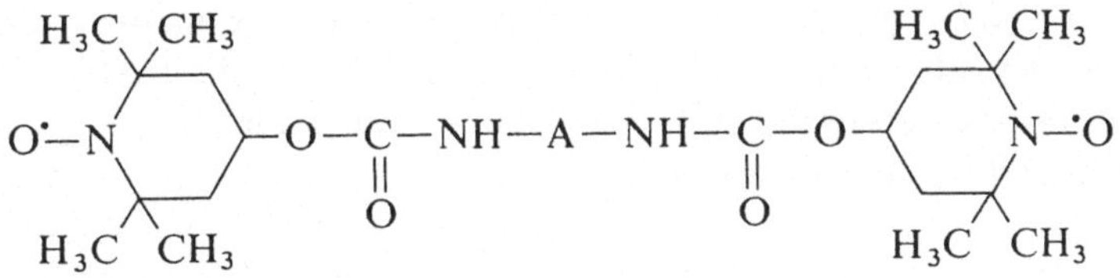

where A = $-(CH_2)_6-$ and $-C_6H_3(CH_3)-$.

A number of individual polyradicals of the type of

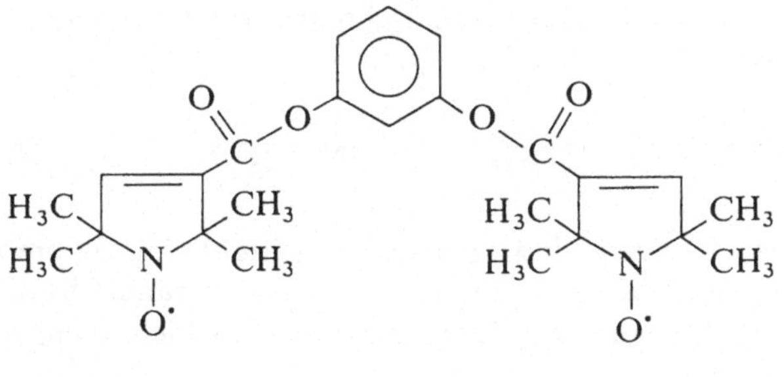

have been synthesized by the reaction of the chloride of 2,2,5,5-tetramethyl-1-oxylpyrroline-3-carboxylic acid with pyridine solutions of resorcinol, catechol, hydroquinone, and phloroglucinol [72].

3. Other Methods

Of the other methods used for the preparative production of nitroxyl radicals, we must first mention the reduction of substituted immonium oxide perchlorates

$$\underset{\underset{Ar}{|}}{\overset{\overset{Ar}{|}}{O \leftarrow N^\oplus ClO_4{}^\ominus}} \quad \xrightarrow{+e} \quad \underset{\underset{Ar}{|}}{\overset{\overset{Ar}{|}}{\cdot N^\oplus - {}^\ominus O}}$$

The initial diarylimmonium oxide perchlorates, many of which were described long ago in the chemical literature, are obtained by Meyer's method [73, 74]:

$$2Ar-H \quad \xrightarrow{HNO_3} \quad \left[\underset{\underset{Ar}{|}}{\overset{\overset{Ar}{|}}{O \leftarrow N - OH}} \right] \quad \xrightarrow{HClO_4} \quad \underset{\underset{Ar}{|}}{\overset{\overset{Ar}{|}}{O \leftarrow N^\oplus ClO_4{}^\ominus}}$$

In this way, dimethoxy-, tetramethoxy-, and hexamethoxydiphenyl-nitroxide and other radicals of this class have been synthesized.

Saturated heterocyclic and acyclic immonium oxides are products of the oxidation of nitroxyl radicals [30]

$$\underset{|}{\overset{|}{\cdot N^\oplus - O^\ominus}} \quad \xrightarrow{Cl^0} \quad \underset{|}{\overset{|}{O \leftarrow N - Cl}} \quad \rightleftarrows \quad \underset{|}{\overset{|}{O \leftarrow N^\oplus}} + Cl^\ominus$$

It has been found that the oxidation potentials of the radicals and salts of the immonium oxides differ fairly considerably, and that it is possible to choose a whole series of compounds capable of selectively reducing the latter to neutral free nitroxyls, even water and many tertiary amines being suitable for this purpose,

$$\underset{|}{\overset{|}{O \leftarrow N^\oplus Cl^\ominus}} + H_2O \quad \longrightarrow \quad \underset{|}{\overset{|}{\cdot N^\oplus - O^\ominus}} + HCl + HO\cdot$$

However, these reducing agents are unsuitable for the preparative production of the nitroxyls, since the process is accompanied by side reactions. The most suitable reducing agents are metals, to which the nitroxyl radicals are apparently inert.

Nevertheless, this method still has little preparative importance, since the immonium salts can in any case be obtained most simply from the nitroxyl radicals corresponding to them.

Stable acyclic nitroxyls of the general formula

$$R_1 - \underset{\underset{H_3C}{|}}{\overset{\overset{CH_3}{|}}{C}} - \underset{\underset{O\cdot}{|}}{N} - \underset{\underset{CH_3}{|}}{\overset{\overset{CH_3}{|}}{C}} - R_2$$

where $R_1 = CH_3$, C_2H_5, and $R_2 = CH_3$, C_2H_5, can be obtained by the action of tertiary alkylmagnesium halides on the corresponding tertiary nitroalkanes [75], since these radicals undergo practically no reaction with Grignard compounds [76]. By this method, Brunel *et al.* [77] have obtained optically active tert-butylcamphenylnitroxyl.

The equilibrium formation of stable radicals has been recorded in the reaction of Grignard compounds with nitrosobenzene [78]:

$$(C_6H_5)_2NO \cdot + \cdot MgBr(C_6H_5MgBr)_n \rightleftharpoons$$

$$[(C_6H_5)_2N-O-MgBr](C_6H_5MgBr)_n$$

The formation of a mixture of stable and unstable radicals in the reduction of nitrosobenzene or nitrobenzene with lithium aluminum hydride has been shown by the EPR method [79].

An elegant method of synthesizing stable radicals has recently been proposed by the Japanese workers Iwamura and Inamoto [80], who, by heating an acidic solution of azo-1,1-isobutyronitrile with tert-butylphenyl nitrone, obtained the first individual acyclic nitroxyl possessing an α-hydrogen atom:

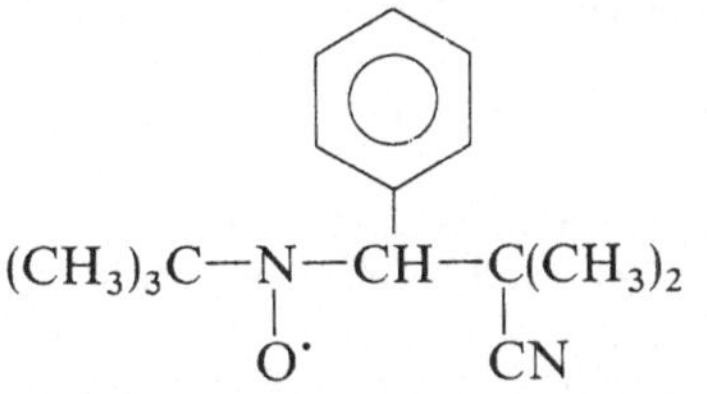

$$(CH_3)_3C-\underset{\underset{O \cdot}{|}}{N}-CH-\underset{\underset{CN}{|}}{C}(CH_3)_2$$

It was previously impossible to isolate compounds of this class in the pure state, because of their ready disproportionation. The only exceptions are bicyclic nitroxyls [81, 82] in which disproportionation is forbidden by Bredt's rule.

Aliphatic-aromatic nitrogen oxides can be obtained in solution by the condensation of aromatic nitroso compounds with branched olefins, the mechanism of the formation of radicals apparently including the stage of a pseudo-Diels–Alder addition [83].

References

1. P. Pascal, Nouveau Traité de Chimie Minérale, Masson et Cie, Paris (1959).
2. F. Chauveau, *Bull. Soc. Chim. France* **1960**, 819.
3. E. Richardson, *J. Less Common Metals* **2**, 360 (1960).
4. G. A. Bogdanov, *Zh. Fiz. Khim.* **25**, 49 (1951).
5. M. Rogers, *Chem. Ind.* (*London*) **1953**, 1033.

6. N. I. Kobozev, *Uch. Zap. Mosk. Gos. Univ.* **174**, 125 (1955).
7. L. Horner and H. Junkermann, *Ann. Chem.* **591**, 53 (1955).
8. German Patent 951,933 (1956).
9. A. A. Nikolaev, *Zh. Fiz. Khim.* **31**, 1185 (1957).
10. A. Hoffman, A. Feldman, E. Gelblum, and W. Hodgson, *J. Am. Chem. Soc.* **86**, 639 (1964).
11. O. L. Lebedev and S. N. Kazarnovskii, *Tr. po Khim. i Khim. Tekhnol.* **1959**, 649.
12. É. G. Rozantsev and R. S. Burmistrova, *Dokl. Akad. Nauk SSSR* **166**, 135 (1966).
13. É. G. Rozantzev and L. A. Krinitzkaya, *Tetrahedron* **21**, 491 (1965).
14. L. A. Krinitskaya, É. G. Rozantsev, and M. B. Neiman, *Izv. Akad. Nauk SSSR, Ser. Khim.* **1965**, 115.
15. É. G. Rozantsev, L. A. Krinitskaya, and M. B. Neiman, *Izv. Akad. Nauk SSSR, Ser. Khim.* **1966**, 675.
16. É. G. Rozantsev, V. A. Golubev, and M. B. Neiman, *Izv. Akad. Nauk SSSR, Ser. Khim.* **1965**, 391.
17. V. A. Golubev and É. G. Rozantsev, *Izv. Akad. Nauk SSSR, Ser. Khim.* **1965**, 716.
18. É. G. Rozantsev and R. A. Papko, *Izv. Akad. Nauk SSSR, Ser. Khim.* **1962**, 2254.
19. É. G. Rozantsev and A. B. Shapiro, *Izv. Akad. Nauk SSSR, Ser. Khim.* **1964**, 1123.
20. É. G. Rozantsev, A. B. Shapiro, and N. N. Komzolova, *Izv. Akad. Nauk SSSR, Ser. Khim.* **1965**, 1100.
21. L. L. Yazina, A. B. Shapiro, and É. G. Rozantsev, *Plast. Massy* **1966**(6), 37.
22. A. B. Shapiro, Thesis, Moscow (1966).
23. R. Briere, H. Lemaire, and A. Rassat, *Bull. Soc. Chim. France* **1965**, 3273.
24. G. Chapelet-Letourneux, H. Lemaire, and A. Rassat, *Bull. Soc. Chim. France* **1965**, 3283.
25. É. G. Rozantsev, V. A. Golubev, and M. B. Neiman, *Izv. Akad. Nauk SSSR, Ser. Khim.* **1965**, 391.
26. É. G. Rozantsev, V. A. Golubev, and M. B. Neiman, *Izv. Akad. Nauk SSSR, Ser. Khim.* **1965**, 393.
27. É. G. Rozantsev, *Teor. i Éksperim. Khim.* **2**, 415 (1966).
28. É. G. Rozantsev, V. A. Golubev, M. B. Neiman, and Yu. V. Kokhanov, *Izv. Akad. Nauk SSSR, Ser. Khim.* **1965**, 572.
29. É. G. Rozantsev and V. A. Golubev, *Izv. Akad. Nauk SSSR, Ser. Khim.* **1966**, 891.
30. V. A. Golubev, É. G. Rozantsev, and M. B. Neiman, *Izv. Akad. Nauk SSSR, Ser. Khim.* **1965**, 1927.
31. C. Harries and F. Lehman, *Ber.* **30**, 230 (1897).
32. C. Harries and F. Lehman, *Ber.* **30**, 2726 (1897).
33. A. B. Shapiro, A. L. Buchachenko, A. A. Medzhidov, and É. G. Rozantsev, *Zh. Strukt. Khim.* **7**, 187 (1966).
34. A. B. Shapiro and É. G. Rozantsev, *Izv. Akad. Nauk SSSR, Ser. Khim.* **1966**, 1650.
35. A. B. Shapiro, A. A. Medzhidov, and É. G. Rozantsev, *Zh. Org. Khim.* **2**, 1873 (1966).
36. L. A. Kalashnikova, M. B. Neiman, É. G. Rozantsev, and L. A. Skripko, *Zh. Org. Khim.* **2**, 1529 (1966).
37. É. G. Rozantsev and Yu. V. Kokhanov, *Izv. Akad. Nauk SSSR, Ser. Khim.* **1966**, 1477.
38. É. G. Rozantsev and M. B. Neiman, *Tetrahedron* **20**, 131 (1964).
39. W. Johnson and G. Daub, Organic Reactions, John Wiley and Sons, New York, Vol. 6 (1953), p. 1.
40. C. Overberger and C. Roberts, *J. Am. Chem. Soc.* **71**, 3618 (1949).
41. E. Müller, Neuere Anschauungen der organischen Chemie, Springer Verlag, Berlin, 2nd ed. (1956).
42. H. Becker, Einführung in die Elektronentheorie Organisch-chemischer Reaktionen, Berlin (1964).

43. M. Newman and B. Magerlein, Organic Reactions, John Wiley and Sons, New York, Vol. 5 (1951), p. 413.
44. V. A. Golubev, M. B. Neiman, and É. G. Rozantsev, *Izv. Akad. Nauk SSSR, Ser. Khim.* **1966**, 343.
45. E. Erlenmeyer, *Ber.* **33**, 3001 (1901).
46. W. Johnson, J. Belew, L. Chinn, and R. Hunt, *J. Am. Chem. Soc.* **75**, 4995 (1953).
47. L. Bellamy, Infrared Spectra of Complex Molecules, Methuen, London, 1st ed. (1954).
48. R. Briere, H. Lemaire, and A. Rassat, *Tetrahedron Letters* **1964**, 1775.
49. T. Campbell, S. Linden, S. Godhalk, and W. Young, *J. Am. Chem. Soc.* **69**, 880 (1947).
50. J. Heilbron, A. Johnson, E. Jones, and A. Spinks, *J. Chem. Soc.* **1942**, 727.
51. I. L. Knunyants, N. P. Gambaryan, and E. M. Rokhlin, *Usp. Khim.* **27**, 1361 (1958).
52. M. Ballester and D. Perez-Blanco, *J. Org. Chem.* **23**, 652 (1958).
53. A. Surrey, Name Reactions in Organic Chemistry, Academic Press, New York (1961).
54. É. G. Rozantsev, *Izv. Akad. Nauk SSSR, Ser. Khim.* **1966**, 770.
55. É. G. Rozantsev, *Izv. Akad. Nauk SSSR, Ser. Khim.* **1963**, 1669.
56. É. G. Rozantsev, G. F. Pavelko, A. L. Buchachenko, and M. B. Neiman, *Izv. Akad. Nauk SSSR, Ser. Khim.* **1967**, 2306.
57. A. A. Medzhidov, É. G. Rozantsev, and M. B. Neiman, *Dokl. Akad. Nauk SSSR* **168**, 348 (1966).
58. É. G. Rozantsev, A. A. Medzhidov, M. N. Neiman, and L. A. Skripko, Omagiu Raluca Ripan [Homage to Raluca Ripan], Bucharest (1966), p. 503.
59. V. V. Gur'yanova, B. M. Kovarskaya, L. A. Krinitskaya, M. B. Neiman, and É. G. Rozantsev, *Vysokomolekul. Soedin.* **7**, 1515 (1965).
60. M. B. Neiman, A. A. Medzhidov, É. G. Rozantsev, and L. A. Skripko, *Dokl. Akad. Nauk SSSR* **154**, 387 (1964).
61. E. N. Aleksandrova, S. N. Zelenin, P. A. Kaikaris, D. D. Mozzhukin, and M. L. Khidekel', *Dokl. Akad. Nauk SSSR* **167**, 1291 (1966).
62. E. Kosower and E. Poziomek, *J. Am. Chem. Soc.* **85**, 2035 (1963).
63. E. Kosower and I. Coffer, *J. Am. Chem. Soc.* **86**, 5524 (1964).
64. É. G. Rozantsev and L. A. Krinitskaya, *Izv. Akad. Nauk SSSR, Ser. Khim.* **1964**, 1573.
65. É. G. Rozantsev, V. A. Golubev, and M. B. Neiman, USSR Authors' Certificate No. 175504, October 24, 1964.
66. M. B. Neiman, É. G. Rozantsev, and V. A. Golubev, *Izv. Akad. Nauk SSSR, Ser. Khim.* **1965**, 548.
67. É. G. Rozantsev and V. A. Golubev, *Izv. Akad. Nauk SSSR, Ser. Khim.* **1965**, 718.
68. É. G. Rozantsev, V. A. Golubev, V. N. Suskina, and M. B. Neiman, *Zh. Organ. Khim.* **2**, 1075 (1966).
69. V. A. Golubev, Thesis, Moscow (1967).
70. G. F. Pavelko, Thesis, Moscow (1968).
71. V. N. Suskina, Diploma Thesis, Inst. Khim. Fiz., Akad. Nauk SSSR, Moscow (1965).
72. L. A. Krinitskaya, A. L. Buchachenko, and É. G. Rozantsev, *Zh. Organ. Khim.* **2**, 1301 (1966).
73. K. Meyer and H. Gottlieb-Billroth, *Ber.* **52**, 1476 (1919).
74. K. Meyer and W. Reppe, *Ber.* **54**, 327 (1921).
75. French Patent 1,376,740 (1964); *Chem. Abstr.* **62**, 9010 (1965).
76. É. G. Rozantsev, M. B. Neiman, and Yu. G. Mamedova, *Izv. Akad. Nauk SSSR, Ser. Khim.* **1963**, 1509.
77. Y. Brunel, H. Lemaire, and A. Rassat, *Bull. Soc. Chim. France* **1964**, 1895.
78. K. Maruyama, *Bull. Chem. Soc. Japan* **37**, 1013 (1964).
79. H. Lamaire, A. Rassat, and J.-P. Ravet, *Tetrahedron Letters* **1964**, 3507.

80. I. Iwamura and N. Inamoto, *Bull. Chem. Soc. Japan* **40**, 703 (1967).
81. É. G. Rozantsev and V. N. Suskina, *Izv. Akad. Nauk SSSR, Ser. Khim.* **1968**, 2106, 2148.
82. R. Dupeyre and A. Rassat, *J. Am. Chem. Soc.* **88**, 3180 (1968).
83. A. Sullivan, *J. Org. Chem.* **31**, 2811 (1966).

Chapter IV

REACTIONS OF NITROXYL RADICALS WITH SATURATION OF THE FREE VALENCE

In the preceding chapter, we discussed general features of the synthesis of nitroxyl radicals and, to some extent, questions of reactivity connected with their reactions not involving the free valence. The stability of nitroxyl radicals, which enables such reactions to be carried out, is, however, extremely limited. In actual fact, these radicals possess a high chemical activity, and, under suitable conditions, they take part in all reactions characteristic of free radicals.

1. Reduction of Nitroxyl Radicals

The presence of unpaired electrons in the nitroxyls and nitroxides imparts a high degree of reactivity to them. A feature of the reduction of nitroxyl radicals is that in the majority of cases, it must be carried out in the presence of not only an electron donor, but also a proton donor. This apparently explains the high stability of the nitroxyl radicals with respect to alkali metals, with which they react only at high temperatures in solvents capable of transferring electrons [1].

Consequently, the reduction of nitroxyl radicals takes place smoothly only in an acid medium or when the reducing agents used are compounds possessing fairly labile hydrogen. This condition is satisfied, in the first place, by hydrogen activated by means of various catalytic systems. In this case, the nature of the products of the reduction of nitroxyl radicals will depend on the catalyst used.

When Adams platinum catalyst is used, $\frac{1}{2}$ mole of hydrogen is absorbed per mole of radical and, as an analysis of the reduction products shows, the sole hydrogenation products are substituted hydroxylamines [2–4].

The presence of several unpaired electrons in one molecule has no effect on the course of this reaction, and in the presence of a platinum catalyst, nitroxyl polyradicals are also reduced to the corresponding polyhydroxylamines [5].

In the presence of other catalysts—for example, Raney nickel or palladium on carbon—the reduction of the radicals is completed by the formation of amines. In this case, the reactions undoubtedly pass through the stage of the corresponding hydroxylamines, since the latter are readily reduced to amines in the presence of these catalysts.

The hydrogenation of aromatic nitroxides on Raney nickel is used for the synthesis of difficulty accessible diphenylamine derivatives [6]:

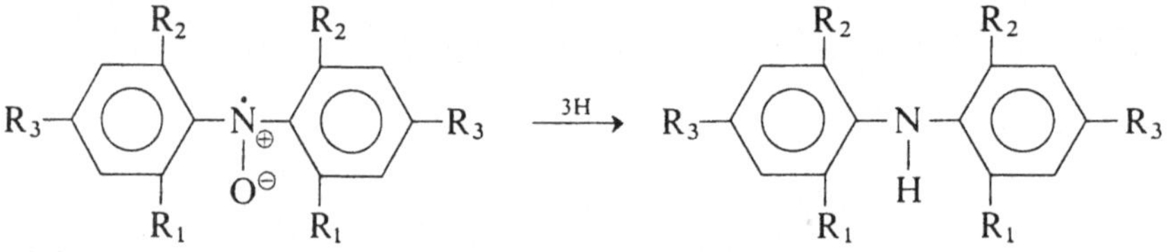

A vigorous reducing agent for nitroxyl radicals is lithium aluminum hydride; with this reagent, the reaction takes place under mild conditions and leads to the formation of the corresponding hydroxylamines [7].

Important from the point of view of the quantitative determination of radicals is the reaction of nitroxides and nitroxyls with potassium iodide in an acid medium. The iodine formed as a result of the reaction does not react with the nitroxyl group, so that the iodometric titration of radicals is very accurate [8].

Organic compounds possessing low oxidation potentials may be used as reducing agents—for example, hydrazine and its derivatives.

The reaction of nitroxyl radicals with an excess of hydrazine forms the corresponding organic analogs of hydroxylamine in quantitative yield and liberates nitrogen, which prevents contact of the hydroxylamines with atmospheric oxygen:

$$4 \ \ce{>N-\dot{O}} + NH_2-NH_2 \ \longrightarrow \ \ce{>N-OH} + N_2$$

The absence of side reactions is a great advantage of this method of reduction, but the hydrogen used as reducing agent is active only at high temperatures.

The action of phenylhydrazine on free nitroxyl radicals also forms the corresponding hydroxylamines. The reaction takes place vigorously even in the cold and is accompanied by the evolution of nitrogen: measurement of the volume of nitrogen evolved has been used for the volumetric determination of nitroxyl radicals [4]:

$$2 \ \ce{>N-\dot{O}} + C_6H_5NHNH_2 \ \longrightarrow \ 2 \ \ce{>N-OH} + N_2 + C_6H_6$$

The reaction of nitroxyl radicals with phenylhydrazine is accompanied by a number of side reactions, apparently explained by the reactions of the primary radical formed in the oxidation of the phenylhydrazine:

$$C_6H_5-NH-NH_2 \longrightarrow C_6H_5-\dot{N}-NH_2$$

$$C_6H_5-\dot{N}-NH_2 \longrightarrow C_6H_5\cdot + N_2 + 2H$$

The main byproducts of such reactions are phenyl ethers of the hydroxylamines, which evidently arise by the recombination of nitroxyl and phenyl radicals [4]:

$$\underset{/}{\overset{\backslash}{}}N-\dot{O} + C_6H_5\cdot \longrightarrow \underset{/}{\overset{\backslash}{}}N-O-C_6H_5$$

By using various methods for reducing nitroxyl radicals and nitroxides, it has been possible to synthesize a large number of hindered hydroxylamine derivatives of hydrogenated pyrroles, pyridines, quinolines, and carbolines, the preparation of which by other methods has been practically impossible until now.

Sterically hindered hydroxylamines are comparatively labile compounds and are gradually oxidized in air to the initial radicals. Their stability with respect to atmospheric oxygen depends to a considerable extent on the nature of the substituent in position 3 or 4 of the heterocycles. In contrast to the rapidly oxidizing hydroxylamines of the type

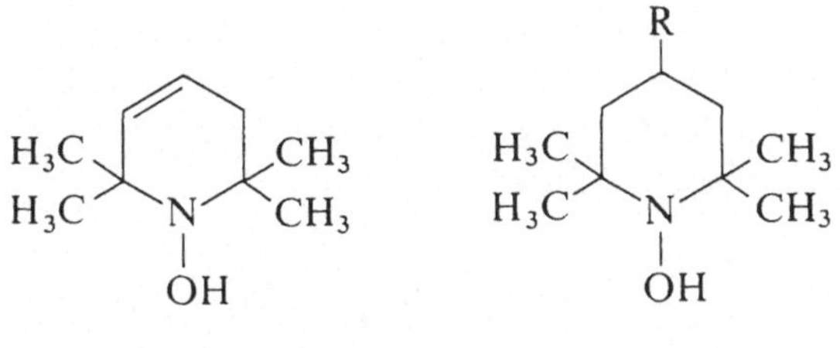

(where R = H, Cl, Br), compounds containing an oxygen atom in the 3 or 4 position of the heterocycle are stable in the crystalline state.

The IR spectra of these compounds show the presence of strong hydrogen bonds of the intra- and intermolecular types. For example, the spectrum of triacetonehydroxylamine has a broad band in the 3320–3390-cm^{-1} region, showing the presence of a polymeric association of the molecules. Dilution of the sample with carbon tetrachloride leads to the rupture of the hydrogen bonds, which is shown in a decreased intensity of the absorption bands relating to the stretching vibrations of bound hydroxy groups and a corresponding increase in the intensity of the absorption of free OH groups. Consequently, the stability of this type of compound to the action of oxygen is apparently due to the capacity of the molecules for forming hydrogen bonds. This can also explain the fact that 1-hydroxy-2,2,6,6-tetramethylpiperidines undergo oxidation far more rapidly in solution than in the crystalline state.

The steric hindrance that partly explains the high stability of the nitroxyls is reflected to a smaller extent in the reactivity of the hydroxylamines.

With carboxylic acid chlorides, arylsulfonyl chlorides, and phenyl isocyanate, they form crystalline derivatives which have been used to establish the structure and to identify the reduction products of the nitroxyl radical.

In contrast to the hydroxylamines themselves, their hydrogen halide salts are completely inert to atmospheric oxygen. This is not unexpected, since the oxidation of the salts would lead to the formation of an unstable protonated form of the nitroxyl radicals. This apparently also explains the high stability of the carboxyl-containing hydroxylamines, which can be stabilized by the formation of intramolecular salts:

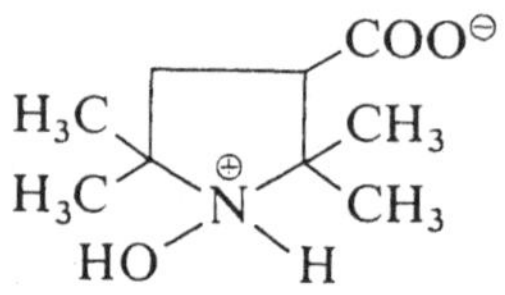

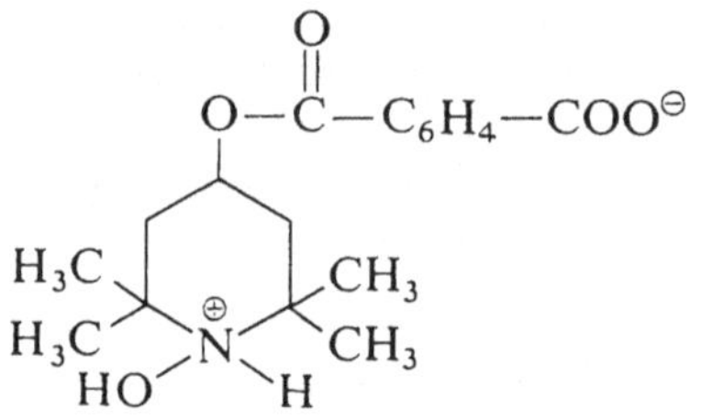

The phenyl ethers of hydroxylamines formed in the reaction of nitroxyl radicals with phenylhydrazine, like the hydroxylamine salts, are stable to atmospheric oxygen. In the presence of mineral acids, they readily hydrolyze to phenol and the corresponding hydroxylamine:

This reaction apparently explains the small amount of radical in samples of phenyl esters stored in air for a long time.

Depending on the nature of the substituent in position 4 of the heterocycle, the phenyl ethers may or may not exist in isomeric forms. Phenyl ethers containing substituents connected with the heterocycle by exocyclic double bonds do not form isomers. Ethers having substituents attached to a piperidine ring by an ordinary bond can be separated into two isomers differing in their physical properties.

The possibility of the hindered conformerism of phenyl ethers of such hydroxylamine derivatives is apparently due to the circumstance that the steric hindrance created by the four geminal methyl groups increases the energy barrier of the inversion transitions.

The unambiguous assignment of the isomers obtained to one spatial configuration or another is difficult; however, on the basis of the empirical Auwers–Skita rule [9, 10], it may be assumed that the isomers with the higher melting point and lower solubility belong to the *trans* forms.

Besides the phenyl ethers, 2-cyanoisopropyl ethers of the heterocyclic analogs of hydroxylamine are known:

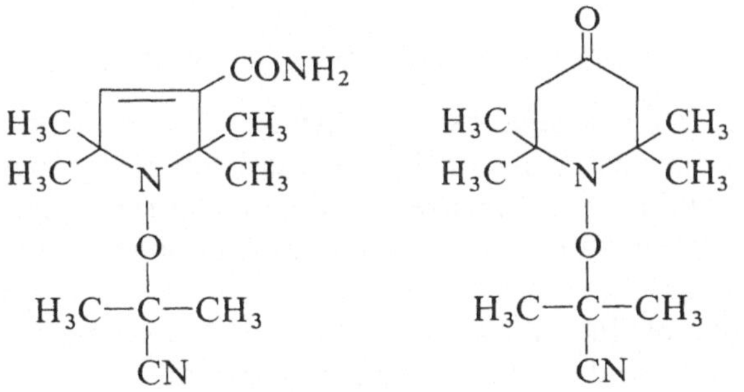

The latter are formed in high yields by heating the corresponding nitroxyl radicals with a toluene solution of azoisobutyronitrile in an atmosphere of argon; in the presence of oxygen, the reaction of the free nitroxyls with the 2-cyanopropyl radical does not take place [11].

Another powerful reducing agent for nitroxyl radicals is hydrazobenzene. The kinetics of this process have been used to explain some features of the reduction of nitroxyl radicals by hydrazobenzene [12]. The reduction was carried out in ethanol, and the course of the process was followed by means of the change in the intensity of the EPR signal of the reaction mixture and by the electronic spectra of the azobenzene formed in the process, which has a sharp absorption maximum at 315 nm. The reaction was carried out in thin tubes placed in the resonator of the radiospectrometer, which enabled it to be followed continuously. The experiments were performed at room temperature and the initial concentration of radicals was made much smaller than the concentration of hydrazobenzenes, so that the latter concentration underwent practically no change during the reaction.

The application of the Arrhenius equation to the results obtained gave the activation energy of the reduction reaction. Assuming the concentration of hydrazobenzene to be constant and equal to $1.08 \times 10^{19} \, \text{cm}^{-3}$ and knowing the rate constants at various temperatures, it is possible to calculate the true preexponential factors and to evaluate the steric factors.

The reaction of free nitroxyl with hydrazobenzene can be written in the form of the following scheme:

$$R_2NO\cdot + C_6H_5-NH-NH-C_6H_5 \xrightarrow{K_1}$$

$$C_6H_5-\dot{N}-NH-C_6H_5 + R_2NOH$$

$$R_2NO\cdot + C_6H_5-\dot{N}-NH-C_6H_5 \xrightarrow{K_2}$$

$$C_6H_5-N=N-C_6H_5 + R_2NOH$$

If it is assumed that $K_2 \gg K_1$, the reaction is first order with respect to the radical and the rate of disappearance of the free nitroxyl is described by the expression:

$$d(R_2NO\cdot)/dt = 2K_1[R_2NO\cdot][C_6H_5NHNHC_6H_5]$$

According to this scheme, two nitroxyls are consumed for each molecule of azobenzene formed. This conclusion was confirmed experimentally by comparing the amounts of azobenzene formed and nitroxyl radical consumed.

The fact that the same activation energy is obtained in the reduction of various nitroxyls is apparently a consequence of the strong solvation of the radicals in ethanol. In this case, the different transannular effects of the substituents are leveled out to a considerable extent.

In order to explain the mechanism of the reduction and the influence of the polar properties of the medium on this reaction, a number of experiments were carried out on the reduction of free nitroxyls at a dropping mercury electrode [13].

From an analysis of the results obtained, it becomes clear that free nitroxyl radicals can be used in chemistry as mild one-electron oxidizing agents. In actual fact, the reaction of free nitroxyls with substituted p-phenylenediamines in an acid medium leads to the formation of stable Wurster's salts.

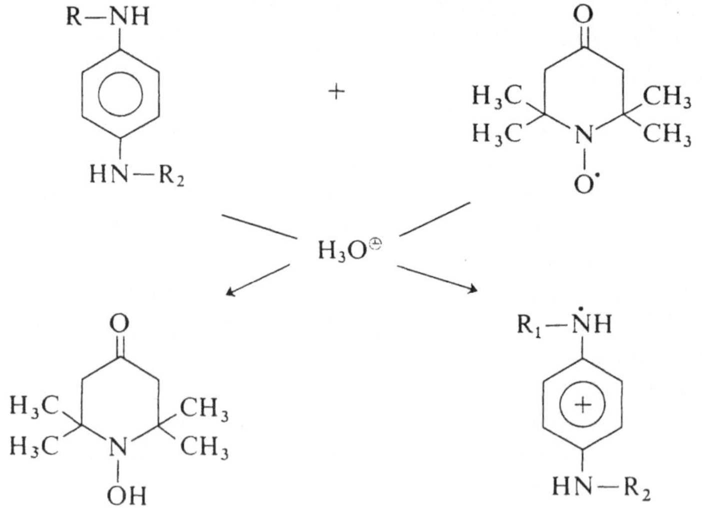

Cation-radicals previously described in the literature, the so-called Wurster's red and blue, have been obtained by this method [14, 15]:

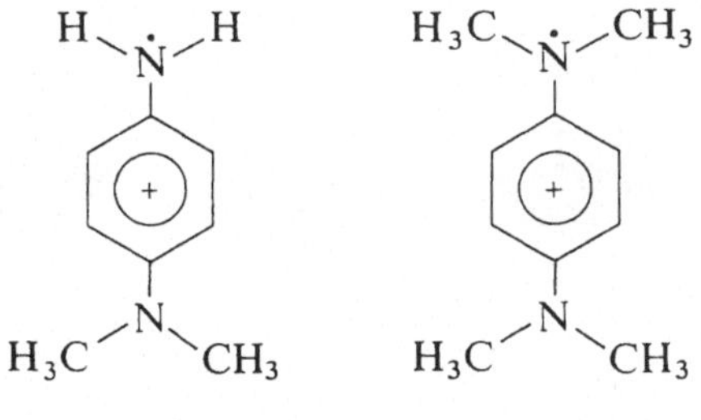

It is noteworthy that according to Michaelis *et al.* [16], stable Wurster's salts can be formed only when the molecules of the initial *p*-phenylene-diamines are highly symmetrical.

Nevertheless, all 20 of the cation-radicals obtained by the new method proved to be fairly stable in acid alcoholic solutions. The most stable solutions of cation-radicals are obtained on using a sesquimolar excess of the *p*-phenylenediamine derivative over the amount of stable radical taken. When equivalent amounts of diamine and radical are used, the cation-radical formed is an intermediate and is destroyed through more far-reaching oxidation.

The recording and analysis of the cation-radicals obtained was carried out by the radiospectroscopic method, and it was found that with a decrease in the pH of the medium, the rate of the reduction of the nitroxyls rises, the process practically ceasing in neutral and alkaline media.

The process of "growth" of the cation-radical in the reaction of 2,2,6,6-tetramethyl-4-oxopiperidine-1-oxyl with N,N'-di-*n*-octyl-*p*-phenylene-diamine is shown in Fig. 19, from which it can be seen that the triplet spectrum of the nitroxyl is replaced by the multiplet spectrum corresponding to the cation-radical in the course of a few minutes.

2. Oxidation of Nitroxyl Radicals

Nitroxyl radicals can not only add electrons, thereby being reduced to the corresponding hydroxylamines or amines, but may also release them.

The possibility of oxidation reactions of nitroxyl radicals is apparently explained by the fact that the removal of the unpaired electron leads to a considerable strengthening of the nitrogen–oxygen bond [17]. There is no doubt that this reaction will take place with all oxidizing agents possessing a sufficiently high oxidation potential.

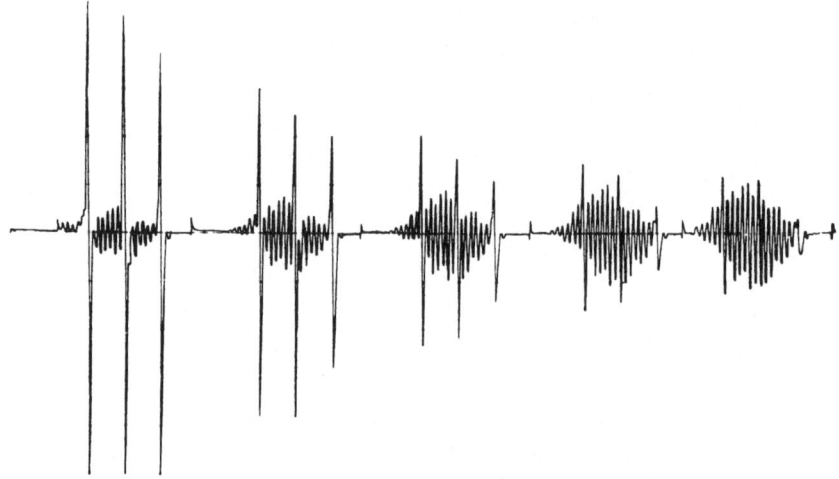

Fig. 19. Appearance on the cation-radical of N,N'-di-n-octyl-p-phenylenediamine and the disappearance of the free nitroxyl.

It is possible to avoid undesirable reactions of the immonium salts formed by oxidizing the radicals in an inert solvent, using bromine or chlorine as oxidizing agent.

The reaction of the radicals with chlorine or bromine gives quantitative yields of brightly colored diamagnetic immonium oxide salts:

$$
\begin{array}{c}
R \\
H_3C \quad \quad CH_3 \\
H_3C \quad \overset{\oplus}{N} \quad CH_3 \\
O \quad \quad X^{\ominus}
\end{array}
$$

where R = HO, OCO—C_6H_5; and X = Cl, Br.

As was to be expected, these salts are readily soluble in polar solvents and possess high decomposition temperatures, resembling in this respect their aromatic analogs, which have been described previously [18, 19].

As in the case of the nitroxyls, the iodometric titration of the salts leads to the formation of the corresponding hydroxylamines:

$$
\overset{\oplus}{N}=O \xrightarrow{\ e\ } N-\dot{O} \xrightarrow{\ e\ } N-O^{\ominus}
$$

The immonium oxide compounds are characterized by their high chemical activity which, in the majority of cases, complicates their isolation and purification.

The reaction of 4-hydroxy-2,2,6,6-tetramethylpiperidine-1-oxyl chloride with water gives a 50% yield of the initial radical and a number of other reaction products, from which it has been possible to isolate only 2,2,6,6-tetramethyl-4-oxopiperidine-1-oxyl.

The formation of nitroxyl radicals can be explained by the transfer of an electron from a molecule of water to the nitroxyl chloride, as is postulated in the case of quinones and other electron-acceptors [20]:

$$\overset{\oplus}{\diagdown \diagup}N{=}O + H_2O \longrightarrow \diagdown \diagup N{-}\dot{O} + H^{\oplus} + HO\cdot$$

or by the formation of an unstable N-oxide of the corresponding hydroxylamine which decomposes into nitroxyl and hydroxyl radicals:

$$\overset{\oplus}{\diagdown \diagup}N{=}O + H_2O \xrightarrow[-H^{\oplus}]{} \diagdown \diagup N \overset{\overset{\dot{O}}{\diagup \!\!/}}{\diagdown_{OH}} \longrightarrow \diagdown \diagup \dot{N}_{\oplus}{-}_{\ominus}O + HO\cdot$$

Since the existence of metal salts of the N-oxide of N,N-di-tert-butyl-hydroxylamine, which hydrolyze with cleavage into a nitroxyl and a hydroxyl radical, has apparently been proved [1, 21], the above scheme is fairly likely.

The formation of 2,2,6,6-tetramethyl-4-oxopiperidine-1-oxyl and other byproducts can be explained by the reaction of the hydroxyl radicals with the initial nitroxyl [22].

The vigorous reaction of the nitroxyl halides with amines also leads to the formation of nitroxyl radicals, which have been recorded by the EPR method. However, as in the reaction with water, it is impossible to isolate the radicals in quantitative yield. From the mixture formed by the reaction of a nitroxyl chloride with triethylamine, only 26% of the radical was isolated, together with the theoretically expected amount of triethylamine hydrochloride.

Apparently, the first stage of the reaction is the transfer of an electron from the anion to the cation with the formation of a radical and a cation-radical. The latter splits off weakly bound hydrogen from the substrate and is converted into triethylamine hydrohalide:

$$\overset{\oplus}{\diagdown \diagup}N{=}O + N(C_2H_5)_3 \longrightarrow \diagdown \diagup N{-}\dot{O} + (C_2H_5)_3\dot{N}^{\oplus}$$

$$^{\oplus}\dot{N}(C_2H_5)_3 + RH \longrightarrow H\overset{\oplus}{N}(C_2H_5)_3 + R\cdot$$

In the case of N,N,N′,N′-tetramethyl-*p*-phenylenediamine, the cation-radical formed (Wurster's blue) is comparatively stable and it can readily be detected from the color and the EPR spectrum.

The reactions of nitroxyl halides with primary and secondary alcohols give quantitative yields of the corresponding hydroxylamines. Simultaneously, the primary alcohols are oxidized to the corresponding aldehydes and the secondary alcohols to ketones.

The reduction can take place either by a radical mechanism or by an intramolecular cyclic mechanism analogous to the thermal decomposition of trialkylamine N-oxides [22] or of the ethers of the aci forms of nitro compounds [24]:

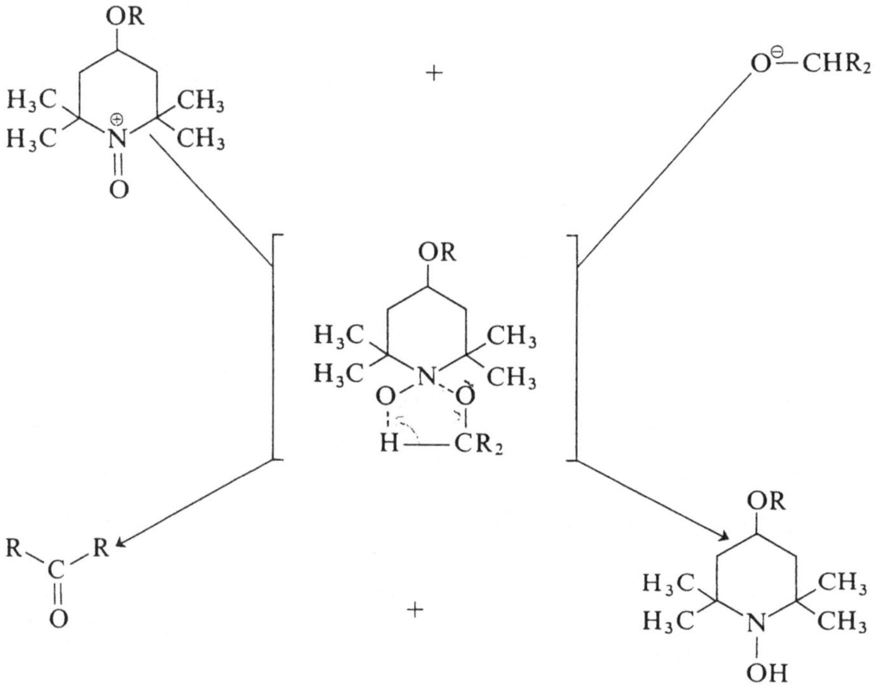

The practically quantitative reduction and the absence of any appreciable amounts of byproducts, the formation of which could have been expected in the case of the radical mechanism, are in favor of the latter scheme. However, in the case of tert-butanol, the formation of an intramolecular ring is impossible and the reaction takes place by the radical mechanism, as is shown by the rapid rise in the concentration of the nitroxyl radical [7].

The high polarity of nitroxyl radicals [2] gives grounds for assuming that they will react with the most reactive electrophilic reagents—for example, proton acids or carbocations. On the basis of this assumption, the mechanism of the interaction of nitroxyl radicals with hydrogen chloride was studied, and this enabled a number of features connected with the reactions of radicals not involving the free valence to be explained.

When dry hydrogen chloride is passed into a solution of a nitroxyl radical in carbon tetrachloride, a colored conglomerate of the nitroxyl chloride and the hydroxylamine hydrochloride is formed:

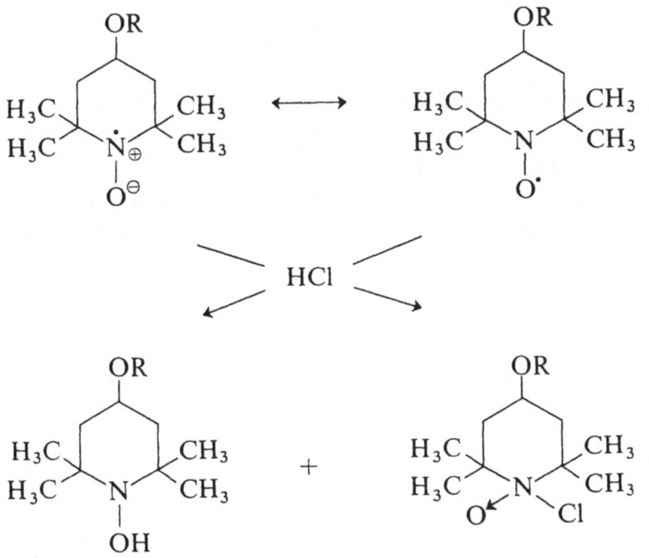

It is impossible to separate the mixture obtained into the individual components, because of the high reactivity of the nitroxyl chlorides; however, the chemical properties of the product unambiguously confirm its nature.

When the product of the reaction of 4-benzoyloxy-2,2,6,6-tetramethyl-piperidine-1-oxyl is dissolved in water, a heterocyclic hydroxylamine and the radical corresponding to it are formed in a ratio of 2:1, which shows the presence of the nitroxyl chloride in the reaction mixture.

When the same conglomerate reacts with ethanol, the hydroxylamine hydrochloride and acetaldehyde are formed in quantitative yield, which also confirms the proposed scheme. The free base obtained from the hydroxylamine hydrochloride has a different melting point from that described previously and is apparently a mixture of stereoisomers. It was possible by fractional crystallization to isolate only the isomer previously described with mp 146.5°C.

The reaction of nitroxyl radicals with hydrochloric acid depends on the concentration of the latter. With concentrated hydrochloric acid, the reaction takes place very vigorously, the solution first becoming deeply colored, which may be ascribed to the formation of the nitroxyl chloride. However, a small concentration of the radical, slowly decreasing with time, remains in solution, its existence being explained by two consecutive reactions:

$$\underset{/}{\overset{\backslash}{N}}\overset{\oplus}{=}O + H_2O \xrightarrow[-H^{\oplus}]{} \underset{/}{\overset{\backslash}{N}}\underset{OH}{\overset{\overset{O}{\diagdown\!\!\!\!/}}{}} \rightleftarrows \underset{/}{\overset{\backslash}{N}}-\dot{O} + HO\cdot$$

With a decrease in the concentration of hydrochloric acid, the initial period of the reaction can be detected by means of EPR, and in highly dilute acid (~ 0.1 N), the radical is destroyed very slowly.

The kinetic results in the initial period correspond to a second-order reaction, and in an aqueous medium, the reaction apparently takes place by a mechanism involving the rapid protonation of the radical and its slow disproportionation:

$$\underset{/}{\overset{\backslash}{\dot{N}}}_{\oplus}-_{\ominus}O \underset{-H^{\ominus}}{\overset{+H^{\oplus}}{\rightleftarrows}} \underset{/}{\overset{\backslash}{\dot{N}}}_{\oplus}-OH$$

$$\underset{/}{\overset{\backslash}{N}}\overset{\oplus}{=}O \quad + \quad \underset{/}{\overset{\backslash}{\ddot{N}}}-OH$$

When hydrogen chloride is passed into an alcoholic solution of the radical, in addition to the hydroxylamine, the nitroxyl chloride is formed, which is shown by the dark red coloration of the reaction mixture. After some time, the color disappears, and when the solvent is removed, the hydrochloride of the corresponding hydroxylamine remains. The quantitative reduction of nitroxyl radicals with hydrogen chloride becomes understandable if one takes into account the interaction of the nitroxyl chloride with the alcohol.

Hydrogen chloride reacts similarly with biradicals. The reaction product formed when hydrogen chloride is passed through a solution of di(2,2,6,6-tetramethyl-1-oxyl-4-piperidyl) phthalate in carbon tetrachloride is a mixture of the nitroxyl chloride and the corresponding hydroxylamine. When the reaction mixture was treated with ethanol, the corresponding dihydroxyl-amine was isolated, and when the mixture was treated with water, the initial biradical, dihydroxylamine, and the monoradical were obtained, as well as other products.

3. Inhibition of Radical Reactions

As is well known, all free-radical processes are hindered in the presence of small amounts of additives capable of readily reacting with free radicals. For example, benzaldehyde containing 1 part by weight of hydroquinone per 135 parts of aldehyde absorbs the same amount of oxygen in one year as it absorbs in 1 min in the absence of an inhibitor [25].

The action of ordinary inhibitors is apparently always connected with the intermediate appearance of a stable radical which, in contrast to the radical of the chain, does not react with the initial substance undergoing oxidation (greater activation energy) and disappears without the regeneration of an active center—for example [26, 27],

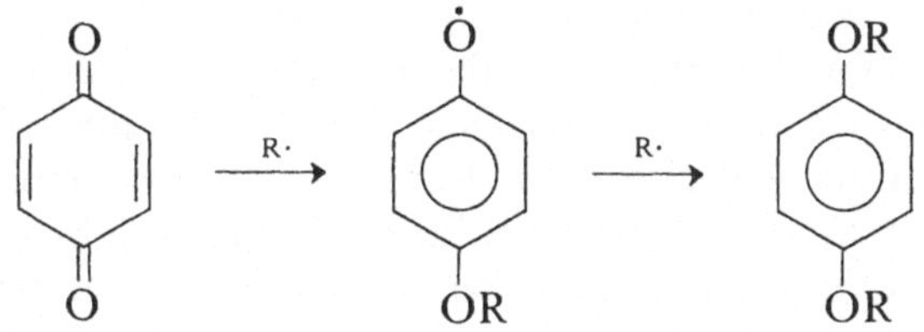

The inhibiting action of polyphenols and amines was previously explained exclusively by their capacity for reacting with free radicals, breaking the polymerization chain [28]. However, it follows from the work of Dolgoplosk and co-workers [29, 30] that the inhibiting action of polyphenols in the presence of oxygen is connected with the formation of the radicals $RO_2\cdot$, because of which, conditions are created for the oxidation of the inhibitors to stable radicals and quinones, which are also chain radical acceptors.

From this, it may be concluded that, in principle, the simplest inhibitors of chain processes should be stable free radicals capable of reacting directly with the radicals of the chain [31].

a. Oxygen. Being a biradical, oxygen readily reacts with many active radicals, and, for this reason, it frequently inhibits polymerization reactions. In such cases, the polymerization process begins only after all the oxygen present in the system has reacted [32, 33].

A classical example of the inhibition of a chain process is the action of oxygen on the photochemical reaction of hydrogen with chlorine

$$H_2 + Cl_2 \xrightarrow{h\nu} 2HCl$$

where the inhibiting action is connected with the fact that the active chain radical H, by reacting with oxygen, gives the relatively inactive radical $HO_2\cdot$, which cannot regenerate H atoms at low temperature.

A study of the kinetics of the reaction has shown [34] that in the presence of 1% of oxygen, the ratio of the rates of the propagation reactions, $H\cdot + Cl_2 \rightarrow HCl + Cl\cdot$, and the chain termination reaction, $H\cdot + O_2 + M \rightarrow HO_2\cdot + M$, at atmospheric pressure is several hundred. Since, in the absence of oxygen, the length of the chain is several hundred thousand links, in the presence of 1% of oxygen, the reaction is slowed down by a factor of approximately 1000.

b. Nitric Oxide. The inhibiting effect of nitric oxide was discovered and exhaustively studied by Hinshelwood and his pupils [35–43]. If it is assumed that the minimum rate of thermal decomposition of *n*-butane obtained in the presence of nitric oxide corresponds to the complete suppression of the chain reaction, it is possible to calculate the mean length of the chain of the uninhibited reaction by means of the relation: Length of the chain = (rate in the absence of nitric oxide)/(rate with maximum inhibition). Detailed investigations of the decomposition of nitric oxide in the inhibited thermal decomposition of *n*-butane have been carried out by Stubbs *et al.* [44].

Although various authors have reported different conversion products of nitric oxide, the majority of them tend to the opinion that the first act of the interaction of nitric oxide with a radical is an addition reaction: $R\cdot + \cdot NO \rightarrow R{-}NO$.

Thompson and Meisner [45], studying the absorption spectra of mixtures of decomposing organic compounds and nitric oxide, came to the conclusion that the overall reaction can be expressed by the equation $CH_3\cdot + \cdot NO \rightarrow CO + NH_3$, and that the intermediate products are apparently formaldoxime and unstable nitroso compounds.

Taylor and Bender [46] studied the pyrolysis of formaldoxime and showed that this reaction obeys the first-order kinetic law and its main products are hydrocyanic acid and water, together with small amounts of carbon monoxide, ammonia, hydrogen, and ethane, and traces of unsaturated compounds.

The experiments carried out confirmed the mechanism of the conversion of nitric oxide with the initial formation of a nitroso compound:

$$CH_3\cdot + \cdot NO \longrightarrow CH_3{-}NO$$

$$CH_3{-}NO \longrightarrow CH_2{=}N{-}OH$$

$$CH_2{=}N{-}OH \longrightarrow \text{conversion products}$$

Later, the dimer of nitrosomethane was isolated, and it was established that above 125°C, formaldoxime trimer is formed [47]. Raley et al. [48] identified formaldoxime as a product of the decomposition of tert-butyl peroxide in the presence of nitric oxide.

Finally, from the products of the reaction of isopropyl radicals with nitric oxide, a compound simultaneously containing nitroso and hydroxyl-amino groups has been isolated [49].

Forsyth [50], and later Durham and Steacie [51], investigated in detail the inhibiting properties of nitric oxide by Paneth's method, using a tellurium mirror. In all the experiments, the rate of disappearance of the free radicals obeyed a first-order kinetic law and was proportional to the partial pressure of the nitric oxide. It was also calculated that of each 7300 collisions of methyl radicals with nitric oxide, only one was effective.

c.   *1,1-Diphenyl-2-Picrylhydrazyl.*   1,1-Diphenyl-2-picrylhydrazyl, which is stable to the action of oxygen, is an example of one of the most stable organic radicals [52]. The work of Bartlett and Kwart [53] and of other workers [54, 55], has established that this compound is an ideal free-radical acceptor. It has been used successfully for the total suppression of the initiated polymerization of vinyl acetate and styrene. The capacity of this inhibitor for effectively trapping the radicals of a chain enables the rate of the homolytic decomposition of certain molecules, including peroxides and azo compounds, to be determined [56].

The chemical structure of the adducts of 1,1-diphenyl-2-picrylhydrazyl with various free radicals is known only for isolated cases. It almost always reacts with the transfer of the reaction center into a phenyl nucleus,

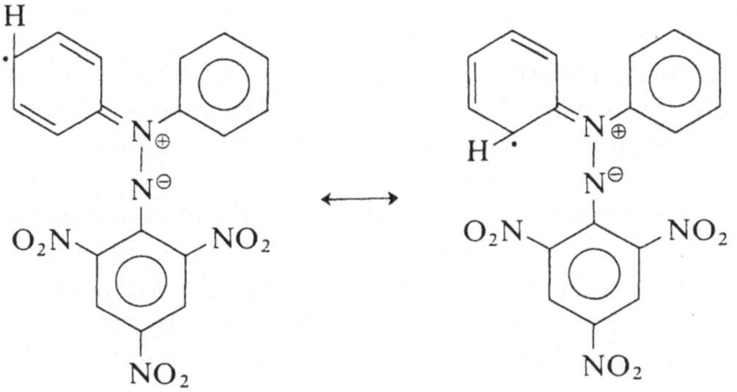

in consequence of which, the products of its recombination with radicals of tervalent carbon have the following structure [52]:

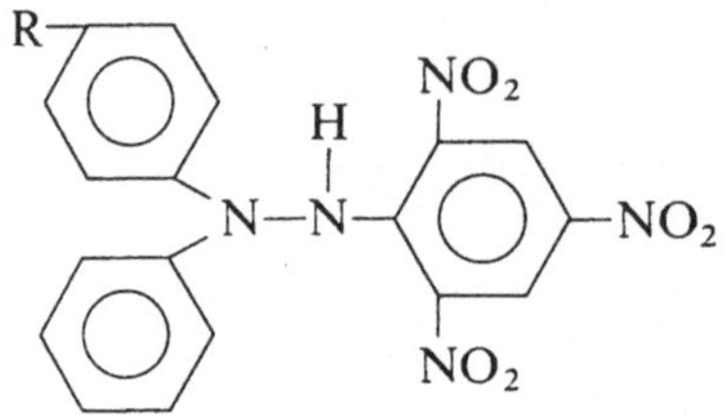

This structure is shown, for example, in the adducts with triphenylmethyl, 2-cyano-2-propyl, and the macroradical of poly(methylmethacrylate) [57].

Subsequently, the work of Hammond *et al.* [58] showed that the reaction of 1,1-diphenyl-2-picrylhydrazyl with free 2-cyano-2-propyl radicals in inert solvents is not stoichiometric and is very sensitive to oxygen. On the basis of an analysis of kinetic data, the authors advanced the hypothesis that the first reaction product may be a π-complex which is capable of reacting both with decomposition and without decomposition of the hydrazyl radical.

Shortly afterward, Möbius and Schneider [59] reported that in the presence of oxygen, 2-cyano-2-propyl radicals react with 1,1-diphenyl-2-picrylhydrazyl with the decomposition of the latter into two new free radicals:

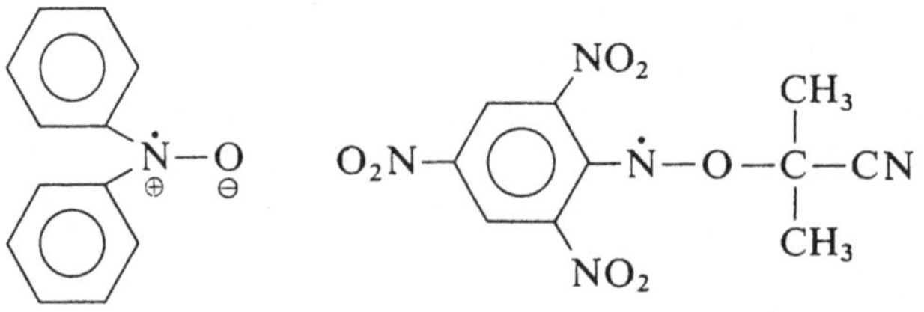

These far from exhaustive examples of the obscurities and complexities of the chemical behavior of 1,1-diphenyl-2-picrylhydrazyl impose serious limitations on the use of this compound for kinetic studies.

d. Other Stable Radicals. The use as inhibitors of less stable radicals is complicated by their considerable capacity for initiating kinetic chains. Thus, for example, triphenylmethyl is in fact capable, within certain ranges of concentration and temperature, of inhibiting the polymerization of styrene [60], but at other concentrations and temperatures, it initiates polymerization.

Still more reactive radicals of this class—for example, benzyl and diphenylmethyl—only initiate the polymerization of acrylonitrile and methyl methacrylate [61, 62].

Apparently, in all cases considered, stable radicals simultaneously exert a retarding and accelerating action, and only the combined effect has a positive or negative sign [63].

One of the most important methods for the direction of chemical reactions is, for example, the action of substances possessing the capacity for trapping free radicals on chain reactions, thus terminating the chain [64].

Inhibitors of the type A—H containing labile hydrogen—for example, phenols, aromatic amines, mercaptans—by reacting with peroxide radicals, terminate the chain process by the reaction

$$RO_2 \cdot + A{-}H \longrightarrow RO_2H + A \cdot$$

As a result of such reactions, instead of active radicals of the type of $RO_2 \cdot$, $RO \cdot$, and $R \cdot$, stable free radicals $A \cdot$ incapable of vigorously propagating the oxidation chain are formed [65].

The appearance of stable free radicals in the oxidation of secondary aromatic amines was observed by Hoskins [66] as early as 1956. Somewhat later, Harle and Thomas [67] reported the accumulation of stable free radicals in the oxidation of octadecene inhibited by phenyl-β-naphthylamine. Thomas [68] succeeded in reliably identifying a stable free radical, diphenyl-nitroxide, in the oxidation products. Thomas and Tolman [69] noted the good inhibiting properties of diphenylnitroxide when it was used as an inhibitor of the oxidation of cumene dissolved in chlorobenzene at 68.5°C.

In contrast to diphenylnitroxide, according to Ziegler and co-workers [70, 71], triarylmethyl free radicals induce autoxidative processes.

Recently, the inhibiting actions of diphenyl- and dianisylnitroxides have been studied in comparison with diphenylamine, taken as a structural model, in the oxidation of isotactic polypropylene at temperatures of 175 and 200°C. The comparative activities were determined by measuring the induction periods in the oxidation of stabilized samples of polypropylene.

The experimental curves of the induction period as a function of the concentration of inhibitor in the oxidation of isotactic polystyrene are given in Fig. 20. The depicted curves show that under these conditions, the greatest induction of the oxidation is found in the presence of di-p-anisylnitroxide.

An analogous feature of the effectiveness of the inhibiting action is also found at 200°C: thus, at an inhibitor concentration of 1.2×10^{-2} mole/kg, the induction period for diphenylamine is 5 min, for diphenylnitroxide 10 min, and for di-p-anisylnitroxide 240 min [72].

A comparison of the curves of the induction period as a function of the concentration of inhibitor leads to the conclusion that the effectiveness of the inhibiting action decreases with a rise in the temperature, apparently as a consequence of an acceleration of the initiation reactions.

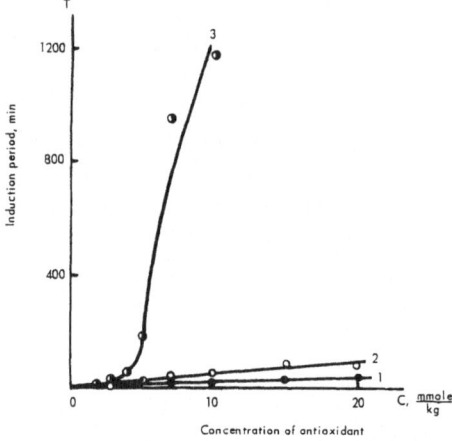

Fig. 20. Induction period in the oxidation of isotactic polypropylene as a function of the concentration of inhibitor at 175°C and an oxygen pressure of 300 mm Hg: (1) diphenylamine; (2) diphenylnitroxide; (3) di-*p*-anisylnitroxide.

The inhibiting action of diansylnitroxide is apparently connected with its participation in chain termination reactions [73]:

In order to ascertain whether nitroxide radicals cause an increase in the probability of chain branching, experiments were carried out on the influence of nitroxide radicals on the decomposition of polypropylene hydroperoxide, which was obtained by oxidizing a film of isotactic polypropylene. The nitroxide radical was introduced into the oxidized film, after which, the film was kept at 130°C, during which time, the kinetics of the decomposition of the hydroperoxide and the kinetics of the disappearance of the nitroxide radicals were measured. The experiments showed that the nitroxide radicals do not accelerate the decomposition of the hydroperoxide by more than 30%. Since, under similar conditions, the addition of aromatic amines accelerates the decomposition of the hydroperoxide by a factor of 30–40, it may be assumed that nitroxide radicals do not appreciably raise the probability of branching by accelerating the decomposition of the hydro-peroxide.

In contrast to the stable di-p-anisylnitroxide, the unstable 2,2',4,4'-tetramethoxydiphenylnitroxide does not participate in the termination of oxidation chains and is rapidly converted into the corresponding amine, which considerably lengthens the induction period of the oxidation of polypropylene at 200°C [74, 75].

Thus, the performed experiments show that at the beginning of the induction period, when the radical has not been consumed, it also inhibits the oxidation reaction. However, even after the complete disappearance of the radical, the induction period is not terminated. There is no doubt that under these conditions, the radicals form new products which are also antioxidants [73].

On comparing the antioxidant activity of 2,2,6,6-tetramethyl-4-oxo-piperidine-1-oxyl with such antioxidants as 2,6-di-tert-butyl-4-methylphenol and diphenylamine, it was found that under otherwise identical conditions, the nitroxyl was not inferior to these typical antioxidants [76, 77] with respect to the induction period of the oxidation of polypropylene. When the antioxidants mentioned were present in isotactic polypropylene in concentrations of 0.04 mole/kg, the inhibiting effect of the free nitroxyl was the greatest (Table 7).

At a temperature of 160°C and an oxygen pressure of 200 mm Hg, Poliamid-68 is oxidized immediately at a high rate, as is shown by the oxygen-absorption curve 1 in Fig. 21. Many antioxidants of the classical type more or less strongly decrease the rate of oxidation. Thus, the addition to Poliamid-68 of 2,6-di-tert-butyl-4-methylphenol ("Ionol") in a concentration of 0.03 mole/kg somewhat lowered the rate of absorption of oxygen, as can be seen from curve 2 of Fig. 21. If, instead of the Ionol, 2,2,6,6-tetramethyl-4-oxo-piperidine-1-oxyl was added to the polymer, a well-defined induction period of 50 min was observed, after which oxygen was absorbed again (curve 3,

TABLE 7

Comparison of the Efficiencies of Some Antioxidants in the Oxidation of Isotactic Polypropylene at 170°C and an Oxygen Pressure of 300 mm Hg

Antioxidant	Induction period, min
2,2,6,6-Tetramethyl-4-oxopiperidine-1-oxyl	210
2,6-Di-tert-butyl-4-methylphenol	135
Diphenylamine	90

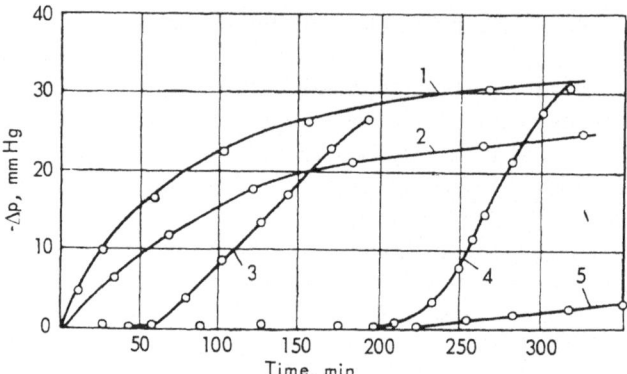

Fig. 21. Fall in the pressure of oxygen as a function of the time of heating of Poliamid-68 at 160°C for an initial oxygen pressure of 200 mm Hg: (1) without a stabilizer; (2) with Ionol; (3) with 2,2,6,6-tetramethyl-4-oxopiperidine-1-oxyl; (4) with 4-ethyl-4-hydroxy-2,2,6,6-tetramethylpiperidine-1-oxyl; (5) with Neozone-D. Concentration of stabilizer in all cases was 0.03 mole/kg.

Fig. 21). When the same concentration of another nitroxyl radical, 4-ethyl-4-hydroxy-2,2,6,6-tetramethylpiperidine-1-oxyl, was added to the Poliamid-68, a considerably longer induction period of 200 min was found. An induction period of the same length was found under otherwise identical conditions when one of the most effective stabilizers, phenyl-β-naphthylamine (Neozone-D), was used.

It must be borne in mind that the nitroxyl radicals studied are comparatively volatile [78], and, during the time of the experiment, about 12% of the added stabilizers volatilized from the polymer without taking part in the inhibition of oxidation. In actual fact, when the concentration of nitroxyl radicals was raised, the induction period of the oxidation of Poliamid-68 lengthened [79]. If the inhibitor is consumed only in chain termination with a rate equal to the rate of initiation, then, at sufficiently high concentrations (greater than the so-called "critical" concentration), the induction period is directly proportional to the initial concentration of inhibitor. In real systems, the dependence of the induction period on the initial concentration of inhibitor is not so simple. At high concentrations of inhibitor, direct proportionality is disturbed [80, 81] and the inhibitor becomes less effective, which is apparently explained by the consumption of the inhibitor in side reactions not connected with chain termination.

The appreciable volatility of the radical inhibitors studied considerably reduces their effectiveness at high temperatures (180–200°C).

The dependence of the amount of radical inhibitor on the time in a plot of $\log[I_0/(I_0 - I)]$ versus t, where I_0 is the initial concentration of inhibitor and

I is the amount of inhibitor consumed in processes not connected with evaporation, is satisfactorily described by a straight line, which shows that the reaction is of the first order [79]. The effective rate constant at 160°C is $1.4 \times 10^{-4} \sec^{-1}$.

Thus, nitroxides and nitroxyls possess a considerable inhibiting action on the thermal oxidation of certain polymers. In contrast to the majority of classical antioxidants, the radical inhibitors completely block the oxidation of a polyamide until they are exhausted, and, after the end of the induction period, the rate of oxidation approximates the rate of the uninhibited oxidation of the polyamide [82].

A similar pattern has been observed in the inhibited oxidation of polyformaldehyde with acetylated terminal groups. When 1.2% of Poliamid-68 and 0.8% of nitroxyl radical were added to the polymer, the rate of its decomposition fell considerably. 4-Ethyl-4-hydroxy-2,2,6,6-tetramethylpiperidine-1-oxyl is approximately as effective as one of the best classical stabilizers "22–46":

It was shown by the EPR method that only after the complete consumption of the nitroxyl radical did vigorous decomposition of the polyformaldehyde begin. With an increase in the concentration of stabilizing mixture, the induction period of the thermal oxidative decomposition of the polyformaldehyde at first rose rapidly, and then a tendency to saturation was observed.

As is known [83], polymerization within a strongly cross-linked ω-polymer of chloroprene leads to the deformation of the polymeric macromolecule and the appearance of localized strains which, in turn, cause the rupture of the polymer chains, preventing the bulk of the chains from assuming the most favorable conformation. The rupture of the individual polymeric chains of an ω-polymer leads to the formation of free alkyl macroradicals capable of continuing ω-polymerization.

Taking the above-mentioned factors into consideration, it appears of interest to consider the process of the growth of an ω-polymer of chloroprene in the presence of stable nitroxyls [84] (Table 8).

The experiments were carried out in sealed glass vessels with a capacity of 250 ml at 25–30°C. The ω-polymerization of the chloroprene was initiated by trimers of the polymer weighing 0.1 ± 0.001 g for each 100 ml of industrial

TABLE 8
Free Stable Nitroxyls

No.	Name	Structural formula	Mp, °C
1	2,2,6,6-Tetramethylpiperidine-1-oxyl	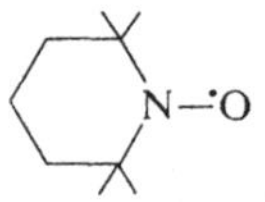	37.5
2	2,2,6,6-Tetramethyl-4-oxopiperidine-1-oxyl	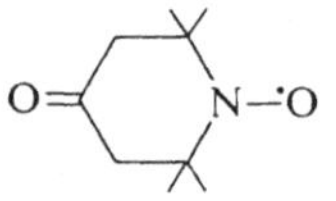	36.6
3	4-Hydroxy-2,2,6,6-tetramethylpiperidine-1-oxyl	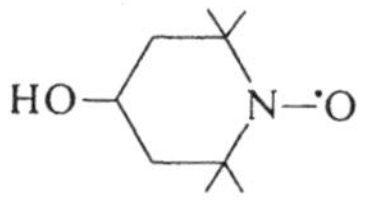	71.5

rectified chloroprene. The amount of ω-polymer formed was determined gravimetrically.

The investigation showed that in the presence of free nitroxyl radicals in concentrations of 0.001–0.01%, the ω-polymerization of the chloroprene took place as rapidly as in the absence of inhibitors.

An appreciable suppression of the polymerization was produced when the nitroxyl was added to the chloroprene in amounts ensuring that not less than 0.05% by weight was present in the polymer. However, the most powerful inhibiting action of free nitroxyls was shown when they were used in concentrations of 0.1% and above. Figure 22 shows the experimental curves of the rate of growth of the ω-polymer as a function of the chemical structure of the nitroxyl radicals when their amount in the chloroprene was 0.1%.

It can be seen from the curves that when the free nitroxyls are present in the chloroprene in a concentration of 0.1%, they completely suppress ω-polymerization for 10–12 days. The relatively high efficiency of radical 1 can apparently be explained by the fact that this compound has the lowest polarity, which enables it to diffuse within the polymer macromolecule most effectively.

The experiments carried out convincingly confirm the free-radical nature of the ω-polymerization of chloroprene, which takes place through the alkyl macroradicals continuously arising in various parts of the polymer. Since

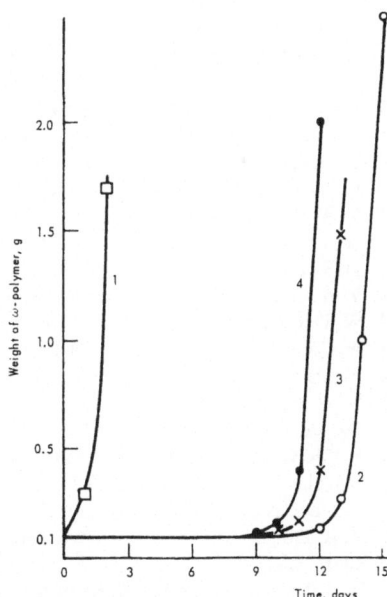

Fig. 22. Rate of formation of an ω-polymer in chloroprene at 25–30°C: (1) in the absence of an inhibitor; (2) in the presence of radical 1 (see Table 8); (3) in the presence of radical 2; (4) with the addition of radical 3. Concentration of inhibitor always 0.1 wt.%.

chloroprene molecules can easily diffuse into the depths of the polymer, the free radicals arising are apparently rapidly consumed in processes of structure formation and those of the "cross-linking" of the newly-formed polymer molecules to the polymer network.

References

1. A. Hoffman, A. Feldman, E. Gelblum, and W. Hodgson, *J. Am. Chem. Soc.* **86**, 639 (1964).
2. É. G. Rozantsev and E. N. Gur'yanova, *Izv. Akad. Nauk SSSR, Ser. Khim.* **1966**, 979.
3. É. G. Rozantsev and A. B. Shapiro, *Izv. Akad. Nauk SSSR, Ser. Khim.* **1964**, 1123.
4. É. G. Rozantsev and V. A. Golubev, *Izv. Akad. Nauk SSSR, Ser. Khim.* **1966**, 891.
5. É. G. Rozantsev, V. A. Golubev, and M. B. Neiman, *Izv. Akad. Nauk SSSR, Ser. Khim.* **1965**, 393.
6. M. N. Volkotrub, É. G. Rozantsev, and L. A. Skripko, *Izv. Akad. Nauk SSSR, Ser. Khim.* **1967**, 179.
7. V. A. Golubev, É. G. Rozantsev, and M. B. Neiman, *Izv. Akad. Nauk SSSR, Ser. Khim.* **1965**, 1927.
8. V. A. Golubev, M. B. Neiman, and É. G. Rozantsev, *Izv. Akad. Nauk SSSR, Ser. Khim.* **1966**, 343.
9. K. Auwers, *Ann. Chem.* **420**, 84 (1920).
10. A. Skita, *Ber.* **56**, 1014 (1923).
11. M. B. Neiman and É. G. Rozantsev, *Izv. Akad. Nauk SSSR, Ser. Khim.* **1964**, 1178.
12. V. V. Gur'yanova, B. M. Kovarskaya, L. A. Krinitskaya, M. B. Neiman, and É. G. Rozantsev, *Vysokomolekul. Soedin.* **7**, 1515 (1965).

13. M. B. Neiman, S. G. Mairanovskii, B. M. Kovarskaya, É. G. Rozantsev, and É. G. Gintsberg, *Izv. Akad. Nauk SSSR, Ser. Khim.* **1964**, 1518.

14. C. Wurster and R. Sendtner, *Ber.* **12**, 1803 (1879).

15. C. Wurster and E. Schobig, *Ber.* **12**, 1807 (1879).

16. L. Michaelis, M. Schubert, and S. Granick, *J. Am. Chem. Soc.* **61**, 1981 (1939).

17. C. Coulson, Valence, Oxford University Press, Oxford, 2nd ed. (1961).

18. K. Meyer and H. Gottlieb-Billroth, *Ber.* **52**, 1476 (1919).

19. K. Meyer and W. Reppe, *Ber.* **54**, 327 (1921).

20. G. V. Fomin, L. A. Blyummenfel'd, and B. N. Sokhorukov, *Dokl. Akad. Nauk SSSR* **157**, 1199 (1964).

21. V. A. Golubev, Thesis, Moscow (1967).

22. W. Dixon and R. Norman, *J. Chem. Soc.* **1964**, 4850.

23. A. Cope, D. McLean, and N. Nelson, *J. Am. Chem. Soc.* **77**, 1628 (1955).

24. F. Arndt and J. Rose, *J. Chem. Soc.* **1956**, 6.

25. N. N. Semenov, Some Problems of Chemical Kinetics and Reactivity, Pergamon Press, Oxford (1958–1959).

26. H. Melville and W. Watson, *Trans. Faraday Soc.* **44**, 886 (1948).

27. S. Cohen, *J. Am. Chem. Soc.* **69**, 1057 (1947).

28. L. Staveley and C. Hinshelwood, *Trans. Faraday Soc.* **35**, 845 (1939).

29. B. A. Dolgoplosk and G. A. Parfenova, *Tr. Vses. Nauchn.-Issled. Inst. Sintetich. Kauchuka* **1951**, 224.

30. B. A. Dolgoplosk and D. S. Korotkina, *Tr. Vses. Nauchn.-Issled. Inst. Sintetich. Kauchuka* **1951**, 198.

31. L. Staveley, *Proc. Roy. Soc.* **A162**, 557 (1937).

32. F. Bovey and J. Kolthoff, *J. Am. Chem. Soc.* **69**, 2142 (1947).

33. Ts. Teidzi, Reactions for the Preparation of Synthetic Polymers [Russian translation], IL, Moscow (1963).

34. M. Bodenstein and W. Unger, *Z. Phys. Chem.* **B11**, 253 (1930).

35. C. Hinshelwood, *Z. Elektrochem.* **42**, 445 (1936).

36. J. Hobbs, *Proc. Roy. Soc.* **A167**, 456 (1938).

37. J. Hobbs and C. Hinshelwood, *Proc. Roy. Soc.* **A167**, 439 (1938).

38. J. Hobbs and C. Hinshelwood, *Proc. Roy. Soc.* **A167**, 447 (1938).

39. L. Staveley and C. Hinshelwood, *Nature* **137**, 29 (1936).

40. L. Staveley and C. Hinshelwood, *J. Chem. Soc.* **1936**, 812.

41. L. Staveley and C. Hinshelwood, *J. Chem. Soc.* **1937**, 1568.

42. L. Staveley and C. Hinshelwood, *Proc. Roy. Soc.* **A154**, 335 (1936).

43. L. Staveley and C. Hinshelwood, *Proc. Roy. Soc.* **A159**, 192 (1937).

44. F. Stubbs, K. Ingold, B. Spall, C. Danby, and C. Hinshelwood, *Proc. Roy. Soc.* **A214**, 20 (1952).

45. H. Thompson and M. Meisner, *Nature* **139**, 1018 (1937).

46. H. Taylor and H. Bender, *J. Chem. Phys.* **9**, 761 (1941).

47. C. Coe and T. Doumani, *J. Am. Chem. Soc.* **70**, 1516 (1948).

48. H. Raley, F. Rust, and W. Vaughan, *J. Am. Chem. Soc.* **70**, 88 (1948).

49. H. Chilton and B. Gowenlock, *Nature* **172**, 73 (1953).

50. J. Forsyth, *Trans. Faraday Soc.* **37**, 312 (1941).

51. R. Durham and E. Steacie, *J. Chem. Phys.* **20**, 582 (1952).

52. S. Goldschmidt and K. Renn, *Ber.* **55**, 628 (1922).

53. P. Bartlett and H. Kwart, *J. Am. Chem. Soc.* **72**, 1051 (1950).

54. M. Matheson, E. Auer, B. Bevilacqua, and E. Hart, *J. Am. Chem. Soc.* **73**, 1700 (1951).

55. D. Verdin, *Trans. Faraday Soc.* **56**, 823 (1960).

56. C. Bawn and S. Mellish, *Trans. Faraday Soc.* **47**, 1216 (1951).
57. A. Henglein, *Makromol. Chem.* **15**, 188 (1955).
58. G. Hammond, J. Senn, and C. Boozer, *J. Am. Chem. Soc.* **77**, 3244 (1955).
59. K. Möbius and F. Schneider, *Z. Naturforsch.* **18a**, 428 (1963).
60. F. Mayo and R. Gregg, *J. Am. Chem. Soc.* **70**, 1284 (1948).
61. W. Theilacker, H. Schulz, U. Baumgarte, H.-G. Drössler, W. Rohde, F. Thater, and H. Uffmann, *Angew. Chem.* **69**, 322 (1957).
62. K. Ziegler, W. Deparade, and H. Kuhlhorn, *Ann. Chem.* **567**, 151 (1950).
63. V. I. Gol'danskii, *Usp. Khim.* **15**, 63 (1946).
64. N. M. Émanuél', Chemistry and Physics, Izd. Znanie, Moscow (1963).
65. E. T. Denisov and N. M. Émanuél', *Usp. Khim.* **27**, 365 (1958).
66. R. Hoskins, *J. Chem. Phys.* **25**, 788 (1956).
67. O. Harle and J. Thomas, *J. Am. Chem. Soc.* **79**, 2973 (1957).
68. J. Thomas, *J. Am. Chem. Soc.* **82**, 5955 (1960).
69. J. Thomas and C. Tolman, *J. Am. Chem. Soc.* **84**, 2930 (1962).
70. K. Ziegler, L. Ewald, and A. Sieb, *Ann. Chem.* **564**, 182 (1933).
71. K. Ziegler and K. Ganicke, *Ann. Chem.* **551**, 213 (1942).
72. É. G. Rozantsev, L. A. Kalashnikova, and M. B. Neiman, *Zh. Prikl. Khim.* **38**, 702 (1965).
73. L. A. Kalashnikova, V. B. Miller, M. B. Neiman, É. G. Rozantsev, and L. A. Skripko, *Plast. Massy* **7**, 10 (1966).
74. L. L. Yasina, A. B. Shapiro, and É. G. Rozantsev, *Plast. Massy* **6**, 37 (1966).
75. L. L. Yasina, V. B. Miller, Yu. A. Shlyapnikov, and L. A. Skripko, *Izv. Akad. Nauk SSSR, Ser. Khim.* **1965**, 1481.
76. É. G. Rozantsev, M. B. Neiman, and G. I. Likhtenshtein, USSR Authors' Certificate No. 166133, May 29, 1962.
77. É. G. Rozantsev, L. A. Krinitskaya, and L. A. Troitskaya, *Khim. Prom.* **3**, 180 (1964).
78. L. A. Kalashnikova, É. G. Rozantsev, and A. M. Chaikin, *Izv. Akad. Nauk SSSR, Ser. Khim.* **1965**, 800.
79. I. I. Levantovskaya, B. M. Kovarskaya, M. B. Neiman, É. G. Rozantsev, and M. P. Yazvikova, *Plast. Massy* **3**, 14 (1964).
80. D. Ryshavy, *Vysokomolekul. Soedin.* **3**, 464 (1961).
81. L. G. Angert and A. S. Kuz'minskii, in: International Symposium on Macromolecular Chemistry [Russian version], Izd. Akad. Nauk SSSR, Vol. 3 (1960), p. 423.
82. M. B. Neiman, L. A. Krinitskaya, and É. G. Rozantsev, *Izv. Akad. Nauk SSSR, Ser. Khim.* **1965**, 2055.
83. A. N. Pravednikov and S. S. Medvedev, *Dokl. Akad. Nauk SSSR* **109**, 579 (1956).
84. M. B. Neiman, N. G. Karapetyan, A. S. Tarkhanyan, A. N. Dyubimova, and É. G. Rozantsev, *Vysokomolekul. Soedin.* **8**, 1237 (1966).

Chapter V

PHYSICAL PROPERTIES AND STRUCTURE OF INDIVIDUAL RADICALS

1. Saturated Vapor Pressure

The vapor pressure of the individual free radicals is their most important thermodynamic characteristic, having not only theoretical interest, but also great practical importance. Until very recently, this aspect has been investigated for only a few radicals [1]. Table 9 gives almost all the nitroxides and nitroxyls and the corresponding hydrides which have been studied.

The vapor pressures of these compounds were measured by the classical Knudsen method [2] with Clausing's correction for the parameters of the membrane aperture. This method presupposes that the saturated vapor pressure is close to the equilibrium value [3], and does not depend on the condensation coefficient α when $A \gg S$; it gives $P = P_{\text{meas}}[1 + (S/\alpha A)]$, where A is the surface of evaporation and S is the cross-sectional area of the aperture in the membrane.

The amount of substance passing through the aperture of the membrane during the time of the experiment was determined by gravimetric, spectrophotometric, and polarographic methods, and the vapor pressure of the

TABLE 9
Constants of Nitroxide and Nitroxyl Radicals and the Corresponding Hydrides

No.	Compound	Mp, °C
I	4,4'-Dimethoxydiphenylnitroxide	150
II	2,2',4,4'-Tetramethoxydiphenylnitroxide	178
III	1,4-Dihydroxy-2,2,6,6-tetramethylpiperidine	158
IV	4-Hydroxy-2,2,6,6-tetramethylpiperidine-1-oxyl	71
V	1-Hydroxy-2,2,6,6-tetramethyl-4-oxopiperidine	95
VI	2,2,6,6-Tetramethyl-1-oxopiperidine-1-oxyl	36

compounds studied was calculated from the formula $P = (0.286G/KSt)$ $(T/M)^{1/2}$, where S is the section of the membrane aperture in cm^2, K is Clausing's coefficient, t is the time in minutes, T is the temperature in degrees Kelvin, M is the molecular weight, and G is the weight of substance passing through the membrane in grams.

It is worth noting that the calculated heats of sublimation for 1,4-dihydroxy-2,2,6,6-tetramethylpiperidine and 2,2,6,6-tetramethyl-4-oxo-piperidine-1-oxyl differ by 4.5 kcal/mole. The difference is apparently due to the tendency of the first substance to associate through "intermolecular" hydrogen bonds.

The aromatic nitroxides proved to be considerably less volatile than the nitroxyls, and the latter more volatile than the corresponding hydroxyl-amines. An equation of the form $\log P = A + (B \times 10^3/T)$ was obtained by the method of least squares, with the following values of A and B for the six compounds given in Table 9: (I) $A = 10.38$, $B = -5.26$; (II) $A = 16.08$, $B = -7.53$; (III) $A = 13.03$, $B = -5.24$; (IV) $A = 14.46$, $B = -5.30$; (V) $A = 11.93$, $B = -4.18$; (VI) $A = 11.93$, $B = -4.35$. Note: all B values are negative.

The saturated vapor pressures and the heats of sublimation of the compounds studied are given in Table 10.

2. Reasons for the Stability of Nitroxyl Radicals

In contrast to the paramagnetic nitroxides, the stability of which is connected with the marked delocalization of the unpaired electron over the aromatic nuclei, the reason for the high stability of the free nitroxyls is less obvious.

To answer this question, the heats of combustion of some nitroxyl radicals and the hydrides corresponding to them were determined and their heats of formation were calculated [4]. The substances studied were carefully purified by repeated recrystallization from various solvents, were chromato-graphed on alumina, and were sublimed in a high vacuum until constant and sharp melting points were obtained.*

Table 11 shows the data on the heats of combustion Q_{comb} of the substances studied, the change in internal energy ΔE_0, and the standard enthalpy of combustion $\Delta H°$, as well as the calculated enthalpies of formation $\Delta H_f°$ of the compounds for the solid and gaseous states under standard conditions. The values of the heats of sublimation were used to calculate the enthalpies of formation of the compounds in the gaseous state. The heat of sublimation of 2,2,6,6-tetramethyl-4-oxopiperidine was determined by

---

*The operations connected with the purification of the hydroxylamine derivatives were carried out in an atmosphere of argon.

TABLE 10
Saturated Vapor Pressures and Heats of Sublimation

No.	Compound	T, °C	$P \times 10^{-5}$, mm	ΔH, kcal/T
I	4,4'-Dimethoxydiphenylnitroxide	55	0.185	
		60	0.332	
		70	1.103	
		80.5	2.565	24.07 ± 0.63
		85	4.260	
		80	8.000	
II	2,2',4,4'-Tetramethoxydiphenylnitroxide	60	0.027	
		67	0.073	
		70	0.123	
		80	0.717	34.47 ± 2.73
		85.0	0.678	
		87.5	1.405	
		90.0	2.707	
III	1,4-Dihydroxy-2,2,6,6-tetramethylpiperidine	40	17.78	
		50.5	59.40	
		60	215.10	
		65	330.25	24.0 ± 0.12
		70	538.00	
		75	892.50	
		75	935.00	
IV	4-Hydroxy-2,2,6,6-tetramethylpiperidine-1-oxyl	20	22.7	
		30	95.4	
		35	171.3	24.26 ± 0.39
		40	321.5	
		45	610.0	
V	1-Hydroxy-2,2,6,6-tetramethyl-4-oxopiperidine	15	194.5	
		30	1503.0	
		35	2400.0	
		40	4100.0	19.15 ± 1.02
		45	5727.0	
		50	7720.0	
		55	18850.0	
VI	2,2,6,6-Tetramethyl-4-oxopiperidine-1-oxyl	2	118.2	
		10	270.0	
		13	443.0	
		15	580.25	
		20	1169.0	19.91 ± 0.39
		25	1450.0	
		28	2490.0	
		30	3285.0	

TABLE 11

Thermochemical Properties of Free Nitroxyls and Their Hydrides

No.	Compound	Mp, °C	Q_{comb}, kcal/mole	$-\Delta E^\circ$, kcal/mole	$-\Delta H^\circ$, kcal/mole	$-\Delta H^\circ_{f,solid}$, kcal/mole	$-\Delta H^\circ_{f,gas}$, kcal/mole
I	1,4-Dihydroxy-2,2,6,6-tetramethylpiperidine	156.5	1387.64 ± 0.33	1387.06 ± 0.33	1388.98 ± 0.33	106.52 ± 0.33	82.52 ± 0.45
II	4-Hydroxy-2,2,6,6-tetramethylpiperidine-1-oxyl	71.5	1366.27 ± 1.82	1365.62 ± 1.82	1367.45 ± 1.82	93.82 ± 1.82	69.61 ± 2.21
III	1-Hydroxy-2,2,6,6-tetramethyl-4-oxopiperidine	90.5	1335.76 ± 0.1	1335.14 ± 0.1	1336.77 ± 0.1	90.39 ± 0.1	71.24 ± 1.12
IV	2,2,6,6-Tetramethyl-4-oxopiperidine-1-oxyl	36.0	1320.79 ± 1.55	1320.15 ± 1.55	1321.63 ± 1.55	71.36 ± 1.55	51.45 ± 1.85
V	2,2,6,6-Tetramethyl-4-oxopiperidine	135.5	1345.92 ± 0.82	1345.37 ± 0.82	1347.26 ± 0.82	79.90 ± 0.82	65.37 ± 1.47

the effusion method from the dependence of the vapor pressure on the temperature. Treatment of the experimental data by the method of least squares gives the equation $\log P_i = 9.567 - (3.175 \times 10^3/T_i)$, from which it follows that the heat of sublimation of triacetoneamine is 14.53 ± 0.65 kcal/mole.

The heat of formation of the radicals is fairly large, which confirms their relatively high chemical stability.

The results of the thermochemical study of the free nitroxyls and the hydrides corresponding to them enable some features of their chemical behavior to be explained [5].

In an analysis of the experimental data, it may be assumed that the sum of the mean thermochemical bond energies is equal to the heat of atomization of the substance: $\sum \sigma = -\Delta H_{gas}^{atom}$.

As the thermochemical practice of the last few years has shown, such an approach is fairly reliable, since a deviation from a constant value of the bond energies may show structural features of the substances under investigation [6].

The mean thermochemical bond energies (in kcal) are, according to Cottrell [7]: C—C, 82.6; C—H, 98.7; C—N, 72.8; N—H, 93.4; O—H, 110.6; C—O, 175.0; C—O, 85.5; and N—O, 53.0.

Since the experimental value of the enthalpy of formation of piperidine is 11.67 kcal, it is easy to evaluate the energy of cyclization. In this case, it may be assumed that the energy of cyclization does not change greatly when symmetrical geminal substituents [8] or hydroxy or carbonyl groups are introduced into the heterocycle, since in these circumstances, the energy state of the nitrogen and carbon atoms vary very slightly on an average.

Taking the heats of atomization of the molecules of hydrogen, nitrogen, and oxygen as 52.10, 112.98, and 59.56 kcal, respectively, and the heat of sublimation of carbon as 170.91 kcal/mole [7], we obtain for the atomization of piperidine $\Delta H_a = 1552.3$ kcal, and $\Delta H_{cyclization} = -\sum \sigma + \Delta H_{atom} = -3.1$ kcal.

The calculated heats of atomization ΔH_a of these compounds are given in Table 12 together with the values of $\sum \sigma_{n-1}$ and $\sum \sigma_{n-2}$, while Table 13 gives the mean energies of the O—H, N—O, and N—H bonds calculated from the formula $\sigma_{bond} = \Delta H_a - \sum \sigma_{n-1}$.

Thus, the values for the mean energies of the nitrogen–oxygen bond of nitroxyl can be regarded as satisfactorily coincident within the error of measurement, although a somewhat smaller value of σ_{bond} in the case of compound I may be the result of a transannular nucleophilic interaction of the nitrogen atom with the carbonyl oxygen.

Nevertheless, we may, with no disadvantage, take the mean value of the energy of the nitroxyl nitrogen–oxygen bond as 99.4 kcal. This value coincides

TABLE 12
Thermochemical Properties of Some Nitroxyls and Their Hydrides

No.	Substance	$-\Delta H_f^0$, kcal/mole	ΔH_a, kcal/mole	$\sum \sigma_{n-1}$	$\sum \sigma_{n-2}$
I	2,2,6,6-Tetramethyl-4-oxopiperidine-1-oxyl	51.45 ± 1.85	2655.34	2557.5	—
II	4-Hydroxy-2,2,6,6-tetramethylpiperidine-1-oxyl	69.61 ± 2.21	2777.7	2676.7	—
III	2,2,6,6-Tetramethyl-4-oxopiperidine	65.37 ± 1.47	2661.77	2557.5	104.2
IV	1-Hydroxy-2,2,6,6-tetramethyl-4-oxopiperidine	71.24 ± 1.12	2727.23	2610.5	—
V	1,4-Dihydroxy-2,2,6,6-tetramethylpiperidine	82.52 ± 0.45	2842.71	2730.3	—

almost exactly with the arithmetic mean of the energies of two- and four-electron bonds determined experimentally by Cottrell (Table 14).

An analysis of the data of Table 13 leads to the conclusion that the energy of the NO—H bond in compound V is 4.3 cal less than in 1-hydroxy-2,2,6,6-tetramethyl-4-oxopiperidine, in which there is transannular 1,4-interaction. The somewhat higher value of the energy of the N—H bond in triacetoneamine as compared with that expected theoretically [6] can apparently be explained in the same way.

The following equation was used to obtain bond-dissociation energies of the compounds studied:

$$D_{R-A} = \Delta H_f^\circ(R \text{ gas}) + \Delta H_f^\circ(A \text{ gas}) - \Delta H_f^\circ(R-A \text{ gas})$$

from which it is easy to find for substance V

$$D_{NO-H} = \Delta H_f^\circ(II) + \Delta H_f^\circ(H) - \Delta H_f^\circ(V) = -69.61 + 52.12 + 82.52$$
$$= 65.03 \text{ kcal}$$

TABLE 13
Mean Thermochemical Bond Energies
For Substances of Table 12

Substance	Bond	V_{bond}, kcal
I	N$\doteq$O	97.8
II	N$\doteq$O	101.0
III	N—H	104.2
IV	NO—H	116.7
V	NO—H	112.4

TABLE 14

**Mean Energies of Various Nitrogen–
Oxygen Bonds**

Type of bond	Bond energy, kcal	Ref.
N—O	53	7
N=O	145	7
N$\doteq$O	99.4	5

A similar calculation for substance IV leads to a dissociation energy of the NO—H bond of 71.91 kcal.

According to Mortimer's data, the dissociation energy of the N—H bond of secondary amines is 107.0 kcal. If we adopt this value for triacetone-amine (III), on deducting the heat of formation of the fragment*

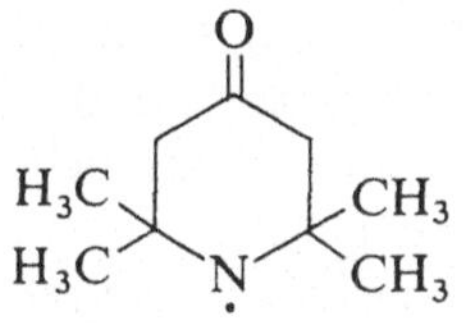

O
‖

H_3C ⟨ring⟩ CH_3
H_3C ⟍ N ⟋ CH_3
      ·

we obtain a value of 10.5 kcal, from which it is not difficult to determine the dissociation energy of the nitrogen–oxygen bond in compounds IV and V, $D_{N-OH} = 70.1$ kcal, and the energy of cleavage of the nitrogen–oxygen bond in compounds I and II, $D_{N\doteq O} = 100.5$ kcal.

The values given above are sufficient for an evaluation of the gain in energy of delocalization of the electron in nitroxyl radicals: $\varepsilon_{deloc} = D_{N\doteq O} - D_{N-OH} = 30.4$ kcal. Such a high delocalization energy of the unpaired electron correlates with the mean thermochemical value of the energy of the N$\doteq$O bond and is one of the reasons for the high stability of the free nitroxyls.

3. Dipole Moments

An important factor for the reactivity of those radicals which contain other atoms besides carbon and hydrogen atoms is the degree of their electronegativity. For example, the electronegativity of a halogen atom is greater than that of nitrogen, and therefore the former more readily attacks the part

*Radicals of this class, possessing an unpaired electron which is completely localized on the nitrogen atom, are extremely unstable [9].

of the molecule with the higher electron density. It is possible that this property is connected with the polarity of the radicals.

The Japanese scientist Furukawa has proposed that the reactivity of free radicals be expressed by means of a resonance factor I and a polarity factor e. Radicals with a high value of I possess high reactivity. The greater the difference between the polarities of the radical and the other reactants, the more easily does it react. The heat effect of the recombination reaction $A\cdot + B\cdot \rightarrow$ $A—B$ is expressed by the equation [10]: $\Delta H = I_A + I_B - e_A e_B$.

In this connection, the dipole moments of the free radicals are of definite interest. In individual cases, information on the distribution of the charge density in the radical can be used in a consideration of the distribution of the spin density of the unpaired electron.

In the case of aromatic radicals, calculation of the molar refraction by the additive method is apparently ineffective, and the experimental determination of the refraction is difficult because of the deep coloration of these compounds.

The dipole moments of the compounds given in Table 15 were determined at 25°C in benzene and n-octane solutions, the dielectric permeability and the density of solutions under study being measured in each case at several different concentrations [11].

The dependence of the density and the refractive index for the sodium D line of benzene solutions of 2,2,6,6-tetramethylpiperidine-1-oxyl on the concentration was studied over a wide range of concentrations at 20°C.

The molecular refraction of this radical, obtained by extrapolating the experimental values to infinite dilution of the solution, came to 45.6. On calculating the group refraction of the hydrocarbon fragment of the radical

TABLE 15
Dipole Moments of Free Radicals and Their Hydrides*

Compound	Mp, °C	P_∞	R_D	$\mu(D)$
1-Hydroxy-2,2,6,6-tetramethylpiperidine	39–40	110.7	47.0	1.76
2,2,6,6-Tetramethylpiperidine-1-oxyl	36–47	248.6	45.6	3.14
		(252.7)		(3.17)
2,2,6,6-Tetramethyl-4-oxopiperidine-1-oxyl	36	84.1	45.6	1.36
4-Hydroxy-2,2,6,6-tetramethylpiperidine-1-oxyl	70.5	248.2	47.2	3.12
Diphenylnitroxide	63–64	240.8	55.8	3.00
Diphenylhydroxylamine	60–61	91.8	56.9	1.30

*Here P_∞ is the molar polarization at infinite dilution; R_D is the molar refraction for the sodium D line; and μ is the dipole moment (in Debyes), calculated from the formula $\mu = 0.22(\sqrt{P_\infty - R_D})^{1/2}$.

(by Eisenlohr's method), the increment of the $N \doteq O$ group came to 4.07, close to the sum of the atomic refractions of nitrogen and oxygen.

Thus, the presence of an unpaired electron in the nitrogen–oxygen group of a free nitroxyl does not cause an appreciable increase of the molecular refraction. The molecular refractions of other free nitroxyl radicals, including the hydrogenated pyrrole series, are also in good agreement with the additive scheme [12]. Thanks to this circumstance, the refraction of the radicals studied was calculated by the additive method, taking the increment for the $N \doteq O$ group that had been found.

In contrast to 1,1-diphenyl-2-picrylhydrazyl, free nitroxyls do not form π-complexes with benzene, as is shown by the constancy of the dipole moments when the measurements are carried out in different solvents. Thus, measurements of the dipole moments of 2,2,6,6-tetramethylpiperidine-1-oxyl in benzene and in n-octane gave practically the same values: 3.14 and 3.17 D, respectively.

If we neglect the dipole moment of the hydrocarbon moiety of 2,2,6,6-tetramethylpiperidine-1-oxyl, the value of 3.14 D can be ascribed completely to the moment of the $C-NO\cdot-C$ group of the free radical.

Measurement of the dipole moment of 1-hydroxy-2,2,6,6-tetramethyl-piperidine enables us to determine the moment of the $C-NO-C$ group of a diamagnetic molecule. Since the dipole moment of the OH bond can be taken as 1.5 D and the valence angle of NOH as 105°, from the experimental value of 1.76 D the contribution of the $C-NO-C$ group of the molecule must be taken as 1.4 D.

Consequently, the presence of an unpaired electron increases the dipole moment of the nitroxyl group by 1.7 D as compared with the moment of the group in the corresponding molecule.

A similar phenomenon has been observed in measurements of the dipole moments of diphenylnitroxide* and diphenylhydroxylamine.

The increase in the dipole moment of the $N \doteq O$ bond in the free radical as compared with its hydride can be explained to a first approximation (without taking into account the atomic dipoles) by the interaction of the unshared electron pair of the nitrogen atom (present to a considerable extent in the p state in such compounds) with the partially-free p orbital of the oxygen atom.

The displacement of one of the electrons of the unshared pair of the nitrogen atom to the oxygen atom explains the increase in the dipole moment of the nitrogen–oxygen bond and the appearance of spin density on the nitrogen atom:

*The dipole moment of the radical is 2.3 ± 0.2 D according to Cambi [13, 14].

By using the obtained value (1.7 D) for the increase in the dipole moment of the nitrogen–oxygen group of the radical as compared with its hydride and assuming an interatomic distance on the N—O group of 1.3 Å, we can evaluate the distribution of the electron spin density in the free nitroxyls which proves to be 27% on the nitrogen atom and 73% on the oxygen atom. According to Berthier *et al.* [15], in the nitroxyl radical, the spin density on the nitrogen atom is close to 0.3.

If we start from the idea of the complete localization of the unpaired electron on the nitrogen atom [16], the calculated dipole moment [17] does not correspond to the experimental value.

Information on the electronic structure of the nitroxyl radicals obtained from an analysis of dipole moments cannot, of course, be considered unambiguous, and, in most cases, correlates poorly with the distribution of spin density obtained by means of EPR. However, in the case of the free nitroxyl radicals of the piperidine series containing a polar substituent in position 4 of the heterocycle, certain questions of the structural configuration of these compounds can be answered with the aid of dipole moment data [11]. By analogy with cyclohexane, three conformations are possible for the heterocycles under consideration: the "chair" form, the "twisted" form, and the "boat" form.

The comparatively large value of the dipole moment of 2,2,6,6-tetramethyl-4-oxopiperidine-1-oxyl (1.36 D) agrees with the calculated value of the dipole moment (1.4 D) for the twisted form [18, 19] of the radical ($\alpha = 152$–$156°$):

$$O{\ominus}{-}\overset{\oplus}{N}\diagup\diagdown\cdots\overset{\alpha}{\diagup}\cdots\overset{\oplus}{}{\ominus}O$$

Cyclohexane-1,4-dione possesses a similar conformation [20].

In the case of 4-hydroxy-2,2,6,6-tetramethylpiperidine-1-oxyl (Table 15), the appearance of both an intramolecular and an "intermolecular" hydrogen bond is possible. A cryoscopic determination of the molecular weight of this substance in benzene gives the following figures: Found mol. wt., 168, 177. $C_9H_{18}NO_2$. Calculated mol. wt., 172.

Thus, in dilute solutions 4-hydroxy-2,2,6,6-tetramethylpiperidine-1-oxyl does not give strong intermolecular associates, and in measuring the dipole moment, we are dealing with the monomeric form of this radical; the dipole moment (3.12 D) correlates best with the "boat" conformation:

Apparently, the latter conformation is energetically more favored than the "chair" form because of the possibility of the formation of an "intra-molecular" hydrogen bond. This conclusion is in agreement with the results of infrared spectroscopy [21].

References

1. L. A. Kalashnikova, É. G. Rozantsev, and A. M. Chaikin, *Izv. Akad. Nauk SSSR, Ser. Khim.* **1965**, 800.
2. S. Dushman, Scientific Foundations of Vacuum Technology, John Wiley and Sons, New York (1949).
3. M. Rossman and J. Yarwood, *J. Chem. Phys.* **21**, 1406 (1953).
4. Yu. A. Lebedev, É. G. Rozantsev, L. A. Kalashnikova, V. P. Lebedev, M. B. Neiman, and A. Ya. Apin, *Dokl. Akad. Nauk SSSR* **168**, 104 (1966).
5. Yu. A. Lebedev, É. G. Rozantsev, M. B. Neiman, and Z. Ya. Apin, *Zh. Fiz. Khim.* **40**, 2340 (1966).
6. C. Mortimer, Reaction Heats and Bond Strengths, Pergamon, Oxford (1962).
7. T. Cottrell, The Strengths of Chemical Bonds, Butterworths, London, 2nd ed. (1958).
8. M. Balasubramanian, *Chem. Rev.* **62**, 691 (1962).
9. V. D. Sholle, É. G. Rozantsev, A. A. Prokof'ev, and S. P. Solodovnikov, *Izv. Akad. Nauk SSSR, Ser. Khim.* **1967**, 2628.
10. Ts. Téidzi, Reactions for Preparing Synthetic Polymers, GNTI, Moscow (1963).
11. É. G. Rozantsev and E. N. Gur'yanova, *Izv. Akad. Nauk SSSR, Ser. Khim.* **1966**, 979.
12. É. G. Rozantsev, L. A. Krinitskaya, and M. B. Neiman, *Izv. Akad. Nauk SSSR, Ser. Khim.* **1966**, 675.
13. L. Cambi, *Gazz. Chim. Ital.* **63**, 579 (1933).
14. L. Cambi and L. Szegö, *Ber.* **64**, 2591 (1931).
15. G. Berthier, H. Lemaire, A. Rassat, and N. Viellard, *Theor. Chim. Acta* **3**, 213 (1965).
16. Yu. Yu. Samitov and O. L. Lebedev, *Fiz. Problemy Spektroskopii* **2**, 103 (1963).
17. Yu. Yu. Samitov and D. Ibraimov, *Teor. i Éksperim. Khim.* **1965**(3), 387.
18. A. Mossel, C. Romers, and E. Havinga, *Tetrahedron Letters* **1963**, 1247.
19. P. Groth and O. Hassel, *Proc. Chem. Soc.* **1963**, 218.
20. M. Bhatt, G. Srinivasan, and P. Neelakantan, *Tetrahedron* **21**, 291 (1965).
21. É. G. Rozantsev, *Izv. Akad. Nauk SSSR, Ser. Khim.* **1964**, 2187.

Chapter VI

ELECTRON PARAMAGNETIC RESONANCE IN NITROXIDES AND NITROXYLS

1. Monoradicals

All stable individual nitroxyls possess a very simple EPR spectrum similar to that of 2,2,6,6-tetramethyl-4-oxopiperidine-1-oxyl (see Fig. 12). However, as already mentioned, when the substance is dissolved in a suitable solvent, the singlet curve is transformed into a group of three lines with a separation between the components of about 15 G (see, for example, Fig. 13).

The theory of electron paramagnetic resonance has been discussed quite extensively in many publications [1–6], and, consequently, before discussing the spectra, we will dwell only briefly on the elementary principles of the analysis of the experimental data.*

The EPR method is based on the Zeeman effect, which consists of the fact that when a paramagnetic particle with a quantum number S is introduced into a constant magnetic field, its ground energy level is split into two $S + 1$ sublevels.

In the simplest case, when the unpaired electron in the free radical does not interact with the nuclear magnetic moments, all the spins and magnetic moments of the unpaired electrons have a chaotic orientation and the same energy. If a sample of such a substance is placed in a constant magnetic field, the spins and the magnetic moments of the electrons oriented parallel and antiparallel to the direction of the lines of force of the applied field. All the intermediate orientations are forbidden by quantum conditions, since the spin of the electron S can assume only two values.

Consequently, the superposition of an external magnetic field separates the electrons into two groups possessing equal energies. Electrons with a parallel orientation of the spins have an energy $\frac{1}{2} g\beta H$ less, and electrons with

---

*A popular account of the essence of the phenomenon is given in a monograph by Solodovnikov [7].

the antiparallel orientation an energy $\frac{1}{2}g\beta H$ greater, than the electrons in a zero field.

The g-factor is a measure of the contribution of the spin moment to the field moment of the electron, and the Bohr magneton β is equal to $eh/4\pi mc$.

When an alternating magnetic field of frequency v satisfying the condition $hv = g\beta H$ is superposed on a sample with oriented spins, electronic transitions will be induced between the two energy levels with equal probability. Because of the fact that the lower energy level is "populated" somewhat more highly than the upper one, high-frequency energy will be absorbed in such transitions, and this is recorded in the form of an EPR signal.

In the majority of free radicals, the unpaired electron is present in an orbital covering several atoms. If even one of these atoms has a nuclear magnetic moment, then, as a result of the interaction between the nuclei and the electron spins, the energy levels of the unpaired electron undergo additional splitting, and hyperfine structure (HFS) appears in the EPR spectrum. The nuclear interaction which causes the splitting of one electron level into several sublevels is called the hyperfine interaction, and the corresponding splitting of the spectral lines, the hyperfine splitting.

Let us consider the particular case of the interaction of an unpaired electron with the nuclear spin of nitrogen ($S = 1$) which takes place in the nitroxyl radicals.

Since the nuclear spin of nitrogen in a constant magnetic field can assume three values ($2S + 1$), i.e., $+1, 0, -1$, with which the magnetic moments of the electron will interact, each electron level is split into three sublevels. When an alternating magnetic field with the resonance frequency is applied, the electron spins change their orientation and absorb the energy of this field, while the nuclear spins do not change their orientation during the time of the electronic transition.

Thus, electron transitions may take place between levels with the same projections of nuclear spin. The corresponding selection rule can be written in the form $\Delta m = 0$, and, in this case, the HFS of EPR spectra will consist of three equidistant lines of the same intensity (Fig. 23).

In accordance with what has been said above, the EPR spectra of nitroxyl radicals consist of a system of three equidistant lines of the same intensity (Fig. 4).

With good resolution, interesting features can be detected in the spectra of the free nitroxyls. Thus, in the spectrum of 2,2,6,6-tetramethyl-4-oxopiperidine-1-oxyl recorded in the absence of oxygen with a high signal-to-noise ratio, a system of symmetrical satellites appears because of the splitting of the spin of the unpaired electron on the nuclei of the natural isotopes of nitrogen ^{15}N ($S = \frac{1}{2}$) and carbon ^{13}C ($S = \frac{1}{2}$).

An analysis of the intensities of the components of the carbon satellite

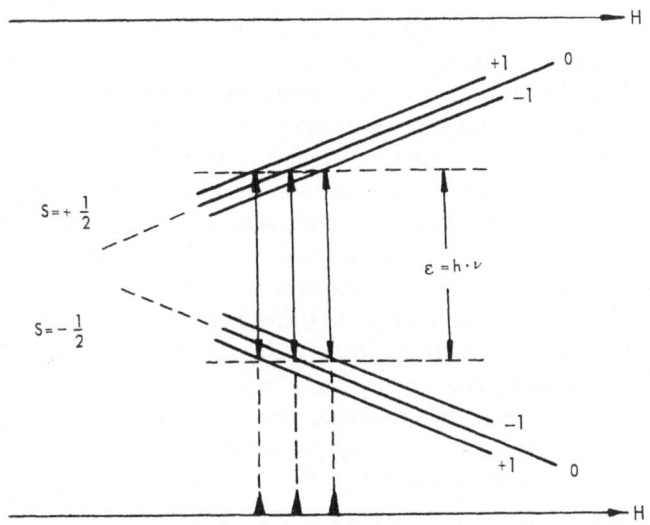

Fig. 23. Hyperfine splitting of the energy levels of an unpaired electron with a nitrogen nucleus.

[8, 9] leads to the conclusion that the unpaired electron interacts with the nuclei of the carbon atoms of the methyl groups (Fig. 24):

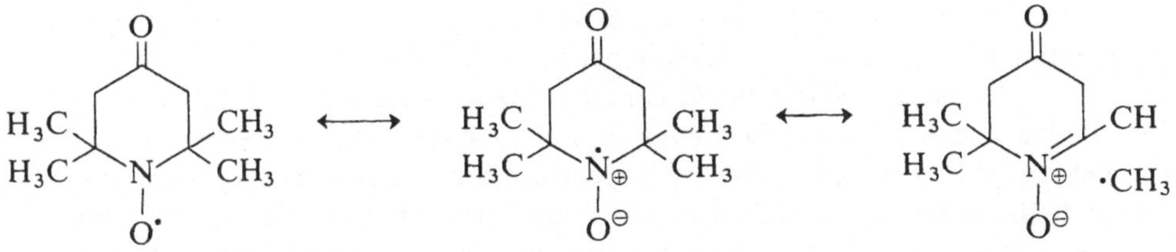

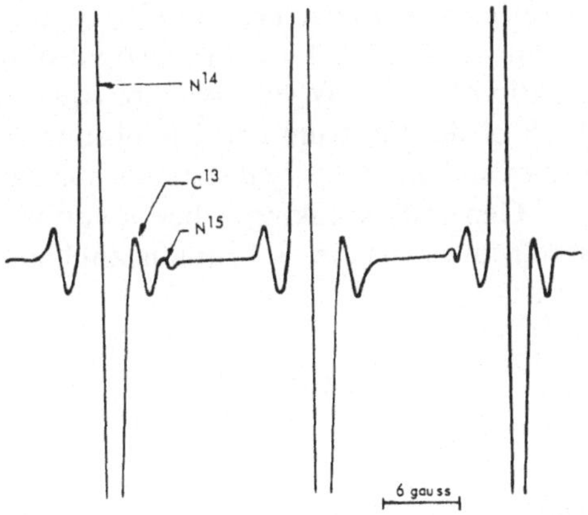

Fig. 24. Appearance in the HFS of the EPR spectrum of 2,2,6,6-tetramethyl-4-oxopiperidine-1-oxyl of a system of satellite lines due to the presence of the natural isotopes of nitrogen ^{15}N and carbon ^{13}C.

Still weaker effects of the interaction of an unpaired electron with protons in the nitroxyl radicals can be observed in some cases with a slow sweep of the magnetic field and a low degree of high-frequency modulation. In these cases, additional splitting of the components of the nitrogen triplet is observed [10].

The presence of such hyperfine structure in the EPR spectra of other radicals studied [11–13] apparently shows the nonzero probability that the unpaired electron will be found beyond the limits of the nitroxyl group of the radical. However, the small value of the hyperfine splitting (about 0.35 G) means that the unpaired electron is almost completely localized in the nitroxyl group and its transfer along σ bonds in the radical rapidly dies away [14].

A similar effect has also been observed by Stock and Suzuki [15] and by Scheffler [16] in a study of some aromatic radicals.

In aliphatic-aromatic, heterocyclic, and aromatic nitrogen oxides, the unpaired electron moves over a strongly delocalized orbital which may comprise several nuclei possessing magnetic moments. Each of these nuclei interacts with the unpaired electron, leading to hyperfine splitting, and the resulting spectra may prove to be fairly complex (see, for example, Figs. 1–3).

There are two cases in which the spectrum can be very easily analyzed:

1. In the case where the unpaired electron interacts more strongly with one of the nuclei ($S = j_1$) than with the other ($S = j_2$): under these conditions, the stronger interaction splits the single absorption line into $2j_1 + 1$ components and the weaker interaction causes an additional splitting of each of these components into $2j_2 + 1$ subcomponents.

2. In the case where the unpaired electron possesses equal links with n similar nuclei ($S = j$), when overlapping of lines with the same distance between them takes place: the EPR spectrum consists of $2nj + 1$ components, with the maximum intensity in the center and intensities falling off symmetrically on either side. Thus, for example, in the hyperfine interaction of the electron with two protons ($S = \frac{1}{2}$), the hyperfine structure of the spectrum comprises $2(2S) + 1$, i.e., three, equidistant lines the intensities of which are in a ratio of $1:2:1$. When there are equal interactions with three protons, the HFS of the spectrum consists of four components with the same splitting values and an intensity distribution in the ratio $1:3:3:1$.

Generally speaking, when n equivalent nuclei ($S = \frac{1}{2}$) are present, the spectrum will have $n + 1$ equidistant lines with an intensity distribution in the form:

$$1:n:\frac{n(n-1)}{2}:\cdots:\frac{n!}{K!(n-K)!}:\cdots:\frac{n(n-1)}{2}:n:1$$

where K assumes values from 1 to $n - 1$.

While the spectrum of a solution of tert-butyl-3,3-dimethylpropynyl-nitroxyl consists of an ordinary triplet, its isopropyl homolog [17], because of the additional interaction of the unpaired electron with one proton ($S = \frac{1}{2}$),

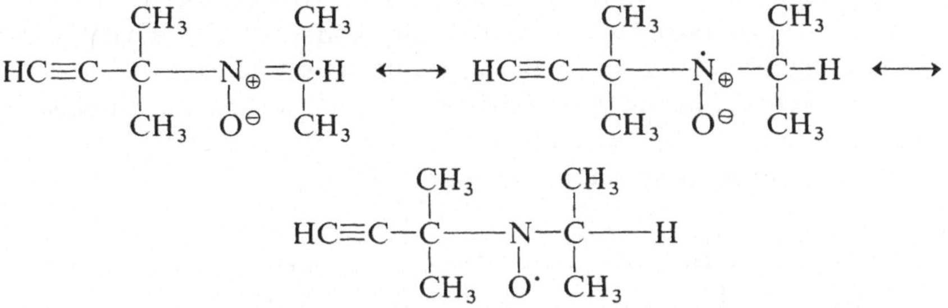

$$
HC\equiv C-\underset{\underset{CH_3}{|}}{\overset{\overset{CH_3}{|}}{C}}----\underset{\underset{CH_3}{|}}{\overset{\overset{O^\ominus}{}}{N_\oplus}}=\underset{\underset{CH_3}{|}}{\overset{\overset{CH_3}{|}}{C}}\cdot H \quad\longleftrightarrow\quad HC\equiv C-\underset{\underset{CH_3}{|}}{\overset{\overset{CH_3}{|}}{C}}----\overset{\cdot}{\underset{\underset{O^\ominus}{}}{N_\oplus}}-\underset{\underset{CH_3}{|}}{\overset{\overset{CH_3}{|}}{C}}-H \quad\longleftrightarrow
$$

$$
HC\equiv C-\underset{\underset{CH_3}{|}}{\overset{\overset{CH_3}{|}}{C}}----\underset{\underset{O^\cdot}{|}}{N}-\underset{\underset{CH_3}{|}}{\overset{\overset{CH_3}{|}}{C}}----H
$$

gives an HFS of the spectrum in which each component of the triplet is split additionally into two lines.

A similar triplet of doublets has been observed in a study of a solution of the pure nitroxide of 6-methoxy-4-methyl-2-spirocyclohexyl-3,4 : 3′,2′-tetra-hydrofurano-1,2,3,4-tetrahydroquinoline [18, 19]:

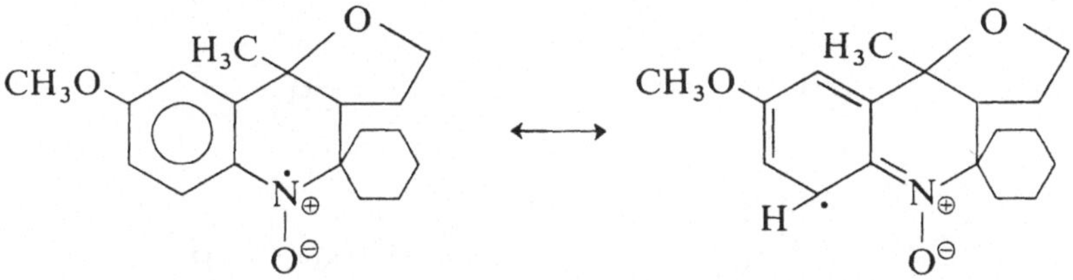

When a benzene solution of 2,2,4-trimethyl-1,2,3,4-tetrahydroquinoline is heated with tert-butyl hydroperoxide in the presence of cobalt stearate, the corresponding paramagnetic nitroxide

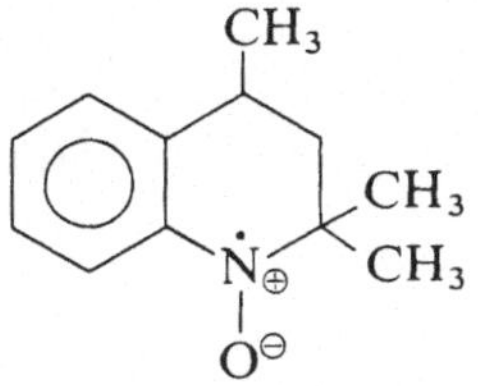

is obtained, and this can be separated by chromatography on alumina in the form of a viscous bright-red oil [20]. The well-resolved spectrum of this radical (Fig. 25) in dilute solution consists of three groups of lines (splitting on the nitrogen nucleus) each of which includes four components with a ratio of the intensities of 1:3:3:1 (approximately equal splitting on the 4, 6, and 8 protons). In turn, each component of the quadruplet is split additionally into three lines, the intensities of which are in a ratio of 1:2:1 (similar weak interaction of the unpaired electron with the 5 and 7 protons).

However, it must be borne in mind that such highly resolved structures of a spectrum are obtained with great difficulty, and we generally have to live with a less highly resolved spectrum consisting of only 12 lines.

In a study of a paramagnetic oxidate [21] and a chromatographically purified free radical [22] it was found that the latter was extremely unstable and rapidly disproportionated in the following way:

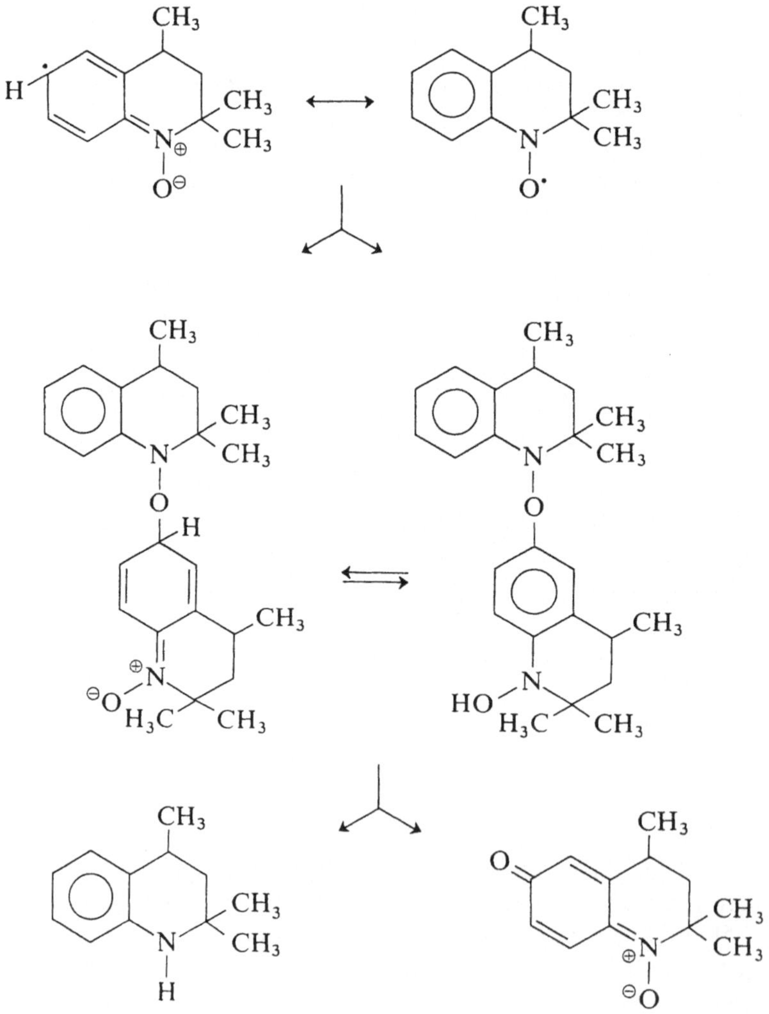

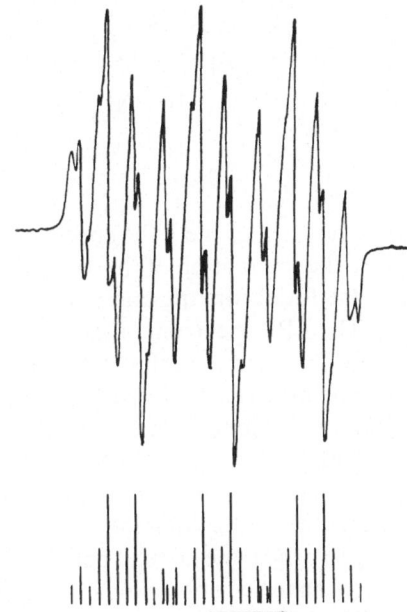

Fig. 25. HFS of the EPR spectrum of a benzene solution of the nitroxide of 2,2,4-trimethyl-1,2,3, 4-tetrahydroquinoline in the absence of oxygen (the HFS of the spectrum calculated theoretically is given below).

By the EPR method, it was established that, in contrast to the nitroxide containing two methyl groups of the geminal type in position 2 of the heterocycle, radicals including the spirocyclohexyl grouping,

(where R = H, OCH$_3$) possess an unusually high stability. Even when such radicals are stored in air for two years, the HFS of their EPR spectra undergoes no change [23].

To explain the reasons for the different stabilities of heterocyclic nitroxides of this type, radicals of different structures were studied in solution. Of all the radicals obtained in solution, only radical III (Table 16) could be isolated in the crystalline state. The bright-red paramagnetic crystals of III were converted in a few minutes into a dark resin for which it was possible to isolate only 2-spirocyclohexyl-1,2-dihydrobenz[5,6]-1,3-oxazine (VI) and to identify the trivial N-oxide of the corresponding quinone imine (VII):

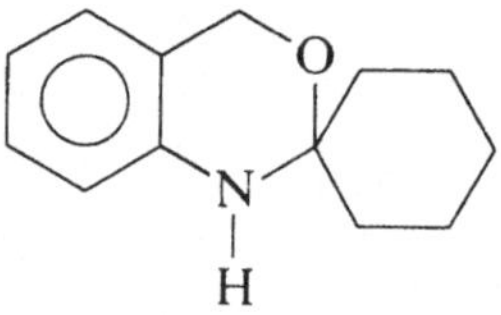

(VI) 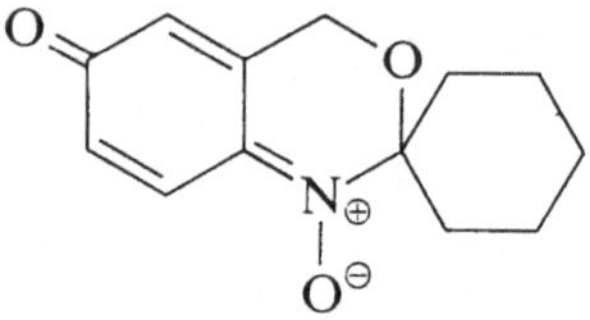   (VII)

The HFS of the EPR spectrum of the nitroxide of 2,2-dimethylindoline shows the interaction of the unpaired electron with the nitrogen nucleus and with the four aromatic and two methylene protons of the radical.

The EPR spectrum of the other radicals listed in Table 16 have approximately the same hyperfine structure.

The majority of the radicals mentioned are destroyed when their dilute solutions are concentrated. However, the rate of destruction of the radicals in vacuum and in the presence of air shows that this process is not connected with oxidative changes. It is a basic fact that during the destruction of these radicals, only the integral intensity of the components of the spectrum changes, while the actual nature of its HFS remains unchanged. Similar features have been observed in a study of pyrryloxyl radicals in solution [24].

The EPR spectra of organomercury nitroxides, solutions of which have been studied recently [25], are peculiar. Thus, the spectrum of p-acetoxymercuriphenyl-tert-butylnitroxide (Fig. 26),

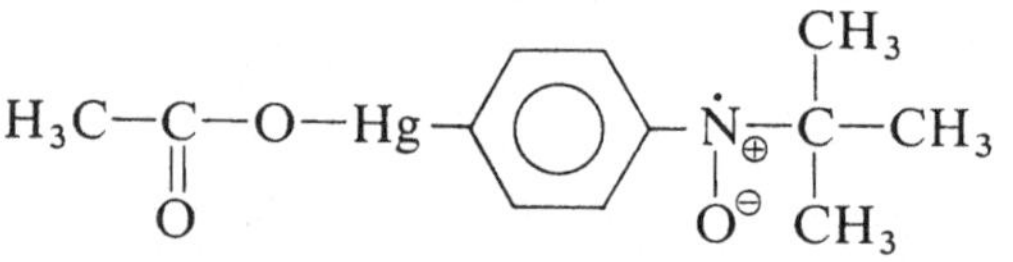

consists of three groups of lines with a Fermi constant of 14.3 ± 0.2 G. The additional splitting is caused by the *ortho* and *meta* protons of the aromatic ring (*ortho*-$a_H = 2.36$, *meta*-$a_H = 1.14$ G, and the spectroscopic splitting factor is 2.0051).

The peripheral groups of lines, each of which copies, as it were, the structure of the nitrogen components of the spectrum, while the splitting between them is also equal to the nitrogen splitting, are clearly separated. There is no doubt that these additional lines are due to the splitting of the electronic Zeeman levels on the nuclei of the magnetic isotopes of mercury with additional subsplitting on the nitrogen nucleus and the protons of the phenylene ring.

The stable isotopes of mercury include only two magnetic ones: ^{199}Hg (spin $\frac{1}{2}$) with a content of 17%, and ^{201}Hg (spin $\frac{3}{2}$) with a content of 13% in natural mercury. Consequently, the line of the ^{199}Hg in the spectrum should be 2.7 times stronger than the line of ^{201}Hg, and the six groups of lines

TABLE 16
Main Parameters of the EPR Spectra of Some Heterocyclic Nitroxides

No.	Nitroxide of	Structure	Splitting constants, G		
			a_N	$a_H^{6,8}$	$a_H^{5,7}$
I	2,2-Dimethyl-1,2-dihydrobenz-[5,6]-1,3-oxazine	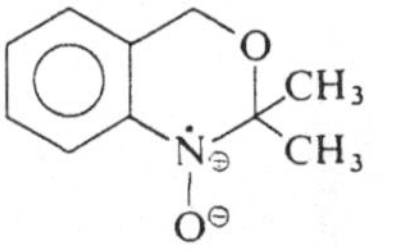	10.8	3.1	0.95
II	2-Ethyl-2-methyl-1,2-dihydrobenz [5,6]-1,3-oxazine	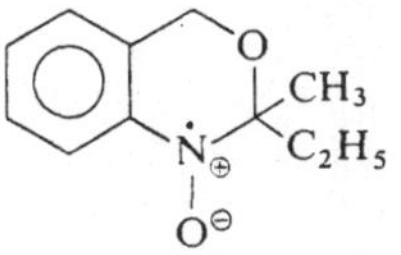	10.6	3.0	1.0
III	2-Spirocyclohexyl-1,2-dihydrobenz [5,6]-1,3-oxazine	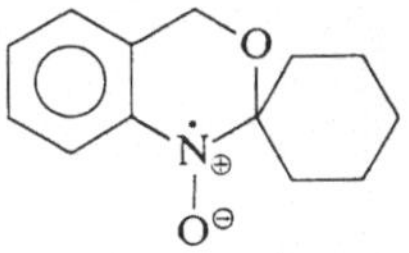	11.0	3.1	—
IV	2,2-Dimethylindoline	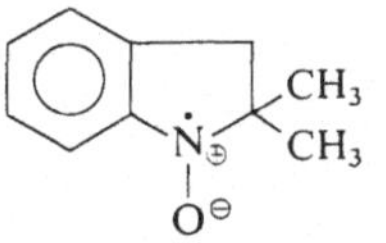	11.26	3.18	0.91
V	1,2,3,4,10,11-Hexahydrocarbazole	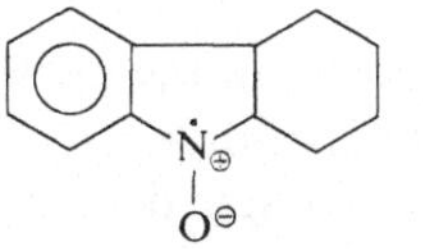	11.48	2.9	—

observed should be due to the ^{199}Hg doublet tripled on the nitrogen nucleus. The hyperfine splitting constant for ^{199}Hg is 44 ± 2 G.

The spectrum of the radical does not show lines due to the natural ^{201}Hg isotope. A theoretical analysis of the expected HFS spectrum shows that the extreme lines of ^{201}Hg coincide with the analogous lines of ^{199}Hg and cannot be observed in the spectrum only if the magnetic moments of the ^{199}Hg and ^{201}Hg nuclei are equal.

If we turn to the literature, we can find extremely contradictory information on the magnetic moments of these elements. The following values (in nuclear magnetons) are given for ^{199}Hg: 0.547, 0.499, 0.5041, and 0.56, and for the ^{201}Hg isotope, correspondingly, -0.607, -0.5591, -0.4438, and -1.110 [26, 27]. However, an analysis of the EPR spectrum leads to the conclusion

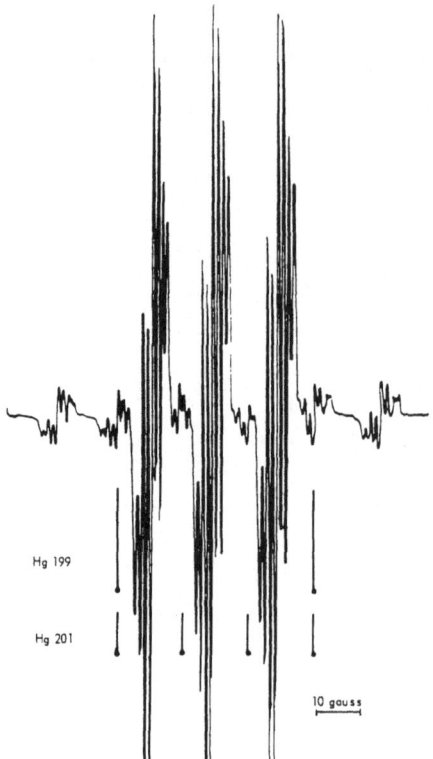

Hg 199

Hg 201

10 gauss

Fig. 26. Hyperfine structure of the EPR spectrum of *p*-acetoxymercuriphenyl-tert-butylnitroxide (in the absence of oxygen).

that the magnetic moments of the nuclei of ^{199}Hg and ^{201}Hg cannot differ essentially in absolute magnitude.

If it is assumed that the spin density in the *ortho* and *para* positions of the phenylene ring are the same, we can obtain for ^{199}Hg a value of the formal McConnell constant (as applied to the case of mercurated nitrogen oxides of this class) of $Q(^{199}\text{Hg})/Q(\text{H}) = 18.7$, where $Q(\text{H})$ is the McConnell constant for a proton.

In the EPR spectrum of the nitroxide of 6,8-diacetyoxymercuri-2,2,4-trimethyl-1,2,3,4-tetrahydroquinoline, each component of the main nitrogen triplet (13.0 ± 0.2 G) is split by the fourth proton of the heterocycle into two lines (4.2 ± 0.2 G). It is noteworthy that in a nonmercurated nitroxide, the same constants have smaller values: 9.7 and 2.8 G, respectively.

An increase in the splittings on the nitrogen nuclei and the protons in the mercurated nitroxides is apparently a general feature for radicals of this class.

The mercury atoms in positions 6 and 8 of the aromatic ring may consist of different isotopes. The possible combinations of the isotopes and the relative intensities of the corresponding lines in the EPR spectrum are given in Table 17.

By making use of the equality $\mu(^{199}\text{Hg}) = \mu(^{201}\text{Hg})$, it is easy to construct the theoretical spectra for each of the combinations given in Table 17.

In comparison with the nonmercurated radicals, the main part of the EPR spectra of the nitrogen oxides of 2,2,4,6-tetramethyl- and 2,2,4,8-tetramethyl-1,2,3,4-tetrahydroquinoline mercurated, respectively, in positions 8 and 6 of the aromatic ring.

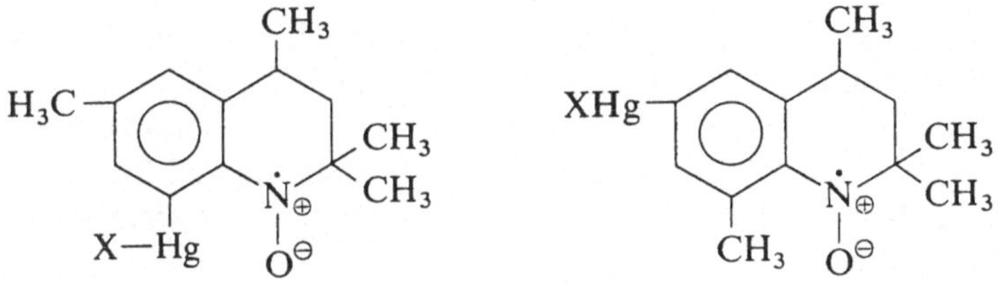

(where $X = \text{OCOCH}_3$, OH, Cl) contain three lines less (substitution of one aromatic proton) and have additional lines in the peripheral fragments of the HFS due to the splitting of the electron levels on the nucleus of ^{199}Hg with additional subsplitting on the nitrogen nucleus and four protons.

From an analysis of the material given above, it may be concluded that in many cases, free radicals can be identified quite unambiguously in solution

TABLE 17

Possible Combinations of Isotopes of Mercury, and Relative Intensities of the Corresponding Lines of the EPR Spectrum of the Nitroxide of 6,8-Diacetoxy-mercuri-2,2,4-trimethyl-1,2,3,4-tetrahydroquinoline

No.	Mercury isotopes		Content, %	Relative intensity of the lines
	Position 6	Position 8		
1	Nonmagnetic	Nonmagnetic	49	100
2	199 Nonmagnetic	Nonmagnetic 199 }	24	24.5
3	201 Nonmagnetic	Nonmagnetic 201 }	18	9.2
4	199 201	201 199 }	4.4	1.1
5	199	199	2.9	1.5
6	201	201	1.7	0.3

from their EPR spectra, particularly if all the atomic nuclei present in the radical are covered by the orbital of the unpaired electron.

Nevertheless, in spite of the enormous possibilities of the method, it cannot always give unambiguous information on the chemical structure of complex organic radicals that have not been isolated in the pure state from a reaction mixture.

Frequently, in order to obtain a paramagnetic oxidate, hydrocarbon solutions of an amine and an organic hydroperoxide are heated in the presence of salts of metals of variable valence [28, 29]. Such oxidation is not infrequently accompanied by side reactions which sometimes change the chemical structure and elementary composition of the radical to a considerable degree. Consequently, in a number of cases, it is impossible solely from the EPR spectra of the unpurified oxidate to make a well-founded choice between several hypothetical structures of a paramagnetic substance obtained in solution [30, 31].

In order to successfully solve problems of the structural chemistry of complex organic free radicals, the main parameters of the HFS of the EPR spectra of *individual* compounds must be studied.

2. Biradicals

As already mentioned, in contrast to free radicals, biradicals possess an even number of electrons, two of which are "free."

In early investigations, the intense color of the substance and its capacity for vigorously adding oxygen, halogens, and nitric oxide were regarded as proof of the biradical state of a molecule. Now it is generally known that the presence of such properties by no means always indicates the biradical state. A bright color* is frequently possessed by valence-saturated systems, and a high chemical activity may be a consequence of the excitation of the molecule into the triplet state.

Beginning in 1935, Müller and his co-workers have made a systematic use of magnetic measurements in the study of biradicals. However, the magnetic method does not permit the reliable identification of a biradical form if it is present in an amount less than 2%. Thus, the violet hydrocarbon obtained in 1907 by A. E. Chichibabin has a very reactive system of bonds, the unusual activity of which was explained by an equilibrium content of a biradical form of the substance:

---

*Even a true biradical may not possess a bright color—for example, oxygen.

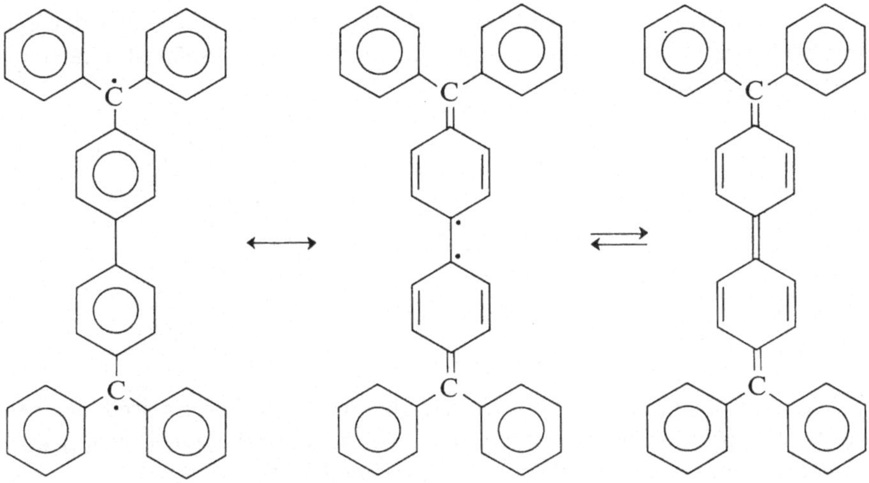

A magnetic study of this hydrocarbon has shown, however, that it is diamagnetic [32], and, consequently, the biradical cannot exist in appreciable concentration.

If in Chichibabin's hydrocarbon, the *ortho* positions of the diphenylene system are occupied by voluminous substituents—for example, chlorine atoms—a genuine biradical, a highly paramagnetic compound, is obtained [33]:

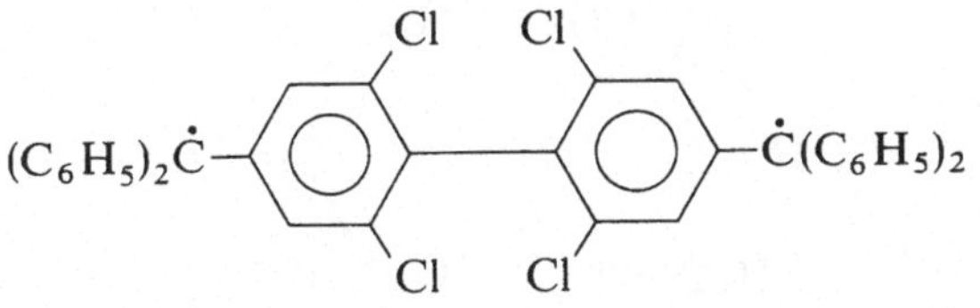

The noncoplanar mutual arrangement of the two phenylene nuclei leads to a considerable independence of the dichlorotriphenylmethyl moieties of the biradical. Thus, when appropriate substituents are introduced into Chichibabin's hydrocarbon, it is possible to obtain in solution (in the absence of oxygen) almost monomeric biradicals with a total magnetic moment $\sqrt{6}\,\mu_B$, agreeing satisfactorily with theory [34].

In spite of the enormous number of studies of biradicals in solution, a number of cardinal questions in the field have remained unanswered. The solution of these problems has been greatly advanced by investigations carried out in the Soviet Union by Dyatkina and Syrkin [35, 36] on the application of the molecular-orbital method to the calculation of the energies of singlet–triplet excitation of various aromatic hydrocarbons.

The main group of biradicals is formed by compounds in which the unpaired electrons are isolated from one another by a system of σ-carbon–carbon bonds—for example,

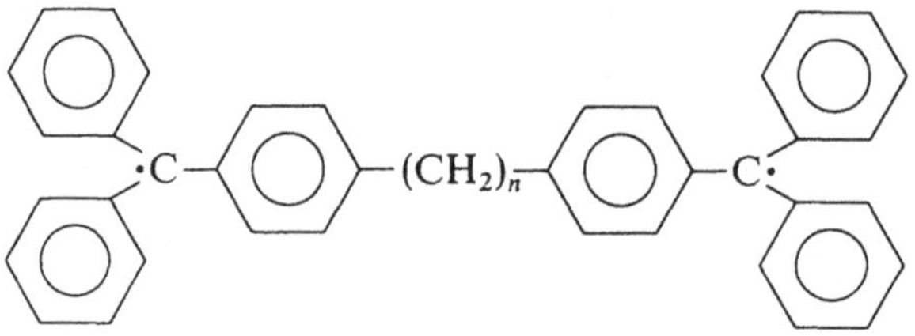

$$\cdot C \text{---} \langle \text{C}_6\text{H}_4 \rangle \text{---} (CH_2)_n \text{---} \langle \text{C}_6\text{H}_4 \rangle \text{---} C\cdot$$

In such triphenylmethyl biradicals, the orbital of the unpaired electron covers the protons of two phenyl groups and one phenylene group in a similar manner to the situation in triphenylmethyl itself.

The presence in the biradical of a bridge of simple σ–σ bonds isolating the free-radical moieties leads to the situation that the latter behave as almost completely independent free radicals and the HFS of their EPR spectra is, according to the results of Jarrett et al. [37], similar to the spectrum of the free triphenylmethyl radical.

In just the same way, the evident independence of the two nitroxyl fragments in biradicals of type

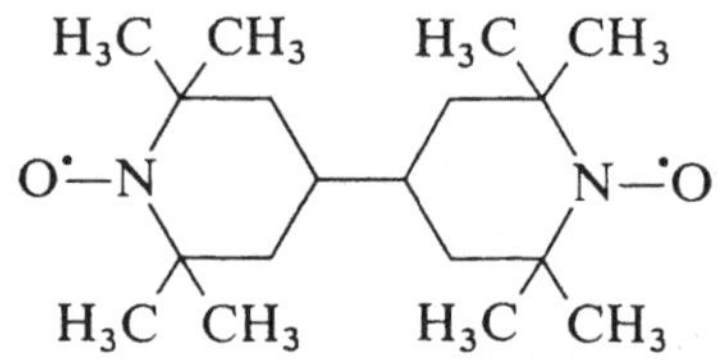

$$O\overset{\cdot}{-}N \langle \rangle \text{---} \langle \rangle N\overset{\cdot}{-}O$$

(with H_3C, CH_3 substituents on each ring)

has given grounds for the assumption that the HFS of their EPR spectra will be similar to the spectra of the corresponding monoradicals. However, in a study of a solution of pure bis(2,2,6,6-tetramethyl-1-oxyl-4-piperidyl) [38], instead of the expected three lines, a five-line EPR spectrum was obtained (Fig. 27).*

The structure of this EPR spectrum is equivalent to the case of the identical interaction of an unpaired electron with both nitrogen atoms (total spin equal to 2) and scarcely differs from the five-component spectrum of 1,1-

---

*This observation, made in 1963, was so unusual that only after numerous discussions was the report of it submitted for publication in the middle of 1964, when the authors had obtained and studied a whole class of this type of biradicals [39].

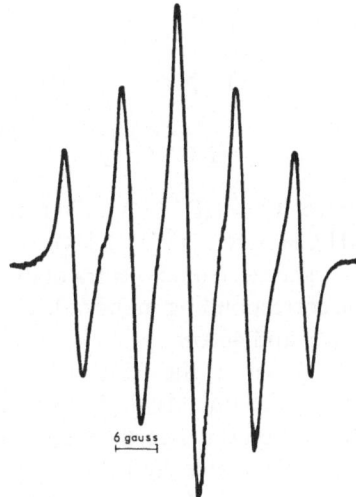

Fig. 27. HFS of the EPR spectrum of a chloroform
solution of bis(2,2,6,6-tetramethyl-1-oxyl-4-piperidyl)
(in the absence of oxygen) [38].

diphenyl-2-picrylhydrazyl, which is explained by the equivalent interaction
of the unpaired electron with the two nitrogen atoms:

Since such interaction cannot exist in the nitroxyl biradical by the migration of the unpaired electron through the system of σ-bonds to the second
nitrogen atom, there is most probably an electronic exchange when the paramagnetic fragments of the biradical approach one another spatially. If this
hypothesis is correct, the length and chemical structure of the connecting
bridge must affect the conformational electronic exchange.

To test the correctness of this hypothesis, a number of individual biradicals with different distances between the nitroxyl groups have been
synthesized and their spectra studied [39, 40]:

$$H_3C \quad CH_3 \qquad\qquad H_3C \quad CH_3$$

O⁻—N〈 〉—O—C—Q—C—O—〈 〉N—⁰O

with C=O (||O) groups beneath, and $H_3C \quad CH_3$ below each ring.

where Q = (II) (—); (III) $(CH_2)_2$; (IV) $(CH_2)_4$; (V) $(C_2H_5)_5$; (VI) $NH(CH_2)_6NH$; (VII) $(CH_2)_8$; (VIII) $p\text{-}C_6H_4$; (IX) $m\text{-}C_6H_4$; (X) $o\text{-}C_6H_4$.

The electronic spectra of these biradicals are identical with the spectra of the corresponding monoradicals and have, as usual, two absorption maxima at 240 and 460 nm.

However, the EPR spectra of the nitroxyl biradicals possess a whole series of characteristic features connected with the effect of conformational electronic exchange through the spatial interaction of the nitroxyl groups. Figure 28 shows the EPR spectra of all ten biradicals in benzene.

The main difference in the EPR spectra consists in the appearance of additional lines between the components of the triplet spectrum characteristic of the mono nitroxyls.

The intensity of the additional lines and their number depend on the structure of the connecting bridge. As already mentioned, in biradical I, where the nitroxyl groups are fairly close to one another (Fig. 29), the HFS of the EPR spectrum contains five lines, with a ratio of the intensities close to that observed when two equivalent nitrogen nuclei take part in the splitting.

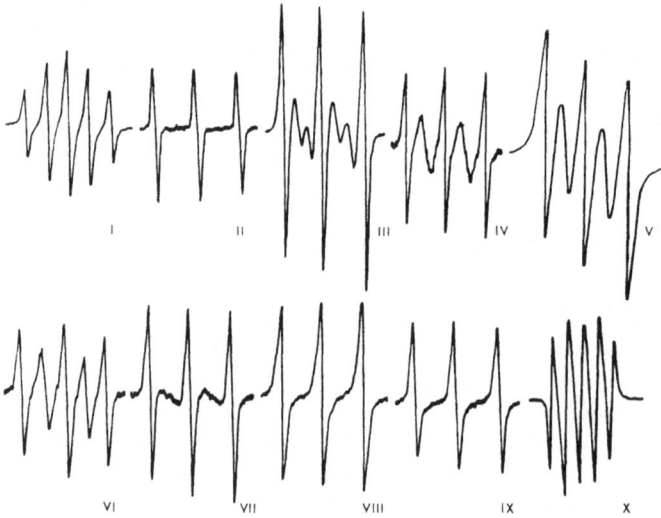

Fig. 28. EPR spectra of the biradicals I–X in benzene at 20°C (the numbers on the spectra correspond to the numbers of the substances).

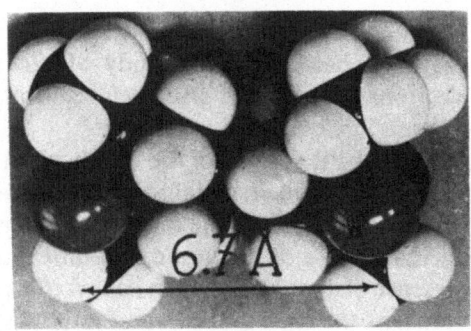

Fig. 29. Stuart–Briegleb structural model of bis(2,2,6,6-tetramethyl-1-oxyl-4-piper-idyl).

In the EPR spectrum of biradical II, where the oxalate bridge, OCOCOO, is incapable of bringing the radical centers together, no additional lines whatever appear, and the spins in this biradical behave independently.

In biradical III, the connecting group increases the freedom of conformational movement, which enables the paramagnetic centers to approach one another fairly closely. This circumstance leads to the appearance in the EPR spectrum of four weak additional lines.

The increase in the length of the connecting bridge in biradicals IV and V favors an increase in the interaction of the spins; however, even in biradicals VI and VII, the mutual separation of the radical moieties begins to play a predominant role, as is shown appropriately in the EPR spectra.

The absence of additional splitting on the nitrogen nuclei of the heterochain bridge in VI is an additional proof of the impossibility of the migration of an unpaired electron through the system of σ–σ bonds of the biradical.

In the biradicals VIII and IX, the electron spins behave independently (triplet EPR spectra), and in the biradical X, where the paramagnetic moieties are in the *ortho* positions of the aromatic ring, spin–spin interaction appears very clearly in the EPR spectrum.

The spectroscopic splitting factors of the biradicals coincide with the *g*-factors of the corresponding monoradicals, and the lines in the spectra are not broadened to any appreciable extent through dipole–dipole interaction. Measurements of the total intensity of the EPR spectra have shown that each paramagnetic particle bears two unpaired electrons, i.e., it is a true biradical [41].

The EPR spectra of the biradicals strongly depend on the temperature. For example, in biradical VIII, the groups are so rigidly connected with one another by the benzene ring that even at 150°C in hexane, spin–spin interaction is very weak and the EPR spectrum remains triplet.

All the temperature changes in the EPR spectra are reversible, and the total number of spins does not change; there is only a change in the ratio of the intensities of the components of the spectra.

Some biradicals of types VI and VII may apparently serve as models of polymer chains, and the study of exchange interactions of the unpaired electrons in them can give useful information on the form and behavior of polymeric molecules in solutions.*

The EPR spectra of the biradicals also depend on the nature of the solvent, and here an essential role may be played not only by the viscosity of the solvent, which affects the intensity of intramolecular motions, but also by the microstructure of the liquid and different solvation effects. In all cases, the presence of dissolved oxygen has no effect on the form of the spectra and only broadens the lines of the hyperfine structure. In no case has it been possible to observe EPR signals with a g-factor of 4, corresponding to the $\Delta m = 2$ transitions. The splitting is 15.6 G in the triplet EPR spectra and 7.8 G in the quintuplet spectra. The latter do not depend on the exchange frequency of the electron spins. This means that electronic exchange does not alter the spin distribution in the nitroxyl fragment.

The observed features of the EPR spectra or biradicals of the nitroxyl series can be explained quantitatively by the assumption of the presence of a weak exchange interaction between the unpaired electrons of the two paramagnetic centers. Since, as was shown above, this interaction cannot take place through the σ-bond system, it is natural to assume that it takes place through the intramolecular approach of the two moieties of the biradical each bearing an unpaired electron. It is obvious that to bring about the exchange conformation effect in the biradicals, a definite energy barrier (of the order of 10 kcal/mole) connected with some deformation of the chemical bonds and the angles between them must be overcome.

In actual fact, in the case of the biradicals VIII, IX, and X,

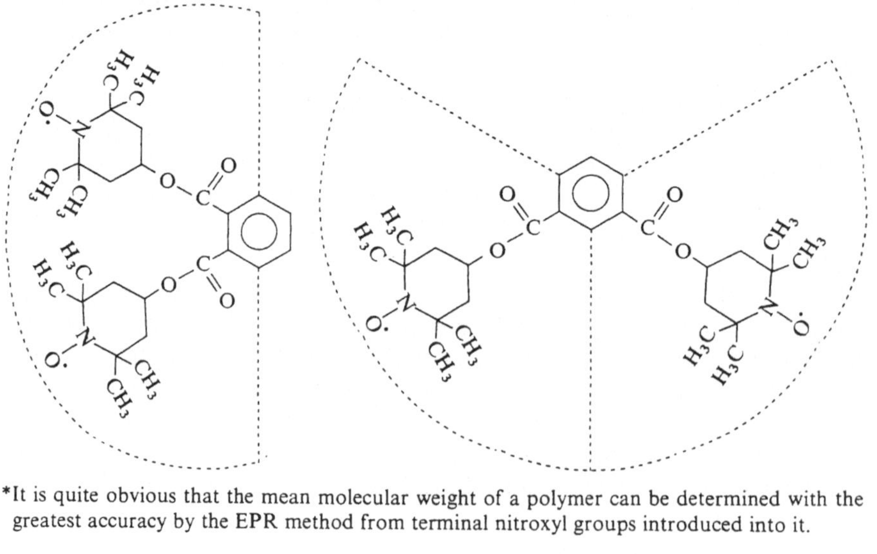

*It is quite obvious that the mean molecular weight of a polymer can be determined with the greatest accuracy by the EPR method from terminal nitroxyl groups introduced into it.

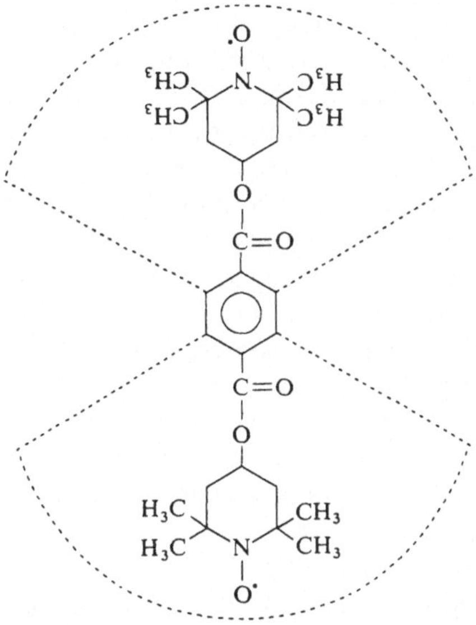

the exchange effect in compound X is found even at room temperature, while for IX the quintuplet spectrum arises only at a temperature of 140°C, and a solution of the biradical VIII does not differ in the nature of the HFS of the EPR spectrum (triplet) from solutions of monoradicals of this class at 150°C [40].

To explain the mechanism of such a peculiar transfer of excitation from one paramagnetic center to another, other biradicals of very different structures were studied [42–45]—for example,

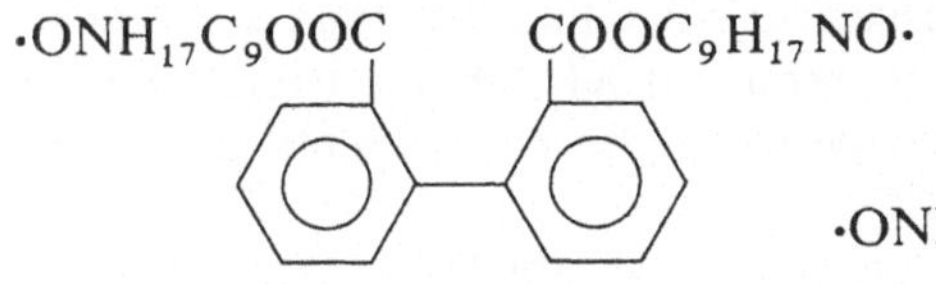

(XI) 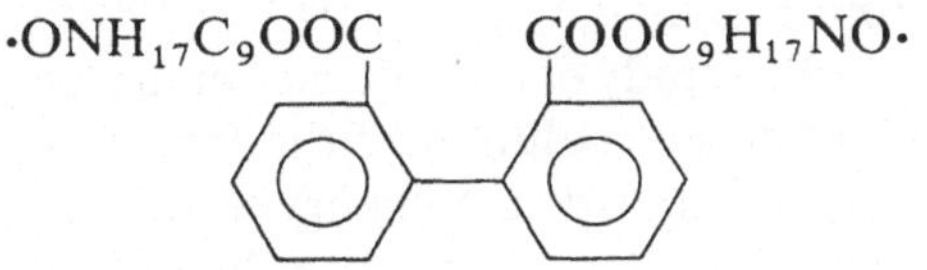   (XII)

where $COOC_9H_{17}NO\cdot$ represents the piperidyl-1-oxyl residue.

In a consideration of the structural formulas of these biradicals, it may be assumed *a priori* that in the case of compound XI, the realization of exchange interaction will be associated with the overcoming of a higher energy barrier than for di(2,2,6,6-tetramethyl-1-oxyl-4-piperidyl) biphenyl-2, 3'-dicarboxylate (XII).

At room temperature, the spectrum of an acetic acid solution of compound XI exhibits three lines with a splitting of 17.8 G. On passing to less polar solvents (benzene, chloroform, dioxane), this spectrum is transformed into a quintuplet with a smaller splitting between the components. In benzene and chloroform, each additional line of the spectrum is split further into a doublet with a distance between its components of about 2.6 G.

If the spectra obtained are compared with those calculated theoretically, it is easy to see that the exchange frequency in the biradical XI is apparently commensurable with the frequency of the hyperfine interaction.

In contrast to the case just considered, the EPR spectra of XII scarcely depend on the polarity of the solvent.

All the spectra consist of five lines and differ only in the ratio of the intensities of the intermediate lines. The splitting between the components of the quintuplet for benzene and dioxane solutions is 8.2 G, and for chloroform and acetic acid solutions, 8.6 G.

While the spectra of the biradicals XI and XII in dioxane solutions are identical, in acetic acid solutions they differ fundamentally. These differences are apparently connected with the specific influence of solvation on the frequency of the electronic exchange and the rate of the conformational transitions.

From an analysis of the temperature dependence of the HFS of the EPR spectra of a chloroform solution of the biradical XI, it may be concluded that the magnitude of the splitting and the ratio of the intensities between the components of the triplet do not change with a change in the temperature. At the same time, a rise in the temperature regularly leads to a decrease in the distance between the intermediate doublet lines, and the intensities of the lines rise.

A similar pattern has been observed for solvents other than acetic acid. In the latter case, the spectrum of the biradical XI remains triplet at high temperatures, and only the distance between the components decreases, from 17.8 (at 22°C) to 17.0 G (at 120°C).

In a study of the influence of the temperature on the HFS of the spectrum of a benzene solution of the biradical XII, it was found that a rise in the temperature led to a change in the ratio of the intensities of both the main triplet and the intermediate lines. It is worthy of note that the spectra of chloroform and acetic acid solutions of the substance do not change with the temperature.

It is an important fact that all the observed changes in the EPR spectra of the radicals studied are reversible and therefore cannot be the result of any chemical conversions.

It can be stated that with a rise in the temperature, the EPR spectrum of a biradical gradually approximates to the theoretically expected quintuplet.

These changes in the spectrum may be connected with an increase in the frequency of electronic exchange between the paramagnetic centers, since the theoretical quintuplet spectrum must correspond to a higher frequency of electronic exchange than the septet spectrum corresponding to the same biradical.

In actual fact, at a high frequency of electronic exchange (greater than the frequency of hyperfine interaction, $2.2 \times 10^7 \text{ sec}^{-1}$), the mean residence times of each electron at the two nuclei will be the same, and the corresponding HFS of the EPR spectrum will consist of five equidistant lines with a ratio of the intensities of $1:2:3:2:1$.

In agreement with the quantum-mechanical calculations [46], at a frequency of the electronic exchange commensurable with the frequency of the hyperfine interaction, the spectrum must contain $(2S + I)^2 + 2$ components (when $S = 1$, a maximum of 11 lines). The observed septet spectra evidently correspond to an intermediate value of the electronic exchange frequency.

Approximately six months after the first report [38], French scientists under the direction of Rassat synthesized a new series of nitroxyl radicals, including some labeled with the isotope ^{15}N, and definitively showed the general nature of the phenomenon discovered [47, 48].

It is worthy of note that nine months later, the Kazan' chemists also submitted a communication on a similar effect in the case of the bishydrazyl radical [49, 50]:

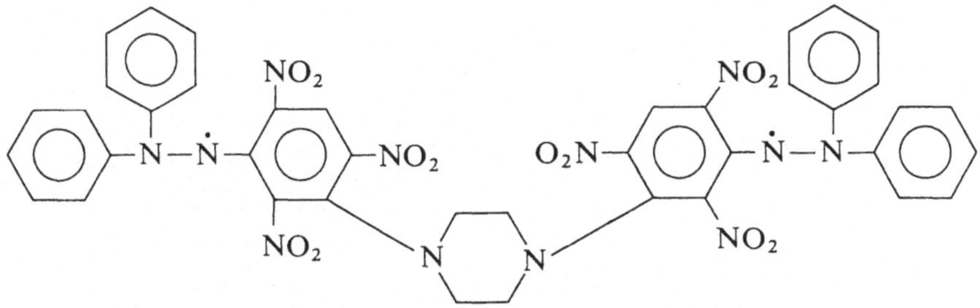

3. Polyradicals

The study of a large number of individual polyradicals first obtained in the Institute of Chemical Physics of the Academy of Sciences of the USSR [40, 51, 52] has also led to extremely interesting observations.

Just as in the case of biradicals, the absorption spectra of tri- and tetraradicals in the physical and ultraviolet region scarcely differ from the corresponding spectra of monoradicals, but their EPR spectra exhibit interesting features [53].

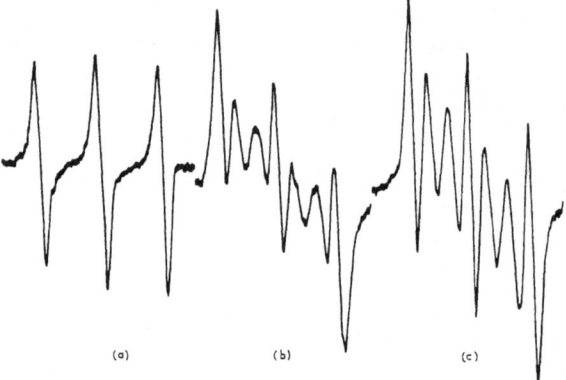

Fig. 30. EPR spectra of tri-radical I in hexane: (a) 20, (b) 130, (c) 150°C.

Figure 30 gives the EPR spectra of a hexane solution of the triradical

(I)

At 20°C, all three spins behave independently (triplet EPR spectra due to the nitrogen nucleus), but with a rise in the temperature, when the intensity of intramolecular motions increases, all the spins begin to participate in exchange, which leads to the appearance in the EPR spectra of four more lines with gradually increasing intensity.

In the triradical II, which possesses the structure of a phosphite,*

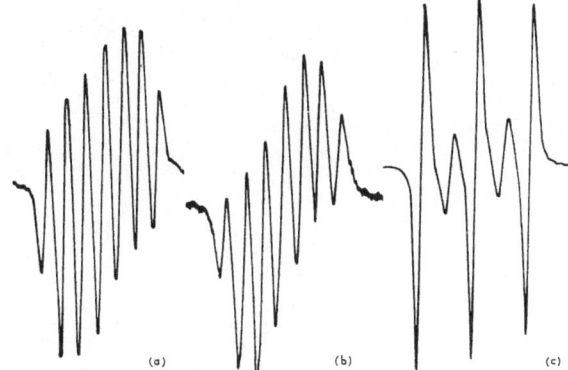

(II)

the intensity of the interactions of the paramagnetic moieties of the radical is so high that it leads to the exchange of all three spins even at 20°C (in ethanol, hexane, and chloroform); consequently, the EPR spectrum consists of seven lines (Fig. 31). However, in acetic acid the spin exchange is greatly retarded, and the EPR spectrum shows only spin exchange through interactions in pairs [Fig. 31(c)]. This dependence of the form of the spectrum on the solvent confirms the characteristics that were found in a study of the biradicals.

Fig. 31. EPR spectra of triradical II at 20°C: (a) chloroform, (b) ethanol, (c) acetic acid [53].

(a)   (b)   (c)

* The same remarks apply in full to the corresponding phosphate.

At 20°C in benzene the EPR spectrum [Fig. 32(a)] of a tetraradical with the pyromellitate structure,

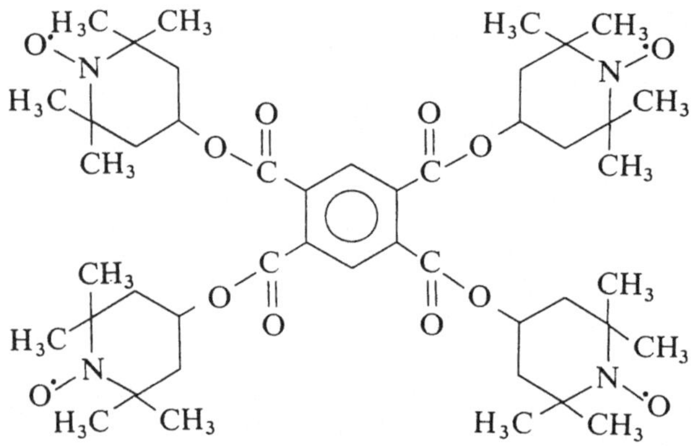

(III)

resembles the spectrum of the corresponding biradical with the phthalate structure. This means that the spins take part in exchange only in pairs with their closest neighbors. Where there are intense "intramolecular" motions (138°C, in benzene), all four spins take part in electronic exchange and nine lines appear in the EPR spectra, the ratio of the intensities being similar to what they should be if an electron was delocalized over four equivalent nitrogen nuclei [Fig. 32(b)].

In a tetraradical with the orthosilicate structure,

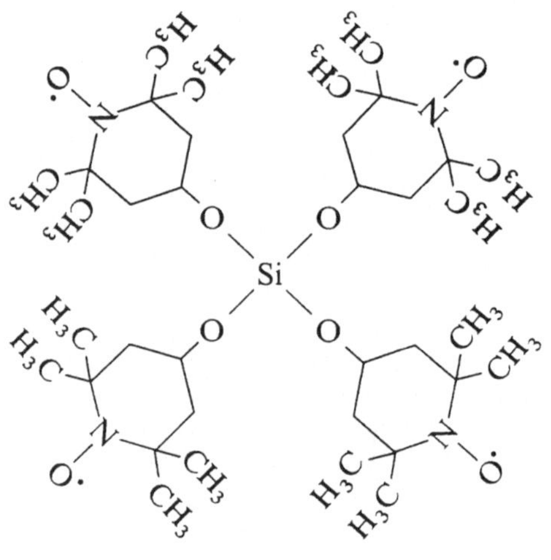

(IV)

even at 20°C, the intensity of the "intramolecular" motions is sufficient for all four spins to take part in exchange, and in this case, the EPR spectrum again contains nine lines [40].

For all the polyradicals, the splitting in the triplet spectra is 15.6 G, in the five-line spectrum, 7.8 G, and in the nine-line spectrum, 4.0 G. The g-factors of the tri- and tetraradicals studied are identical with the g-factors of the corresponding monoradicals. In none of the polyradicals was it possible to detect electronic transitions with $\Delta m > 1$.

To this point, the exchange of only two electrons has been considered. Here, however, cases will be discussed in which the number of unpaired electrons taking part in exchange is greater than two. In view of this, the question arises of the mechanism of the exchange interactions. An analysis of the results obtained excludes the possibility of exchange through ternary, quaternary, and similar interactions of free-radical centers. We must consider only the possibility of a successive "relay" transition of excitation from spin to spin through interactions in pairs [53].

It is desirable to consider from this point of view the EPR spectra of the two tetraradicals V and VI with a biphenyl connecting bridge [54]:

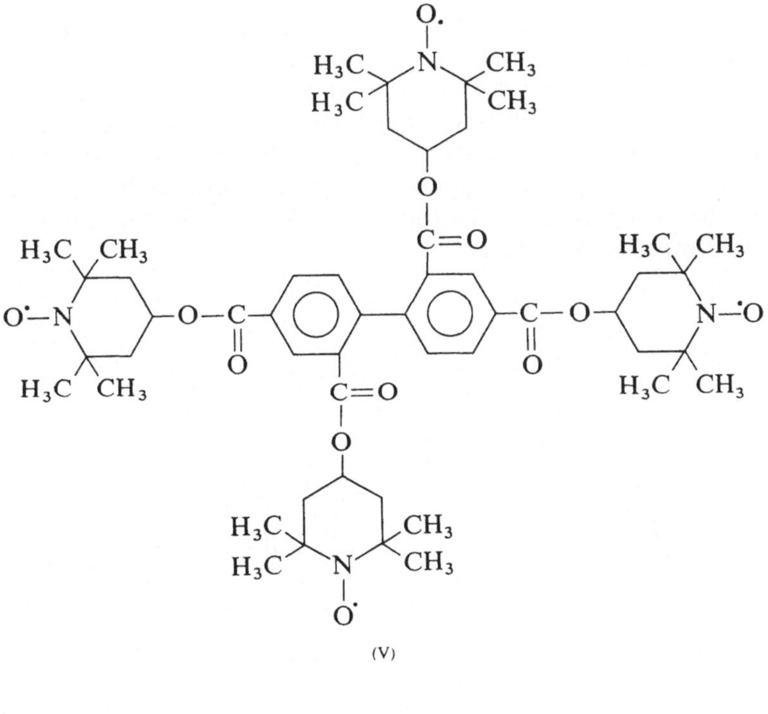

(V)

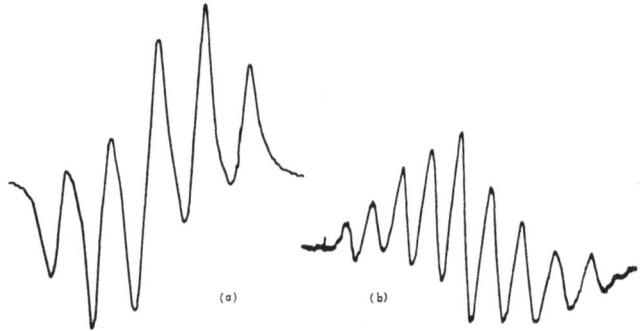

(VI)

As was to be expected, the EPR spectrum of tetra(2,2,6,6-tetramethyl-1-oxyl-4-piperidyl) biphenyl-2,2'-4,4'-tetracarboxylate (V) in acetic acid at room temperature consists of three lines, and in chloroform, dioxane, and benzene, of seven lines. The splitting between the components of the basic triplet in acetic acid solution (17.4 G) is somewhat higher than in the other solvents (16.4 G).

With a rise in the temperature, the HFS of the EPR spectra of the tetraradicals gradually approximate to the theoretically expected ($2nS + 1$, where $S = 1$) nonuplet spectrum corresponding to the very high frequency of electronic exchange between the four paramagnetic substituents.

With a rise in the temperature, the original septet spectrum of compound V in dioxane is gradually transformed into a nonuplet, whereupon the intensity of the intermediate lines increases and the ratio of the intensities of the basic triplet changes correspondingly. It is fundamental that the splitting between the components of the triplet (16.3 G) and the nonuplet (4.07 G) does not depend on the temperature.

Fig. 32. EPR spectra of tetraradical III in benzene: (a) 20, (b) 138°C [53].

Similar features are observed in the case of the second tetraradical studied (VI).

As for biradicals, the widths of the intermediate lines of the HFS in the spectra of radicals V and VI become commensurable with the widths of the triplet lines at sufficiently high temperatures; for example, for solutions of radical V, this characteristic is found at 102°C in acetic acid, benzene, and dioxane.

Similar phenomena occur in the case of a benzene solution of tetra-(2,2,6,6-tetramethyl-1-oxyl-4-piperidyl)   biphenyl-2,2',5,5'-tetracarboxylate. In this case, at high temperatures, the HFS of the EPR spectrum consists of nine well-resolved lines with a separation of 4.05 G.

By comparing the material obtained from a study of mono-, bis-, tris-, and tetrakis-nitroxyl radicals, we can state that the experimental results are in complete qualitative agreement with the theory developed by Sakhun *et al.* [46].

From the practical point of view, the effect of conformational electronic exchange can apparently be used, for example, for solving some special problems of structural chemistry and, in particular, for determining the stereochemical configuration of organic molecules [55]. By using the following compounds, for example [56],

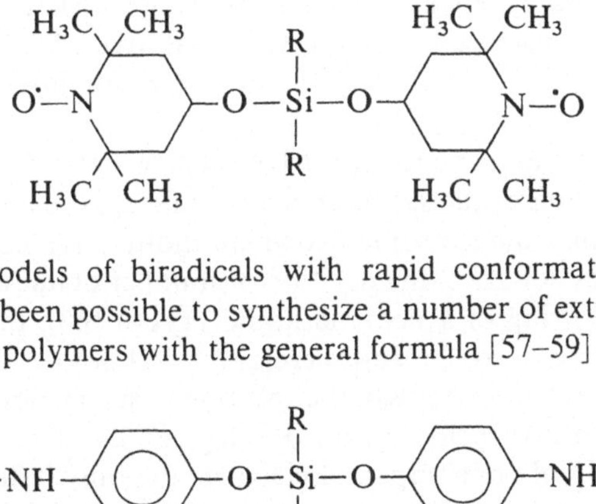

as structural models of biradicals with rapid conformational electronic exchange, it has been possible to synthesize a number of extremely powerful antioxidants for polymers with the general formula [57–59]

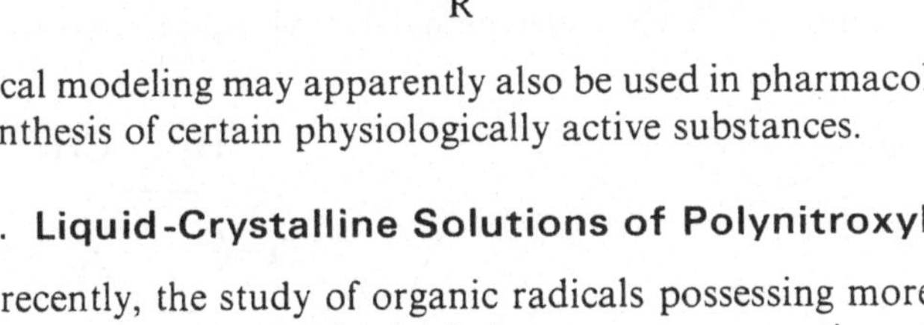

Biradical modeling may apparently also be used in pharmacology in the directed synthesis of certain physiologically active substances.

4. Liquid-Crystalline Solutions of Polynitroxyls

Until recently, the study of organic radicals possessing more than one unpaired electron by the method of electron paramagnetic resonance in solutions was hampered either by the short lives of these molecules or by the strong dipole interaction between the electrons. Now it is possible to create stable biradicals which have small values of the splitting in zero field and therefore give well-resolved spectra in solution. The use of liquid crystals [60, 61] as solvents may prove to be of great help for the identification of such radicals. The detailed development of an elegant method for identifying polyradicals in which the effect of conformational electronic exchange does not appear is due to Luckhurst [66].

The form of the EPR spectrum in a solution of a biradical depends critically on the magnitude of the scalar interaction between the two unpaired electrons, i.e., on the exchange integral I. As an example, let us consider a biradical in which the two electrons can interact with two equivalent nitrogen atoms. When I is much greater than the hyperfine splitting on the nitrogen atom, the spectrum consists of five lines separated by intervals equal to $a_N/2$ with a ratio of intensities of $1:2:3:2:1$, which is characteristic for two equivalent nitrogen atoms.

If I is approximately equal to a_N, the hyperfine structure of the spectrum is more complex and consists of more than five lines [41, 46]. In all cases, the observation of such EPR spectra confirms the presence of two electrons in one molecule.

If I is considerably less than a_N, the spectrum consists of only three lines of equal intensity separated by an interval equal to a_N. This simple spectrum can be obtained from a radical with one electron interacting with one nitrogen atom.

Although the scalar interaction I between two electrons may be small, the dipole interaction or the splitting in a zero field, D, is fairly considerable. In normal solutions, the isotropic Brownian motion averages out D to zero and therefore does not affect the position or number of lines in the spectrum. In a solvent consisting of a nematic liquid crystal [63], the motion of the biradical will not be isotropic, and therefore D will not be zero and will split each line of the spectrum into two: the splitting in such a doublet will depend on the splitting in zero field and on the degree of orderedness [62]. If the radical contained only one unpaired electron, the only influence of a liquid crystal solvent should be a decrease in the hyperfine splitting on the nitrogen atom. For example, di(2,2,6,6-tetramethyl-1-oxyl-4-piperidyl) terephthalate [39]:

in isotropic *p*-azoxyanisole at 140°C (nematic liquid crystal) gives a simple triplet spectrum [39–41]. However, when the temperature is lowered below the point of phase transition (135°C), a nematic mesophase is formed, and the EPR spectrum at 130.4°C consists of a sextet. This doubling of the spectrum clearly shows that di(2,2,6,6-tetramethyl-1-oxyl-4-piperidyl) terephthalate is in fact a biradical. For a polyradical containing *n* unpaired electrons, each component of the isotropic spectrum will be split into *n* lines in a nematic mesophase [64, 65].

Thus, Luckhurst's method has been used in a study of the triradical possessing the structure [39–41]

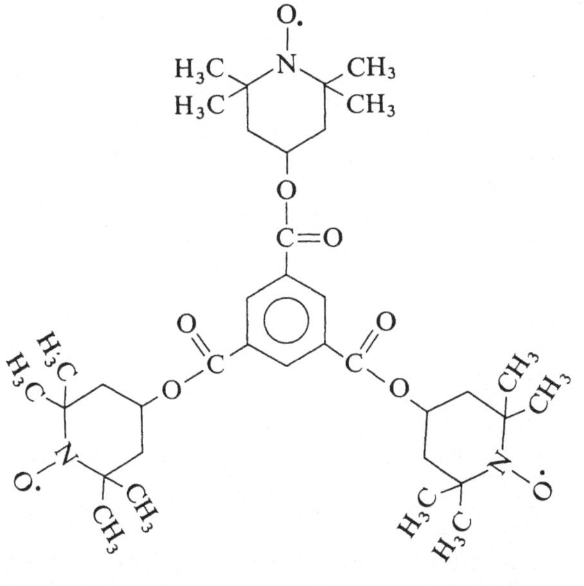

In an attempt to measure the anisotropic spectrum of the triradical in *p*-azoxyanisole, it was possible to obtain only three very broad lines. However, when the *p*-azoxyanisole was replaced by anisaldehyde azine, which possesses a high degree of nematicity at temperatures below 180°C (phase transition), it was possible to obtain a normal HFS, each component of which was split into three lines [66]. This result can be considered as yet another convincing proof of the correctness of the structure of this compound assumed previously, including three unpaired electrons [51].

On taking the features observed into consideration, it may be expected that polyradical models will prove useful in medical and biological investigations, since many materials of vegetable and animal origin possess the structure of liquid crystals of the nematic, smectic, and cholesteric types: myosin, collagen, DNA, mitochondrin, liquid-crystalline precipitates of cholesterol esters, and many polypeptides, enzymes, and lipoids.

References

1. D. Ingram, Free Radicals As Studied by Electron Resonance, Butterworths, London (1958).
2. L. A. Blyummenfel'd, V. V. Voevodskii, and A. G. Semenov, The Use of EPR in Chemistry, Novosibirsk (1962).
3. L. A. Blyummenfel'd and V. V. Voevodskii, *Usp. Fiz. Nauk* **68**, 32 (1959).
4. S. A. Al'tshuler and B. M. Kozyrev, Electron Paramagnetic Resonance. Fiz. Mat. Izd., Moscow (1961).
5. D. Ingram, Spectroscopy at Radio and Microwave Frequencies, Butterworths, London, 2nd ed. (1967).
6. V. Gordy, W. Smith, and R. Trambularo, Microwave Spectroscopy, John Wiley and Sons, New York (1953).
7. S. P. Solodovnikov, Signals from the Micro World, Izd. Akad. Nauk SSSR, Moscow (1963).
8. Yu. G. Mamedova, Thesis, Moscow (1965).
9. H. Lemaire, A. Rassat, and P. Servoz-Gavin, *J. Chim. Phys.* **1962**, 1247.
10. A. L. Buchachenko and É. G. Rozantsev, *Izv. Akad. Nauk SSSR, Ser. Khim.* **1964**, 556.
11. R. Bonnett, V. Clark, and A. Todd, *J. Chem. Soc.* **1959,** 2109.
12. R. Kreilick, *J. Chem. Phys.* **46**, 4260 (1967).
13. V. A. Golubev, Thesis, Moscow (1967).
14. A. L. Buchachenko and A. M. Vasserman, *Zh. Strukt. Khim.* **8**, 27 (1967).
15. L. Stock and I. Suzuki, *Proc. Chem. Soc.* **1962**, 136.
16. K. Scheffler, *Z. Electrochem.* **65**, 439 (1961).
17. É. G. Rozantsev and R. S. Burmistrova, *Dokl. Akad. Nauk SSSR* **166**, 135 (1966).
18. A. B. Shapiro, É. G. Rozantsev, L. S. Povarov, and V. N. Grigos, *Izv. Akad. Nauk SSSR, Ser. Khim.* **1965**, 1102.
19. A. B. Shapiro, Thesis, Moscow (1967).
20. A. A. Medzhidov, A. L. Buchachenko, É. G. Rozantsev, and M. B. Neiman, *Izv. Akad. Nauk SSSR, Ser. Khim.* **1963**, 1713.
21. G. A. Razuvaev and M. L. Khidekel', *Korrelyats. Uravneniya v Organ. Khim.* (*Tartu*) **1**, 365 (1962).
22. A. B. Shapiro, A. A. Medzhidov, and É. G. Rozantsev, *Zh. Organ. Khim.* **2**, 1873 (1966).
23. L. S. Povarov, A. B. Shapiro, and É. G. Rozantsev, *Izv. Akad. Nauk SSSR, Ser. Khim.* **1966**, 339.
24. É. G. Rozantsev, A. A. Medzhidov, and M. B. Neiman, *Izv. Akad. Nauk SSSR, Ser. Khim.* **1963**, 1876.
25. A. B. Shapiro, A. L. Buchachenko, A. A. Medzhidov, and É. G. Rozantsev, *Zh. Strukt. Khim.* **7**, 187 (1966).
26. W. Kunz and I. Schintlmaister, Tabellen der Atomkerne, Vol. 2. Berlin, 1959.
27. G. Konferman, Nuclear Moments [Russian translation], IL, Moscow, 1960.
28. A. L. Buchachenko, *Opt. i Spektroslopiya* **13**, 795 (1962).
29. G. Chapelet-Letourneux, H. Lemaire, and A. Rassat, *Bull. Soc. Chim. France* **1965**, 3283.
30. É. G. Rozantsev and M. B. Neiman, *Zh. Organ. Khim.* **1**, 1337 (1965).
31. L. A. Kalashnikova, M. B. Neiman, É. G. Rozantsev, and L. A. Skripko, *Zh. Organ. Khim.* **2**, 1529 (1966).
32. E. Müller and I. Müller-Rodloff, *Ann. Chem.* **517**, 134 (1935).
33. E. Müller and H. Neuhoff, *Ber.* **72**, 2063 (1939).
34. E. Müller, *Chem. Z.* **77**, 203 (1953).

35. M. E. Dyatkina and Ya. K. Syrkin, *Izv. Akad. Nauk SSSR, Otd. Khim. Nauk* **1965**, 543.
36. M. E. Dyatkina and Ya. K. Syrkin, *Usp. Khim.* **16**, 29 (1947).
37. H. Jarrett, G. Sloan, and W. Vaughan, *J. Chem. Phys.* **25**, 697 (1956).
38. É. G. Rozantsev, V. A. Golubev, and M. B. Neiman, *Izv. Akad. Nauk SSSR, Ser. Khim.* **1965**, 393.
39. É. G. Rozantsev, V. A. Golubev, M. B. Neiman, and Yu. V. Kokhanov, *Izv. Akad. Nauk SSSR, Ser. Khim.* **1965**, 572.
40. É. G. Rozantsev and V. A. Golubev, *Izv. Akad. Nauk SSSR, Ser. Khim.* **1965**, 718.
41. A. L. Buchachenko, V. A. Golubev, A. A. Medzhidov, and É. G. Rozantsev, *Teor. i Éksperim. Khim.* **1**, 250 (1965).
42. L. A. Krinitskaya, A. L. Buchachenko, and É. G. Rozantsev, *Zh. Organ. Khim.* **2**, 1301 (1966).
43. É. G. Rozantsev, *Teor. i Éksperim. Khim.* **2**, 415 (1966).
44. É. G. Rozantsev, V. A. Golubev, V. I. Suskina, and M. B. Neiman, *Zh. Organ. Khim.* **2**, 2235 (1966).
45. É. G. Rozantsev, V. A. Golubev, V. N. Suskina, and M. B. Neiman, *Zh. Organ. Khim.* **2**, 2235 (1966).
46. V. P. Sakhun, A. L. Buchachenko, and I. B. Aleksandrov, *Teor. i Eksperim. Khim.* **2**, 269 (1965).
47. R. Dupeyre, H. Lemaire, and A. Rassat, *J. Am. Chem. Soc.* **81**, 3111 (1965).
48. R. Briere, R.-M. Dupeyre, H. Lemaire, C. Morat, A. Rassat, and P. Rey, *Bull. Soc. Chim. France* **1965**, 3290.
49. Yu. M. Ryzhmanov, Yu. V. Yablokov, B. M. Kozyrev, R. O. Matevosyan, and L. I. Stashkov, *Dokl. Akad. Nauk SSSR* **164**, 1073 (1965).
50. V. V. Korshak (ed.), The Development of Organic Chemistry in the USSR, Nauka, Moscow (1967), p. 371.
51. É. G. Rozantsev, V. A. Golubev, and M. B. Neiman, USSR Authors' Certificate 175504, October 24, 1964; *Byul. Izobretenii* **1965**(20).
52. G. F. Pavelko and É. G. Rozantsev, *Izv. Akad. Nauk SSSR, Ser. Khim.* **1967**, 2466.
53. A. L. Buchachenko, V. A. Golubev, M. B. Neiman, and É. G. Rozantsev, *Dokl. Akad. Nauk SSSR* **163**, 1416 (1965).
54. V. A. Golubev, M. B. Neiman, É. G. Rozantsev, and V. I. Suskina, *Zh. Organ. Khim.* **2**, 1916 (1966).
55. É. G. Rozantsev and L. A. Krinitskaya, *Izv. Akad. Nauk SSSR, Ser. Khim.* **1967**, 1137.
56. É. G. Rozantsev, G. F. Pavelko, A. L. Buchachenko, and M. B. Neiman, *Izv. Akad. Nauk SSSR, Ser. Khim.* **1967**, 2306.
57. É. G. Rozantsev, L. A. Skripko, and Yu. I. Temchin, USSR Authors' Certificate No. 190016, August 26, 1965.
58. É. G. Rozantsev, L. A. Skripko, Z. V. Ivanova, and N. S. Mel'nik, USSR Authors' Certificate No. 202940, August 26, 1965.
59. É. G. Rozantsev, P. I. Levin, and L. A. Skripko, USSR Authors' Certificate No. 197941, May 19, 1966.
60. O. Lehmann, Flüssige, Kristalle, Leipzig, 1904.
61. G. Friedel, *Ann. Phys.* **19**, 273 (1922).
62. H. Falle, G. Luckhurst, H. Lemaire, Y. Marechal, A. Rassat, and P. Rey, *Mol. Phys.* **11**, 49 (1966).
63. I. G. Chistyakov, Liquid Crystals, Nauka, Moscow (1966).
64. G. Luckhurst, *Mol. Cryst.* **2**, 363 (1967).
65. A. Hudson and G. Luckhurst, *J. Chem. Phys.* **46**, 55 (1967).
66. G. R. Luckhurst and É. G. Rozantsev, *Izv. Akad. Nauk SSSR, Ser. Khim.* **1968**, 1708.
67. S. S. Khalatov, Cholesterol Disease, Moscow (1946).

Chapter VII

SOME CHARACTERISTICS OF INDIVIDUAL NITROXYLS

1. Magnetic Properties

The magnetic properties of crystalline undiluted paramagnetic substances in a strong field are determined by the specific interaction between the unpaired electrons of neighboring molecules.

Two types of such interactions are distinguished [1, 2]: (1) long-range, dipole–dipole and (2) short-range, exchange interactions; the latter are due to the overlapping of the wave functions of the unpaired electrons, and fall rapidly with distance. The exchange interaction averages both the dipole interaction between the unpaired electrons and the intramolecular hyperfine interaction of the unpaired electrons with the nuclei [2]. The influence of the exchange interaction on the hyperfine structure of the EPR spectrum leads first (with a weak exchange) to a broadening of each component of the HFS [3]. When the exchange of the components increases, the HFS comes closer to the center of the spectrum and fuses into one line having the Lorentzian form [3] (when $I \gg a$, where a is the hyperfine interaction constant and I is the exchange integral). The exchange interaction in crystalline paramagnetics decreases the dipole–dipole width of the line. In these circumstances, from the observed width of the EPR line, it is possible to evaluate the magnitude of the exchange forces acting between the unpaired electrons. Anderson and Weiss have obtained a formula for the width of an EPR line narrowed by strong interaction,

$$\Delta H = (2/\sqrt{3})H_p^2/H_e \tag{1}$$

Here ΔH is the distance between the maxima of the derived absorption line; H_e is the exchange field, $H_e = h\nu_e/g\beta \approx I'/g\beta$, where ν_e is the exchange frequency; and H_p^2 is the second dipole moment; H_p is the mean-square width due to dipole–dipole interaction, and H_p^2 depends on the mutual location of the molecules in the crystal lattice, which is unknown for the majority of organic paramagnetics. For very rough estimations of H_e and v_e, an isotropic arrangement of the molecules is generally assumed (simple cubic lattice). In this case, for a polycrystal, H_p^2 can be calculated without difficulty [2]:

$$H_p^2 = 5.1(g\beta n)^2 S(S + I) \tag{2}$$

where n is the number of spins in unit volume.

For nitroxyl radicals, the characteristic magnitude $v_e \approx 5 \times 10^9$ Hz ($I'/K \approx hv_e/K \approx 0.2°$K). It has been determined [4, 5] by means of Eqs. (1) and (2) that for the radical 2,2,6,6-tetramethyl-4-oxopiperidine-1-oxyl (II) ($\Delta H = 11.5$ G and $T = 293°$K), $H_p \approx 140$ G at a density of 1.1 g/cm^3. The frequency of electronic exchange v_e in organic paramagnetics in a strong field is determined by the delocalization of the unpaired electron over the molecular skeleton. With an increase in the degree of delocalization, the overlapping of the electron clouds of neighboring radicals is facilitated, which leads to a rise in the exchange forces between the unpaired electrons.

Besides the narrowing of the line, the influence of the exchange leads to an interesting effect of the dependence of the width [2, 6] of the EPR line of the observation frequency. This dependence exists when the resonance frequency v varies in the neighborhood of the exchange frequency. A dependence of the linewidth on the frequency v has in fact been observed for some paramagnetics. The change in the width of the EPR line with a change in v is called the "10/3 effect." This name is explained by the fact that with a completely isotropic distribution of the molecules of a paramagnetic, the linewidth ΔH at $v \ll v_e$ is 10/3 times greater than ΔH at $v \gg v_e$. The 10/3 effect can apparently explain the difference in the width of the EPR lines of some crystalline nitroxyls at low and high resonance frequencies. Table 18 gives the values of the EPR linewidths for the radicals

The values for the linewidths at a frequency $v_0' = 4.5 \times 10^8$ Hz were taken from the literature [7].

The difference from 10/3 of the ratio $\Delta H(v_0')/\Delta H(v_0)$ must be ascribed both to the closeness of the exchange frequency v_e to v_0 and to an isotropic arrangement of the radicals.

TABLE 18
EPR Linewidths of Crystalline Nitroxyls at Different Resonance
Frequencies*

Sample	$\Delta H(v_0)$, G	$\Delta H(v_0')$, G	$\Delta H(v_0)/\Delta H(v_0')$
Polycrystal (I)	10	15	1.5
Polycrystal (II)	11.5	22	1.9
Polycrystal (III)	22.4	50	2.2

*Here, $v_n = 10^9$ Hz, $v_0' = 4.5 \times 10^8$ Hz; $T = 293°$K.

The shape of the EPR lines of some polycrystalline nitroxyls is asymmetrical, the nature of the curves showing the axial anisotropy of the g-factor [8]. As a rule, the asymmetry of the line changes with a change in the temperature. It has been observed [4] that in the monoclinic modification of radical II, a decrease in the temperature from room temperature to liquid-nitrogen temperature leads to an increase in the anisotropy of the g-factor. At temperatures close to room temperature, the EPR line in both mono-crystalline and polycrystalline samples is symmetrical with $g_0 = 2.0066 \pm 0.0003$. At 78°K, the axial anisotropy of the g-factor is characterized by the parameters $g_\parallel = 2.0075 \pm 0.003$ and $g_\perp = 2.0062 \pm 0.003$.

By analyzing the molecular motions in the crystalline radical by the NMR method, Karimov *et al.* [4] came to the conclusion that the appearance of anisotropy of the g-factor with a decrease in the temperature is not caused by prevention of the rotation of the radical as a whole, but is connected with the influence of the high-frequency motions of the molecular groups, although the actual mechanism of the effect is not completely clear.

The features of the exchange interaction in organic paramagnetics are responsible for the extremely specific behavior of the magnetic susceptibility. Figure 33 shows the temperature dependence of the susceptibility of the

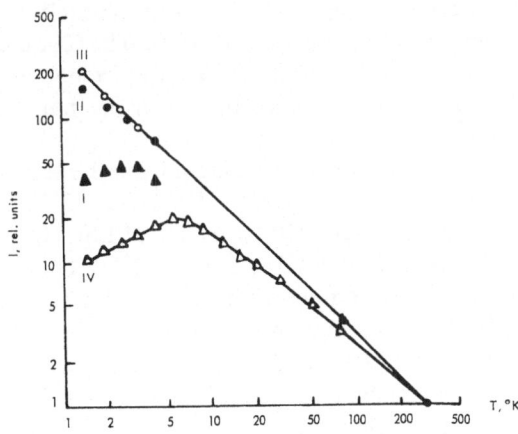

Fig. 33. Intensities of the EPR signals (paramagnetic susceptibilities) of the crystalline nitroxyls I–IV as functions of the temperature.

nitroxyl radicals I, II, and III and the radical 4-hydroxy-2,2,6,6-tetramethyl-piperidine-1-oxyl (IV) measured by the low-frequency EPR method [7]. The curves of the susceptibility of radicals I and IV as a function of the temperature each have a maximum. Fedutin and Stryukov (see [9]), have measured the susceptibility of a number of organosilicon nitroxyl radicals in the range from 1.5 to 300°K and have found that they deviate considerably from Curie's law. Thus, for example, for the pure tetraradical

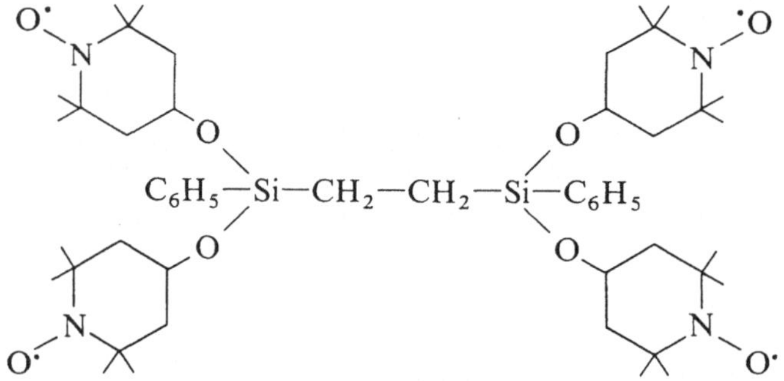

(V)

a well-defined susceptibility maximum was found at 8°K. When the temperature was lower to 2°K, the susceptibility fell by a factor of 10 as compared with the maximum value.

The change in the susceptibility of this radical is described fairly satisfactorily by the well-known formula

$$\chi \sim [T(e^{I/kT} + 3)]^{-1} \tag{3}$$

This equation describes the exchange interactions in pairs of the unpaired electrons. The ground state of such a pair is a singlet with an antiparallel arrangement of spins ($S = 0$), and the excited state is a triplet with a parallel arrangement of the spins ($S = 1$). The distance between the singlet and the triplet states is determined by the exchange integral I. At sufficiently low temperatures ($T < I/k$), the electrons pair off and the susceptibility falls sharply.

The value of I/k calculated from the temperature change of the susceptibility of the tetraradical V [Eq. (3)] is 13°K. At the same time, the value of $I'/k = h\nu_e/k$ calculated by means of Eqs. (1) and (2) from the EPR linewidth at 300°K ($\Delta H = 9$ G) is about 0.1°K. The existence of a maximum in the susceptibility–temperature curve is characteristic for linear spin chains. The position of the maximum is determined by the magnitude of the exchange interactions between the spins. As has been reported by McConnell, because of the asymmetry of the molecules, as a rule, the crystal structure of organic radicals enables linear chains of the most strongly interacting spins to be isolated. The majority of radicals studied can be roughly divided into two classes [10, 11]: The first comprises crystalline compounds in which each spin interacts, basically, with only one neighbor. In these circumstances, the interaction between different pairs, I', is far smaller than the exchange I within each pair ($I \gg I'$). The second class comprises radicals in which the interactions of each spin with two neighbors are similar in magnitude ($I \approx I'$). The radical V that we have studied (and some other polyradicals and individual crystalline nitroxyls) must be assigned to the first class. For this type of compound, the change in susceptibility with temperature is determined mainly by the magnitude of I [Eq. (3)], while the width of the EPR line is determined by the magnitude of I' [Eqs. (1) and (2)]. We must also have $I \gg I'$. At the same time, for the second class of radicals, the value of I, which determines the change in susceptibility, and the magnitude of I', which determines the width of the EPR line, must be similar [11].

A knowledge of crystal structure is very important for an understanding of the paramagnetic properties of a crystal. At the same time, structural data on the nitroxyl radicals are extremely sparse. Leizerovich [12] has made a complete interpretation of the structure of radical IV. He has established that the molecules of the crystalline radical IV are connected by hydrogen bonds into infinite chains. The molecules themselves have a distorted chair shape. We may note that in solution, the radical IV has the boat form with an intermolecular hydrogen bond [13]. Apparently, the intermolecular forces change the configuration of the molecule. Shibaeva has studied the structure of radical II and has found that it has two crystalline modifications: monoclinic [4] with cell parameters $a = 14.2\,\text{Å}$, $b = 6.15\,\text{Å}$, $c = 18.4\,\text{Å}$, $\beta = 98°$, $Z = 6$; and rhombic with parameters $a = 7.94\,\text{Å}$, $b = 11.94\,\text{Å}$, $c = 10.54\,\text{Å}$, $Z = 4$.

2. Nuclear Magnetic Resonance in Nitroxyl Radicals

The basic information on the constants of hyperfine interaction in organic paramagnetics has been obtained by studying their EPR radicals in solution. However, it is impossible in this way to determine the sign of these constants, which we must know in order to study the distribution of the spin density in the radical. This shortcoming does not exist in the nuclear resonance method. The NMR method has recently been applied successfully to the determination of the constants of the hyperfine interaction of unpaired electrons with protons in nitroxyl radicals. The energy of hyperfine interaction of the electron and the nucleus is written in the form of the Hamiltonian

$$\mathscr{H}_{hf}^i = A_i S_z I_{zi} + B_i S_x I_{xi} + C_i S_y I_{yi} \tag{4}$$

The main values of the hyperfine interaction tensor, A, B, and C, are determined [14] by the isotropic contact interaction and the anisotropic dipole interaction of the unpaired electron with the magnetic moment of the nucleus:

$$A_i = A_d^i + a_i \qquad B_i = B_d^i + a_i \qquad C_i = C_d^i + a_i \tag{5}$$

The anisotropic constants A_d^i, B_d^i, and C_d^i depend on the orientation of the radicals in the magnetic field and are connected by the relation [14]

$$A_d^i + B_d^i + C_d^i = 0 \tag{6}$$

The isotropic constant a_i is proportional to the density of the unpaired electron on the ith nucleus. If a paramagnetic molecule is present in solution, because of the unordered motions, an averaging of the anisotropic constants takes place, and in high-resolution NMR spectra (which will be discussed below), only the contact constant a_i appears. The following condition is essential for the observation of NMR in solutions of organic paramagnetics [15]

$$1/T_e \gg a_i \tag{7}$$

where the time T_e characterizes electronic exchange. Since in nitroxyl radicals the constants a_i are comparatively small, condition (7) is easily satisfied in concentrated solutions or in a melt of the radical.

Under these conditions, for each proton of the radical or each group of equivalent protons, a narrow line shifted in the field relative to the resonance field of the same nucleus in a diamagnetic molecule will be observed. The magnitude and sign of the constant can be found from the magnitude and the direction of the shift:

$$a_i = -\frac{\Delta H_i}{(\gamma_e/\gamma_N)} \frac{4kT}{g\beta H_0} \tag{8}$$

where γ_e and γ_N are the hydromagnetic ratios of the electron and the nucleus, respectively. As can be seen from Eq. (8), at a given a_i, the paramagnetic shift is proportional to the external field and the paramagnetic susceptibility.

The possibility of using high-resolution NMR for studying the hyperfine interactions of radicals in the liquid state was shown by Hausser et al. [15]. After this, other authors [16–20] investigated the paramagnetic shifts for solutions of a large number of nitroxyl radicals. Table 19 gives the constants a_i for radicals I and II (Krelick [17]).

The number of groups of equivalent protons is determined by the number of lines in the spectrum. The difference in the constants a_i for the same proton groups in radicals with different substituents is due to the influence of the substituents on the stereochemical configuration of the radicals.

TABLE 19
Constants a_i Measured at 45°C

Compound	Group of protons	a_i, G
I	Methyl	−0.22
	β-Methylene	−0.39
	γ-Methylene	+0.18
II	Methyl	−0.11
	β-Methylene	−0.01

While studies of paramagnetic shifts in dissolved radicals by the high-resolution NMR method have become very widespread only recently, paramagnetic shifts in crystalline radicals have been studied for a fairly long time. It is natural that in the crystalline state the anisotropic constants A, B, and C are not averaged out, and the resulting NMR lines of a polycrystalline paramagnetic are fairly complex. However, the difficulties of interpreting a spectrum are compensated by the possibility of simultaneously evaluating both the isotropic and anisotropic constants of the hyperfine interaction.

Karimov and Makova [21] have studied the polycrystalline radicals I–VI by the low-resolution NMR method. In view of the smallness of the constants a_i, the experiments were carried out at liquid-helium temperatures (4.2 ± 1.5°K). It may be concluded from these experiments that the hyperfine interaction constants for a radical in the solid state differ considerably from the values of the same constants in the liquid state (Table 19), which is apparently due to a change in the conformation of the radicals.

The low value of the constants a_i confirms the conclusion drawn from an analysis of the EPR spectra that, because of the saturated bonds of the heterocycle, the unpaired electron is weakly delocalized over the nitroxyl radical.

The low-resolution NMR method has been used successfully to study internal motions in solids [22]. This method has also been used [4, 5] to study molecular motions in the crystalline radicals I and II in the range of temperatures from 310 to 80°K.

With a change in the temperature, the width of the NMR signal changes over two ranges (Fig. 34). The first range, from the melting point to 250°K, is apparently connected with the freezing of the motion of the radical as a whole. The second range of broadening, from 110°K and below, may relate to the freezing of the high-frequency motions of the CH_3 groups, which rotate around the single C—C bonds. The possibility of observing comparatively narrow NMR lines in crystalline paramagnetics is closely connected with the presence of strong intermolecular exchange interactions between the unpaired electrons [22]. For the nitroxyl radicals, an exchange frequency

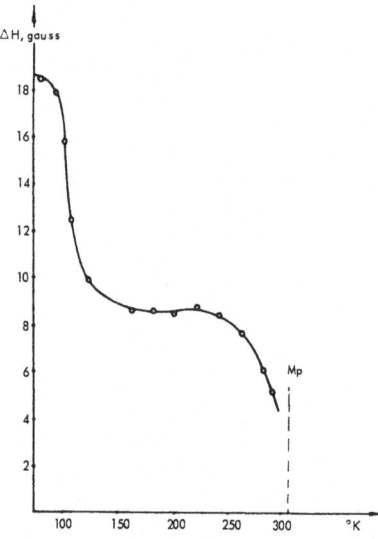

Fig. 34. Width of the NMR signal of the crystalline radical I as a function of temperature.

$v_e \approx 5 \times 10^9$ Hz proved to be sufficient to decrease the instantaneous dipole field of the unpaired electron at the closest methyl proton (~ 600 G) to the average value of ~ 0.3 G. This magnitude is far less than the observed linewidth (Fig. 34), which is therefore due to the dipole–dipole interaction of the protons.

3. Conductivity

Eley *et al.* [23] have studied the conductivity of the crystalline radicals 1,1-diphenyl-2-picrylhydrazine-2-oxyl, di-*p*-anisylnitroxide, di-*p*-nitrophenyl-nitroxide, and the Kenyon–Benfield radical. All the radicals behave as semiconductors and their specific conductivity varies in accordance with the law

$$\sigma = \sigma_0 e^{-\Delta\varepsilon/kT} \tag{9}$$

The parameters $\Delta\varepsilon$ and σ_0 for the radicals are given in Table 20.

TABLE 20
Parameters of the Conductivity of Some Radicals

Parameter	Radical*			
	1	2	3	4
$\Delta\varepsilon$ (eV) (tablet)	2.03	—	2.60	2.31
$\Delta\varepsilon$ (eV) (monocrystal)	—	2.83	—	2.42
$\log \sigma_0 (\Omega^{-1}\text{-cm}^{-1})$ (tablet)	1.61	—	7.46	5.00
$\log \sigma_0 (\Omega^{-1}\text{-cm}^{-1})$ (monocrystal)	—	12.35	—	7.32

*Radicals: (1) 1,1-diphenyl-2-picrylhydrazine-2-oxyl, (2) di-*p*-anisylnitroxide, (3) di-*p*-nitrophenyl-nitroxide, (4) the Kenyon–Benfield radical.

The fairly high threshold and low conductivity lead to the conclusion that the carriers have low mobility. This indicates the strong localization of the unpaired electron within the limits of each radical, which is confirmed by the EPR and NMR data.

4. A Radical in a Diamagnetic Matrix

The possibility of using nitroxyl radicals as mobility detectors in various media is closely connected with the fact that each of the unpaired electrons in these radicals interacts mainly with a single nitrogen nucleus and the resulting EPR spectrum has the form of a simple triplet (special conditions are required for the observation of the HFS from the ring protons [17]). The interaction of an unpaired electron with a ^{14}N nucleus in an external magnetic field is written in the form of the Hamiltonian

$$\mathcal{H} = \beta S \hat{g} H + h(AS_z I_z + BS_x I_x + CS_y I_y) \tag{10}$$

To find the parameters of the tensor $\hat{g}$ (g_{xx}, g_{yy}, g_{zz}) and of the hyperfine interaction tensor T (A, B, C), Griffith *et al.* [24] have studied nitroxyl radicals in a diamagnetic monocrystalline matrix of the dimer of dimethyl-ketene. It was found that the radicals in the diamagnetic crystal had a strictly definite orientation, since, when such a monocrystal was rotated in the magnetic field, the distance between the components of the HFS and the position of the center of the EPR spectrum of the radicals changed. The distance between the components of the HFS determines the constants A, B, and C, while the position of the center of the spectrum determines g_{zz}, g_{xx}, and g_{yy}. It was found that the hyperfine interaction of the unpaired electron with the nitrogen nucleus possesses axial symmetry to a first approximation: $A \approx 90$ MHz, and $B = C = 15$ MHz, while the values of A, B, and C changed only slightly from radical to radical ($\sim 15\%$). The experimental value of the isotropic constant determined in solution ($a \approx 40$ MHz) agrees with the value $a = 1.3(A + B + C)$, which showed that A, B, and C have the same signs. The main values of the tensor $\hat{g}$ have been found for di-tert-butylnitroxyl ($g_{xx} = 2.0089, g_{yy} = 2.0061,$ and $g_{zz} = 2.0027$). The spectra of the radicals studied by Griffith *et al.* [24] resemble one another. This shows that the parameters A, B, C, g_{zz}, g_{xx}, and g_{yy} are similar to one another for various nitroxyls.

The calculated values of the constants A, B, and C and the equality of B and C led Griffith *et al.* to the conclusion that the unpaired electron is mainly localized in the π orbitals of the nitrogen atom ($\rho = 0.9$), apart from which it was concluded that the CNOC fragment is plane. This is apparently true only to a first approximation. Thus, for example, X-ray measurements by Leizerovich* have shown that in the crystalline radical 4-hydroxy-2,2,6,6-

*Private communication.

tetramethylpiperidine-1-oxyl, the N—O bond forms an angle of 17° with the CNC plane. The height of the pyramid (distance of the N atom from the COC plane) is 0.15 Å.

5. Migration of Stable Radicals in Polymers

In a study of the properties of nitroxyl radicals by the EPR method, it was found [25] that some radicals readily diffuse into various polymers. At room temperature, the EPR spectrum of a radical in such a polymer as, for example, polyethylene, consists of a well-resolved triplet and is very similar to the spectrum of the same radical in liquid glycerol. Since the form of the EPR spectrum of the radical in a viscous medium is determined unambiguously by its rotational mobility, it was concluded, surprisingly, that the mobilities of the radical in glycerol and in polyethylene were similar. The form of the EPR spectrum and therefore the mobility of the radical at a given temperature depend on the nature of the polymer (for example, the spectra of the radical in polyethylene and in polystyrene at $T = 293°K$ differ markedly). When the temperature of the polymer samples changes, the form of the EPR spectrum changes, and therefore the mobility of the radicals also changes. With a fall in the temperature, each component of the HFS gradually broadens [5]. In polyethylene, for example, close to liquid-nitrogen temperature, the EPR spectrum of the radical acquires the form characteristic for the radical in a vitrefied liquid. It became obvious that the method of radical migration can be used to study both the radicals themselves and polymers. The symmetry and stereochemical configuration of the radicals also affect their EPR spectra in polymers. Thus, the EPR spectra of the radicals I, II, an IV in polyethylene at the same temperature differ from one another [25]. When high-pressure polyethylene is irreversibly deformed, the components of the HFS of the EPR spectrum of the radical broaden somewhat as compared with the spectrum of the radical in undeformed polyethylene [25]. This shows the retardation of the rotational movements of the radical in the deformed polymer.

It is known [26] that the irreversible deformation of a polymer reduces the number of vibrational degrees of freedom of the polymer molecules, which leads to an increase in their "rigidity." The size of the segments of the macromolecule moving simultaneously increases and their mobility decreases. The fact that the irreversible deformation of the polymer retards the rotational movements of the radical shows the close connection of the rotational mobility of the radical with the segmentary mobility of the macromolecules of the polymer. This purely qualitative result shows that nitroxyl radicals can be used to detect the strains arising in a loaded polymer.

It has been established that when a polymer swells—for example, when poly(vinyl chloride) swells in a solvent—the rotational mobility of a radical increases. As has been shown by NMR [27], the swelling of a polymer in a solvent leads, in the first place, to an increase in the segmentary mobility of the polymer chains without a considerable increase in the free volume of the polymer.

As is well known, the chaotic thermal Brownian movements of a particle (radical) are characterized by two parameters: the coefficients of translational diffusion (D_{tr}) and of rotational diffusion (D_{rot}). The nitroxyl radical migration method enables D_{tr} and D_{rot} to be determined separately [28].

As the basic radical for studying mobility in polymers, the highly stable and nonvolatile radical 2,2,6,6-tetramethylpiperidine-1-oxyl has been selected [5, 25, 28, 29]. Rotational mobility is characterized by the correlation time τ_c, which determines the effective frequency of rotation of the radical, $v = 1/2\pi\tau_c$.

If it is considered that the process of rotational movement is described by the equation of rotational diffusion, the coefficient D_{rot} is connected with τ_c by the relation [22]

$$\tau_c = 1/6D_{rot} \tag{11}$$

The form of the EPR spectrum of the radical in a viscous medium depends on the degree of averaging out of the rotational movements by the anisotropy of the hyperfine interaction and the anisotropy of the g-factor. A complete picture of the changes of the EPR spectra with a change in the rotational mobility of the radical 2,2,6,6-tetramethylpiperidine-1-oxyl in polyethylene (in the temperature range from 370 to 100°K) has been given [5]. Quantitative values of τ_c for a number of polymers have been determined in the so-called region of rapid rotations [5, 28, 29], where the theories of the averaging out of the anisotropic interactions of the rotational movements of a paramagnetic molecule in viscous solutions are valid [30, 31]. These theories are limited by the velocity conditions

$$(\sigma\tau_c)^2 \ll 1 \tag{12}$$

In this case, $\sigma \approx (\Delta\gamma H_0 + b)$, where $\Delta\gamma$ is the parameter of the anisotropy of the g-factor $\Delta\gamma = (|\beta|/\hbar)[g_{zz} - \frac{1}{2}(g_{xx} + g_{yy})]$; b is the parameter of the anisotropy of the hyperfine interaction, $b = (4\pi/3)(A - B)$; and H_0 is the external field. The measurements performed [5, 28, 29] were carried out in a spectrometer of the 3-cm range with $v_0 = 9300$ MHz and $H_0 = 3300$ G. The maximum value of τ_c determined by condition (12) is $\tau_c \approx 10^{-9}$ sec. In the region of rapid rotations, the EPR spectrum consists of three HFS components having different widths. The spectra were treated in accordance with Kivelson's theory [30], which has been used repeatedly to describe the behavior of a paramagnetic molecule in viscous media [32, 33]. The theory leads to the conclusion of the Lorentzian form of the components of the HFS.

The linewidth of each component $1/T_2$ is described by means of the spin–spin, T'_2, and the spin–lattice, T_1, relaxation times, which depend on τ_c and M: $1/T_2 = (1/T'_2) + (1/T_1) + K$, where K does not depend on τ_c and M (for example, unresolved hyperfine structure of protons).

The fact that the measurements were carried out at a sufficiently high frequency, $(2\pi v_0 \tau_c)^2 \gg 1$, enables us to neglect T_1—the nonsecular contribution to the linewidth. Under these conditions, τ_c is found from the equation [33]

$$\frac{T_2(0)}{T_2(M)} = 1 - \frac{4}{15}\tau_c b \Delta\gamma H_0 T_2(0)M + \frac{\tau_c}{8}b^2 T_2(0)M^2 \qquad (13)$$

We may note that an analysis of the EPR spectra of a nitroxyl on the basis of Freed and Frenkel's theory leads to an equation identical with Eq. (13) [31, 34]. The values of the relaxation times $T_2^{(0)}$ and $T_2^{(M)}$ (where $M = \pm 1.0$, the values of the spin projections of the ^{14}N nitrogen nucleus) are found experimentally. For the Lorentzian form of the line, $T_2^{-1} = \pi\sqrt{3}\,\Delta v$, where Δv is the distance between the maxima of the derivatives of each component of the absorption. Using the values of the tensor parameters $\hat{g}$ and $\hat{T}$, we find $\Delta\gamma \approx 4 \times 10^4$ and $b \approx 320$ MHz. At these values of $\Delta\gamma$ and b, for all the polymers studied, the value of the time τ_c determined from the member at M [Eq. (13)] is several times less than the value of τ_c determined from the member at M^2. At the same time, the ratio of the members at M and M^2 did not depend on the temperature. This difference in the values of τ_c can be interpreted as the result of the effective decrease in $\Delta\gamma$ in polymers. In particular, it is shown in a smaller asymmetry of the EPR spectrum of the radical in the polymer as compared with the spectrum in a viscous liquid (Fig. 35). It seems that the apparent effective decrease in $\Delta\gamma$ in polymers may reflect the complexity of the dynamic behavior of the radical in such

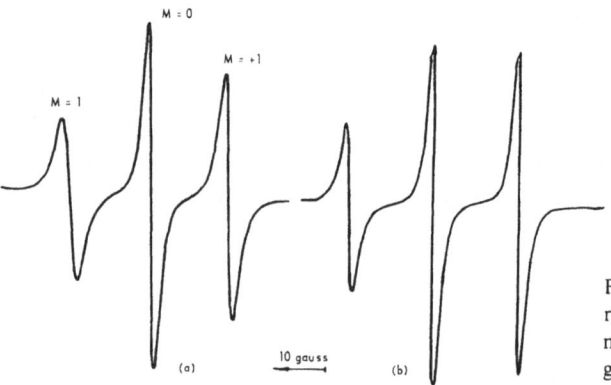

Fig. 35. EPR spectra of radical I: (a) in poly(butyl-methacrylate) (102°C); (b) in glycerol (20°C).

unhomogeneous media as polymers. This question is still under study. For the sake of definiteness, in the papers cited [5, 25, 29], the value of τ_c in polymers was calculated from the member at M^2 [Eq. (13)].

In all the polymers investigated, in the region of rapid rotations (temperature range $T_{min} - T_{min} \approx 50°C$), the temperature changes in the correlation time of the radical obey the law $\tau_c = \tau_0 \exp(E_a/RT)$. In Table 21, the values of E_a, τ_0, and D_{rot}^0 for some polymers are given.

TABLE 21
Parameters of the Rotational Movements of the 2,2,6,6-Tetramethylipiperidine-1-oxyl Radical in Various Polymers

Polymer	τ_0, sec	D_{rot}^0, sec^{-1}	E_a, kcal/mole	$T_{min} - T_{max}$, °K
Polyethylene, $\rho = 0.918$ g/cm^3	1.3×10^{-17}	1.3×10^{16}	10.7	~290–345
Linear polyethylene, $\rho = 0.95$ g/cm^3	6×10^{-17}	2.8×10^{15}	10.4	~310–365
Polyisobutene	1.8×10^{-17}	9.3×10^{15}	11.8	~325–370
Poly(butyl methacrylate)	6×10^{-17}	2.8×10^{15}	12	~365–410
Polystyrene	1.3×10^{-21}	1.3×10^{20}	22.6	~413–450

It has been shown [29] that the minimum temperature at which the theory of rapid rotations is valid for each of the polymers studied considerably exceeds the characteristic temperature of the freezing of the segmentary motions of the polymer molecules. From this, it was concluded that the rotational movement of a radical is determined to a considerable extent by the segmentary motion of the polymer macromolecules. The values found for the activation energies of the radical are close (but not equal) to the activation energies of the segmentary motions of the corresponding polymers measured by the NMR method [27].

In this work [29], no dependence of the correlation time of the rotational motions on the molecular weight of the polyethylene was found ($M = 4 \times 10^5$, 1.5×10^5, 2.4×10^4). This shows that a rotational diffusion of the radical is most greatly affected by the motions of the polymer molecules, the scale of which is comparable with the dimensions of the radical. Such motions must depend weakly on the length of the polymer chains when it varies within wide limits. Hence, apparently, it also follows that the voids in which the radicals are located are distributed (and migrate) through the whole volume of the polymer and are not connected solely with terminal defects of the polymer chains.

It has been found [5] that on passing from linear to high-pressure branched polyethylene, the rotational mobility of the radical rises severalfold. This may be compared with the analogous conclusion made on the basis of

NMR data [27] of the greater intensity of the segmentary motions in high-pressure polyethylene. By using a radical as a peculiar kind of probe, it is possible to determine the effective microscopic viscosity [5], which is connected with the rotational movement of the radical in the polymer

$$\eta_{rot}^{M} = (3kT/4\pi a^{3})\tau_{c} = (kT/8\pi a^{3}D_{rot}) \tag{14}$$

where a is the effective radius of the radical.

In the region of rapid rotations, η_{rot}^{M} amounts to several poises, i.e., it corresponds to viscous liquids and has nothing in common with the rheology of the polymer. This confirms the idea that polymers occupy an intermediate position between solids and liquids, exhibiting a dual nature [26].

The condition (12) limiting the field of rapid rotation enables us to use the theory of the averaging out of the anisotropic interactions in practice for the range of values $\tau_{c} = \sim 10^{-10}$–10^{-9} sec [at $\tau_{c} \approx 10^{-10}$ sec, all three components of the HFS are equal, and it becomes impossible to determine τ_{c} from Eq. (13)].

The temperature range of the region of rapid rotations is $\sim 50°C$, while the whole range of changes in the EPR spectra in polymers covers $\sim 250°C$. Thus, τ_{c} remains undetermined over a large range of temperatures. This circumstance is connected with the fact that the method of perturbations used by Kivelson [30] and by Freed and Frenkel [31] cannot be applied when condition (12) is violated.

The rotational motions of the radical for which the condition $\tau_{c} > 10^{-9}$ sec is not satisfied are called slow. The glass transition temperature of a polymer belongs to the region of slow movements, for example. The shape of the EPR spectrum of a radical with a change in the temperature close to the glass transition point of polystyrene (80°C) has been studied [29]. It was found that there was no jump-like change in the spectrum, nor, therefore, in the time τ_{c}. A change in the temperature dependence of the rotational mobility of the radical close to the glass transition point of the polymer must be expected. The absence of a theory of slow rotations prevents this hypothesis from being checked. In a paper by Itzkowitz [35], the rotational movements of a nitroxyl radical are modeled by arbitrary jumps. The resulting spectra, calculated by machine methods, greatly resemble the EPR spectra. However, Itzkowitz was unable to connect the calculated spectrum with the physical characteristic of the rotational movement, the time τ_{c}. In connection with experiments carried out in the Institute of Chemical Physics of the Academy of Sciences of the USSR [5], a theory of the shape of the EPR signal in the region of slow rotations of a radical has been created which uses the model of Brownian rotational diffusion. The theory has explained, in particular, the effect of the shift of the components of the HFS that have been observed [5]. The theory of slow rotations undoubtedly

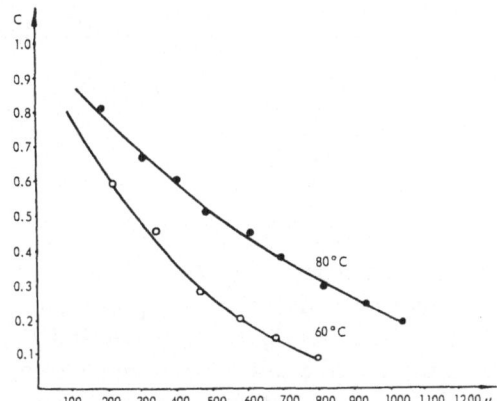

Fig. 36. Relative concentration of 2, 2, 6, 6-tetramethylpiperidine-1-oxyl in sections of polyethylene cut at various distances from the surface of the sample.

considerably broadens the field of application of nitroxyl radicals not only in polymers, but also in the study of the mobility of macrobiomolecules.

The study of the migration of stable radicals in polymers [28] makes it possible to determine D_{tr} accurately by the EPR method. As in the experiments described above, the 2,2,6,6-tetramethylpiperidine-1-oxyl radical was chosen to study the process of translational diffusion. The measurements of D_{tr} were carried out in high-pressure polyethylene with $\rho = 0.918$ g/cm^3.

To determine D_{tr}, a sample of the polymer was placed in the vapor of the radical for some time. Under the experimental conditions, the diffusion of the radical into the sample was uniform. Under these conditions, an integral measurement of the radical absorbed enables D_{tr} to be determined. However, Stryukov [28] has used the more accurate thin-section method, in which the distribution of the concentrations of the radical over the diffusion front is measured.

Figure 36 shows the relative concentration of radicals migrating into polyethylene over 8 hr at 80°C and 60°C. The diffusion curves obtained approximately describe the solution of the one-dimensional diffusion equation [36]:

$$C(x, t) = C_0 \left[1 - (2/\sqrt{\pi}) \int_0^{\lambda x} \exp(-y^2)\, dy \right], \qquad \lambda = (4D_{tr}t)^{1/2}$$

where C_0 is the concentration of radical vapor at the surface of the polymer and t is the time for which diffusion takes place.

In order to exclude C_0, the value of the coefficient D_{tr} was found from the ratio of the concentrations of the radical determined at different distances from the surface of the sample (this method had previously been used in a study of diffusion in metals by the labeled-atom method):

$$\frac{C_1(x_1, t)}{C_2(x_2, t)} = \frac{1 - (2/\sqrt{\pi}) \int_0^{\lambda x_1} \exp(-y^2)\, dy}{1 - (2/\sqrt{\pi}) \int_0^{\lambda x_2} \exp(-y^2)\, dy} = \frac{\text{erfc}(\gamma x_1)}{\text{erfc}(\gamma x_2)} \tag{16}$$

Equation (16) was solved graphically, and the value of D_{tr} was determined for various temperatures [28]. The temperature dependence of the coefficient of translational diffusion is exponential in nature, $D_{tr} = D_{tr}^0 \, e^{E_a/RT}$. In the range 313–353°K, $D_{tr}^0 = 10 \text{ cm}^2/\text{sec}$ and $E_a = 13 \text{ kcal/mole}$.

Just as in the case of rotational motion, the coefficient D_{tr} enables us to make a rough evaluation of the effective microscopic viscosity corresponding to the translational movement of the particle, $\eta_{tr}^M = kT/6\pi D_{tr} a$.

At 80°C, η_{tr}^M amounts to a few poises, i.e., has nothing in common with the macroviscosity of the polymer. Equations (14) and (17) follow from Stokes's law with the microviscosity parameters η_{rot}^M and η_{tr}^M. In the classical case of the Brownian motion with a sphere with radius a in a viscous liquid, η_{rot}^M and η_{tr}^M are equal to the viscosity of the liquid η, and the ratio $D_{tr}/D_{rot} = \frac{4}{3}a^2$ is determined solely by the dimensions of the diffusing particle. For a nitroxyl in polyethylene in the range 40–80°C, the ratio $(D_{tr}/D_{rot}) = (8 \times 10^{-16})e^{-2300/RT} \text{ cm}^2$ is not a constant. Thus, the rotational and translational motions of the radical in the polymer are described by different microviscosity parameters. The closeness of the values E_a^{rot} and E_a^{tr} (10.7 and 13 kcal/mole, respectively) shows the similarity of the processes determining the translational and rotational diffusion of the molecule in the polymer. This process is, to all appearances, the migration of microscopic defects, which is closely connected with the mobility of the polymer molecules.

References

1. G. Pake, Paramagnetic Resonance, An Introductory Monograph, Benjamin, New York (1962).
2. P. Anderson and P. Weiss, *Rev. Mod. Phys.* **25**, 269 (1953).
3. D. Kivelson, *J. Chem. Phys.* **27**, 1087 (1957).
4. Yu. S. Karimov, V. B. Stryukov, and I. F. Shchegolev, *Teor. i Éksperim. Khim.* **1967**, 557.
5. V. B. Stryukov and É. G. Rozantsev, *Vysokomolekul. Soedin.* **10**, 616 (1968).
6. R. Kubo and K. Tomita, *J. Phys. Soc. Japan* **9**, 888 (1954).
7. Yu. S. Karimov and É. G. Rozantsev, *Fiz. Tverd. Tela* **8**, 2727 (1966).
8. L. A. Blyumenfel'd, V. V. Voevodskii, and A. G. Semenov, The Use of Electron Paramagnetic Resonance in Chemistry, Izd. Akad. Nauk SSSR, Otd. Sib., Novosibirsk (1962).
9. G. F. Pavelko and É. G. Rozantsev, *Izv. Akad. Nauk SSSR, Ser. Khim.* **1967**, 2466.
10. P. Nordio, Z. Soos, and H. McConnell, *Ann. Rev. Phys. Chem.* **17**, 237 (1966).
11. Z. Soos, *J. Chem. Phys.* **46**, 4284 (1967).

12. Zh. Leizerovich, in: Abstracts of Papers at the VIIth International Congress of Crystallo-graphers [Russian Version], Moscow (1966), p. 114.
13. É. G. Rozantsev and E. N. Guryanova, *Izv. Akad. Nauk SSSR, Ser. Khim.* **1966**, 979.
14. H. McConnell, C. Heller, T. Cole, and R. Fessenden, *J. Am. Chem. Soc.* **82**, 766 (1960).
15. K. Hausser, H. Brunner, and J. Iochims, *Mol. Phys.* **10**, 253 (1966).
16. R. Krelick, *J. Chem. Phys.* **45**, 1922 (1966).
17. R. Krelick, *J. Chem. Phys.* **46**, 4260 (1966).
18. R. Z. Sagdeev, Yu. N. Molin, and É. G. Rozantsev, *Zh. Strukt. Khim.* **8**, 697 (1967).
19. N. A. Sysoeva, A. L. Buchachenko, V. I. Sheichenko, É. G. Rozantsev, V. F. Bystrov, and M. B. Neiman, *Zh. Strukt. Khim.* **8**, 1094 (1967).
20. A. Rassat, in: VIIIth International Symposium on Free Radicals [Russian abstracts], Novosibirsk (1967).
21. Yu. S. Karimov and M. K. Makova, in: Abstracts of Papers at the 14th All-Union Confer-ence on Low-Temperature Physics, Khar'kov (1967).
22. A. Abragam, The Principles of Nuclear Magnetism, Clarendon Press, Oxford (1961).
23. D. Eley, K. Jones, and M. Willis, *Nature* **212**, 72 (1966).
24. O. Griffith, P. Cornell, and H. McConnell, *J. Chem. Phys.* **43**, 2909 (1965).
25. V. B. Stryukov, Yu. S. Karimov, and É. G. Rozantsev, *Vysokomolekul. Soedin, Ser. B* **7**, 493 (1967).
26. V. A. Kargin and G. L. Slonimskii, Brief Outlines of the Physical Chemistry of Polymers, Khimiya, Moscow (1967).
27. W. Slichter, in: The Physics of Polymers [Russian translation], IL, Moscow (1961), p. 172.
28. V. B. Stryukov, *Dokl. Akad. Nauk SSSR* **179**, 33 (1968).
29. V. B. Stryukov and G. B. Korolev, Thesis, Moscow (1968).
30. D. Kivelson, *J. Chem. Phys.* **33**, 1094 (1960).
31. I. Freed and G. Frenkel, *J. Chem. Phys.* **39**, 326 (1963).
32. R. N. Rogers and G. E. Pake, *J. Chem. Phys.* **33**, 1107 (1960).
33. T. Stone, T. Buckman, P. Nordio, and H. McConnell, *Proc. Natl. Acad. Sci. US* **54**, 100 (1965).
34. A. L. Buchachenko and A. M. Vasserman, *Zh. Strukt. Khim.* **8**, 27 (1967).
35. M. Itzkowitz, *J. Chem. Phys.* **46**, 3047 (1967).
36. R. Barrer, Diffusion in and through Solids, Cambridge University Press (1941).

Chapter VIII

FIELDS OF APPLICATION OF INDIVIDUAL RADICALS

1. Organic Chemistry

The directed synthesis of complex organic compounds frequently leads to the necessity for protecting functional groups. Secondary amino and hydroxyl-amino groups can be protected, for example, by acylation, alkylation, and salt formation. However, by no means do these methods always give the desired result.

In particular cases, a satisfactory method for the protection of secondary amino and hydroxylamino groups is the oxylation reaction—for example [1–3],

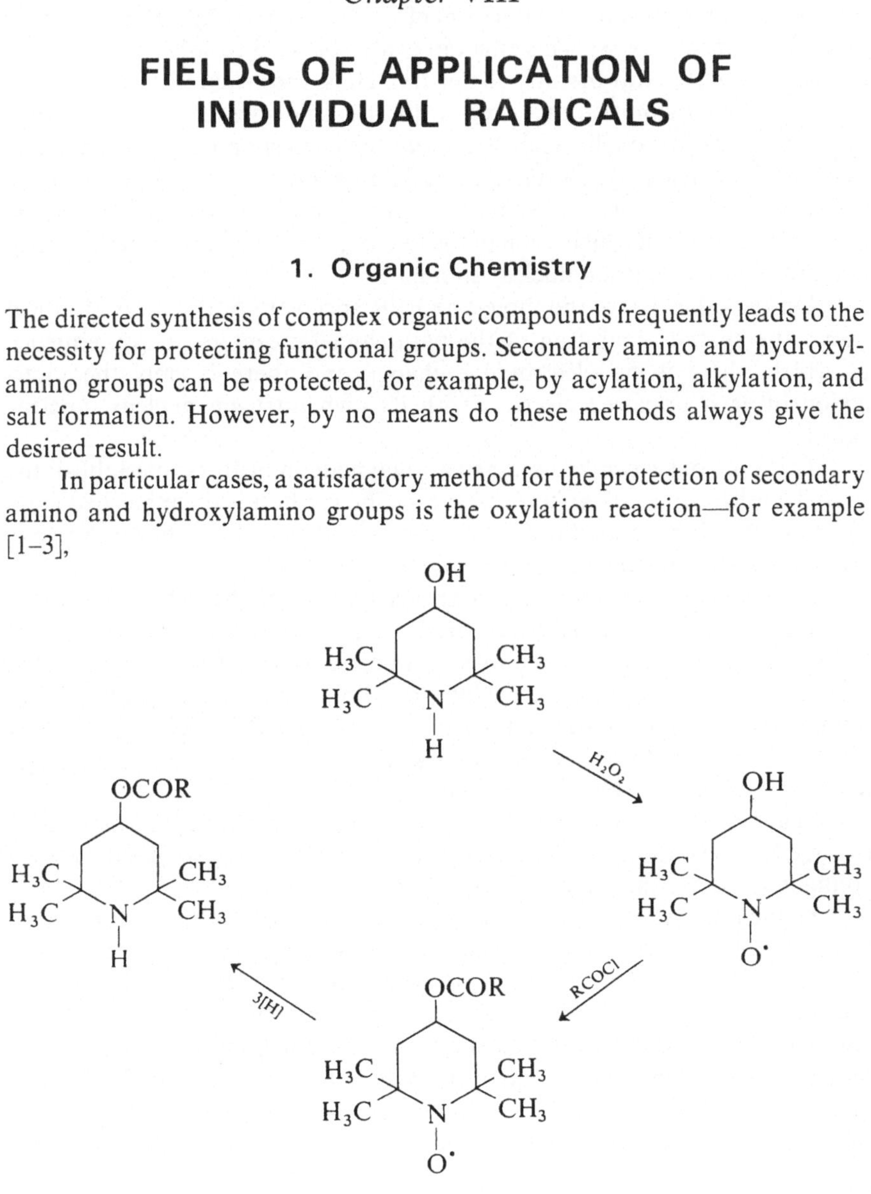

As is well known, the specific reactions of nitroxyls without saturation of the free valence can lead to polyradicals with different spatial arrangements of the paramagnetic fragments with respect to one another.

It has been mentioned previously that when the paramagnetic centers in nitroxyl biradicals are spatially close to one another, electronic exchange takes place, which is shown in the form of five-component hyperfine structure of the EPR spectra [4–8]. This effect may also be used to solve some special problems of structural organic chemistry. Thus, the reaction of 4-hydroxy-2,2,6,6-tetramethylpiperidine-1-oxyl with the chlorides of *cis*- and *trans*-cyclopropanedicarboxylic acids has given the corresponding *cis*- and *trans*-esters of these acids [9]. The *cis*-biradicals possessed a five-component EPR spectrum, which shows the spatial propinquity of the two paramagnetic centers, as a result of which, each of the two unpaired electrons interacts with two nitrogen nuclei (total nuclear spin 2).

On the other hand, the three-line EPR spectrum of the *trans* biradical shows the spatial separation of the paramagnetic centers, in consequence of which, there is no electronic exchange and there is only the interaction of each unpaired electron with its own nitrogen nucleus (nuclear spin 1).

Thus, nitroxyls can be used as paramagnetic indicators to establish the spatial configuration of *cis-trans* isomers. The method is distinguished from the majority of those used previously by the fact that it is sufficient to have only one of the isomers for the determination.

When there is an unfixed spatial orientation of the substituents, this method may also give valuable information on the mutual spatial location of the individual fragments of the molecule in certain cases. For example, in the case of the *cis*- and *trans*-1,2-cyclohexanediol esters of 3-carboxy-2,2,5,5-tetramethylpyrrolidine-1-oxyl, the corresponding EPR spectra possess a five-line hyperfine structure, thereby showing that in both cases, the paramagnetic centers are spatially adjacent.

In actual fact, as has been established by other methods [10], in the case of the 1,2-disubstituted cyclohexanes, the projected valence angles in the two configurations are the same, amounting to about 60°.

Stable nitroxyl radicals are a convenient means of studying the reactivities of various molecules and chemical bonds as functions of their nature and structure. Their convenience is due to the fact that because of their stability, they do not take part in recombination or disproportion reactions with one another, and this considerably facilitates the kinetic analysis. Furthermore, they are easily identified and quantitatively determined by simple physical methods. All this, apart from the convenience of the process, also ensures the reliability and unambiguousness of the results. Finally, the most important fact is that the results obtained relate to elementary steps of the chemical reactions [11].

Recently, Plyusnin and Chirkov [12] have proposed the use of stable nitroxyls to determine the rate constants of elementary steps. They have developed a direct method for determining the concentration of active centers by the introduction of a stable radical at the beginning of and during the process of polymerizing tetrafluoroethylene. From the results obtained, they have calculated the rate constants of the elementary steps.

This far from complete list of cases of the use of nitroxyl radicals in organic chemistry shows that they can serve as a fine instrument for the study of the structure and reactivity of molecules and of energy and steric factors in kinetics [13].

As already mentioned, free nitroxyls may be used as mild one-electron oxidizing agents in the dehydrogenation of phenols and amines. In some cases, nitroxyl radicals can play the part of *cis-trans* isomerization catalysts and promoters of certain oxidation reactions. It is known, for example, that complexes of the bivalent copper ion with phenanthroline catalyze the oxidation of methanol [14]. It has recently been established [15, 16] that di-tert-butylnitroxyl promotes this reaction. In the presence of the nitroxyl radical, the phenanthroline complex of bivalent copper is readily reduced by methanol with the formation of the equivalent amount of formaldehyde.* In this process, the stable radical is regenerated, and the rate of this reaction is proportional to the concentration of the nitroxyl.

The study shows that a transition complex including phenanthroline and the Cu^{++} ion in a ratio of 1:1 takes part in the limiting stage of the reaction. With a different stoichiometry of the components of the complex, the rate of the reaction falls.

It has been established by the EPR mthod that in the promotion process, the free nitroxyl radical takes part in an oxidation–reduction cycle with the corresponding hydroxylamine,

$$(H_3C)_3C-\underset{\underset{O^\cdot}{|}}{N}-C(CH_3)_3 \quad \underset{-H}{\overset{H}{\rightleftarrows}} \quad (H_3C)_3C-\underset{\underset{OH}{|}}{N}-C(CH_3)_3$$

and the presence of a base is necessary to neutralize the acid formed in the reaction

$$2Cu^{\oplus\oplus} + CH_3OH \longrightarrow 2Cu^{\oplus} + 2H^{\oplus} + CH_2O$$

---

*The oxidation of alcohols to the corresponding carbonyl compounds with the aid of nitroxyl radicals has been studied by Golubev [17] and Krinitskaya [18].

Two transition complexes are possible in which bivalent copper passes into the univalent state

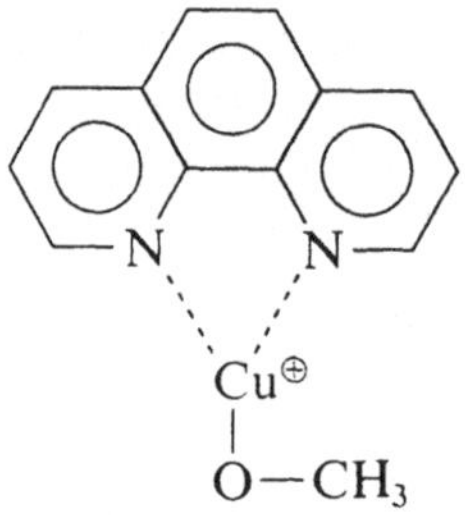

(I) 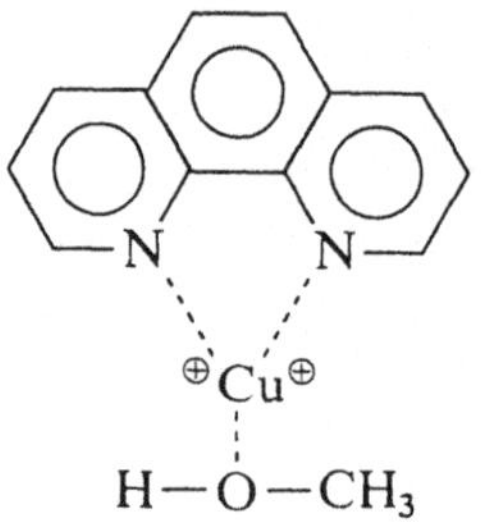   (II)

In principle, the two complexes play the same role. In this situation, the free nitroxyl radical may be regarded as an electron acceptor or a mild dehydrogenating agent:

$$Cu^{++} + \overset{\diagdown}{\underset{\diagup}{N}}-O \longrightarrow Cu^{+}\cdot\Phi + CH_2O + H^{+} + \overset{\diagdown}{\underset{\diagup}{N}}-OH$$

where Φ represents phenanthroline.

The complexly bound copper ion activates the molecule of coordinately bound methanol:

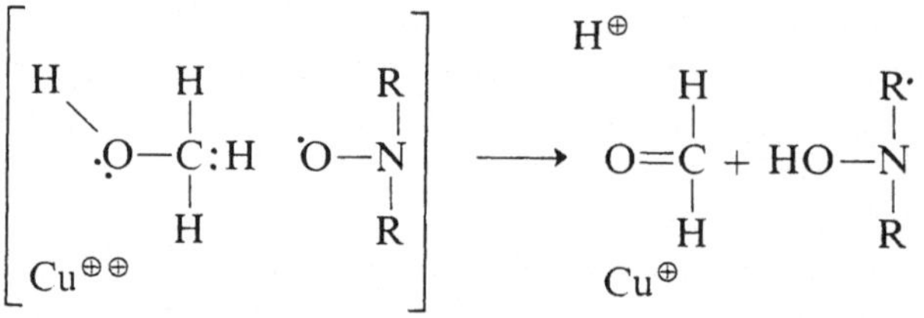

The overall mechanism of the catalytic oxidation of methanol in the presence of the nitroxyl radical can be represented by the following diagram:

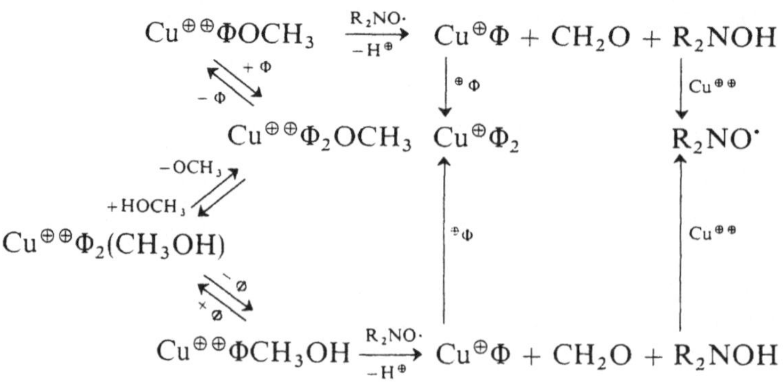

In the case of aerobic oxidation, the nitroxyl also catalyzes the reaction of methanol with the complexly bound $Cu^{\oplus\oplus}$ ion, the hydroxylamine so produced regenerating the radical by the reaction:

$$R_2NOH + Cu^{\oplus\oplus} \longrightarrow R_2NO\cdot + H^{\oplus} + Cu^{\oplus}$$

In contrast to the first case analyzed, the univalent copper ion reacts further with oxygen, reducing the latter to $O_2^{\ominus}\cdot$, which also oxidizes the methanol:

$$Cu^{\oplus}\Phi + O_2 \longrightarrow Cu^{\oplus\oplus}\Phi + O_2^{\ominus}\cdot$$

$$O_2^{\ominus}\cdot + CH_3OH\cdot Cu^{\oplus\oplus}\Phi \longrightarrow HO_2^{\ominus} + CH_2O + Cu^{\oplus}\Phi + H^{\oplus}$$

$$HO_2^{\ominus} + Cu^{\oplus}\Phi \longrightarrow Cu^{\oplus\oplus}\Phi O^{\ominus}\cdot + HO^{\ominus}$$

$$Cu^{\oplus}\Phi + O_2 + CH_3OHCu^{\oplus\oplus}\Phi \longrightarrow Cu^{\oplus\oplus}\Phi + Cu^{\oplus\oplus}\Phi O^{\ominus}\cdot + CH_2O + H_2O$$

The intermediate labile complex $Cu^{\oplus\oplus}\Phi O^{\ominus}\cdot$ then passes into a stable complex in accordance with the scheme:

$$2Cu^{\oplus\oplus}\Phi O^{\ominus}\cdot + CH_3OH \longrightarrow 2Cu^{\oplus\oplus}\Phi + CH_2O + HO^{\ominus}$$

2. Biochemical Investigations

In addition to the wide use of nitroxyls in the study of chemical and physical–chemical processes, these radicals are now beginning to find ever-increasing use in biochemical studies.

Thus, the capability of nitric oxide to increase the sensitivity of living organisms to the action of X rays [19] has led to the study of a number of stable nitroxyls in this field. Emmerson and Howard-Flanders [20, 21] have discovered the radiosensitization of the colibacillus *Escherichia coli B/r* in the absence of oxygen under the action of di-tert-butylnitroxyl, tert-butyl-2,6-dimethoxyphenylnitroxide, and 2,2,6,6-tetramethylpiperidine-1-oxyl.

An analysis of the curves of the survival of the bacteria under anaerobic conditions in the presence of nitroxyls led the authors to the conclusion that with an increase in the concentration of radicals, the sensitivity of the organisms to X rays rose considerably. The phenomenon can apparently be used in the therapy of malignant tumors. In actual fact, the blood supply is almost always disturbed in malignant tumors, and therefore individual colonies of cells are found in them in which the concentration of oxygen is extremely low. On irradiation, these cells are not destroyed, and subsequently may serve as active centers for the growth of the tumor. There is no doubt that an increase in the sensitivity of the cells to X-radiation is an urgent problem.

In this connection, investigations devoted to the toxicity of nitroxyls with respect to various types of cells appear not less important. According to Emmerson and Howard-Flanders [21], these radicals do not affect the rate of growth of *Escherichia coli*. However, they are toxic to tumor cells. Thus, the influence of di-tert-butylnitroxyl on the reproduction and growth of cultures of tumor cells *in vitro* has been studied [22]. In a special manner, the cultured tumor cells were subjected to the action of a radical for 40 min and for 11 days. It was found that in the case of the 40-min exposure, the radical had no effect on the reproductive capacity of the cells at concentrations of less than 5 μM. At a concentration of 10 μM, 68% of the cells lost their capacity for reproduction, and at 40 μM, only 1 7% of the cells retained this capacity. A study of the toxicological effect with the 11-day exposure showed that a concentration of radicals below 0.5 μM had no effect; a concentration of 1 μM brought about the destruction of 20% of the cells, and at 5 μM, the whole population underwent reproductive destruction.

Konovalova *et al.* [23] have studied the antitumoral activity of nitroxyls *in vivo*. The investigation was carried out on mice of the strain *C 57 BL* with a transplanted leukosis of strain *La*. The antileukemic activity of the substances was characterized by the authors by means of the inhibition coefficients κ showing how many fewer times the tumoral process developed in the presence of the radical as compared with a control experiment.

The nitroxyl radicals studied showed very slight toxicity, which permitted their carcinolytic activity to be investigated over a wide range of concentrations.

An attempt to study stable nitroxyls as chemical mutagens deserves attention [24]. We investigated the influence of 2,2,6,6-tetramethyl-4-oxo-piperidine-1-oxyl (P-1), 2,2,6,6-tetramethylpiperidine-1-oxyl (P-2), and 4-hydroxy-2,2,6,6-tetramethylpiperidine-1-oxyl (P-3) on the mutation frequency of the fruit fly *Drosophila melanogaster*. The trials were carried out on adult males and larvae of Drosophila of strain U^{3p}, the radicals being introduced either in the gas phase or in admixture with the nutritional medium. The results obtained in the two cases agreed well. The males subjected to the treatment were crossed with females of the strain *ClB* and the recessive features and partially visible mutations in the resulting progeny were counted.

The experiments showed that the radical P-1 exhibits considerable selectivity of its effect, causing hereditary changes in the maturing sex cells 2–3 times more frequently than in the mature sperms of the adult males.

The radical P-2 proved to be the most effective mutagen, although its mutagenic activity was masked to a considerable extent by its high toxicity. This compound has a strong narcotic action on the fly, which permitted the authors to use only small doses and exposures; nevertheless, P-2 gave a high percentage of mutations in all cases.

A study of the hypotensive action of di-tert-butylnitroxyl, 2,2,6,6-tetramethylpiperidine-1-oxyl, and a number of compounds related to them is of interest [25, 26]. The substances studied were administered intravenously and *per os* to rats, dogs, and cats. The number of contractions of the urinary bladder and the cardiac muscle and also the pressure of the arterial blood in the experimental animals were recorded.

The authors believe that the presence of an unpaired electron causes no specific pharmacological effect and that the ganglion-blocking activity of the radicals studied is explained exclusively by the steric screening of the nitrogen atom.

The previously mentioned property of stable nitroxyls of effectively trapping active radicals [12, 18] has been used successfully by Corker *et al.* [27] in a study of the primary free-radical products of photosynthesis. On studying the photochemical reaction of di-tert-butylnitroxyl with spinach chloroplast, the authors found that the nitroxyl recombined with the photo-induced radicals, the nature of which was analyzed subsequently by means of radiocarbon, ^{14}C. However, when the radical 3-carbamoyl-2,2,5,5-tetramethylpyrroline-1-oxyl [28] was added to a culture of *Chlamidomonas* [29], the EPR signal similarly decreased with time, but this had no effect whatever on the *Chlamidomonas* itself. The rate of disappearance of the radicals increased greatly when the substrate was irradiated with visible light. The authors assume that the disappearance of the EPR signal is connected with the reaction of the nitroxyl and plastoquinone, which, in this case, apparently acts as a reducing agent.

Thus, both the first studies on the use of stable radicals in biochemical investigations [30, 31] and all the subsequent studies have been based on the reactions of free radicals with the participation of the unpaired electron. As far as the reaction of radicals without the participation of the free valence is concerned, their use in biology is connected entirely with the name of the American scientist McConnell, and these uses will be discussed next.

3. Molecular Biology

As already mentioned, the discovery [32–34] and the detailed study [35–37] of free-radical reactions without the involvement of the unpaired electron have opened a wide range of applications and have enabled a new class of unusually stable radicals with strongly localized free valences to be created (Scheme 1).

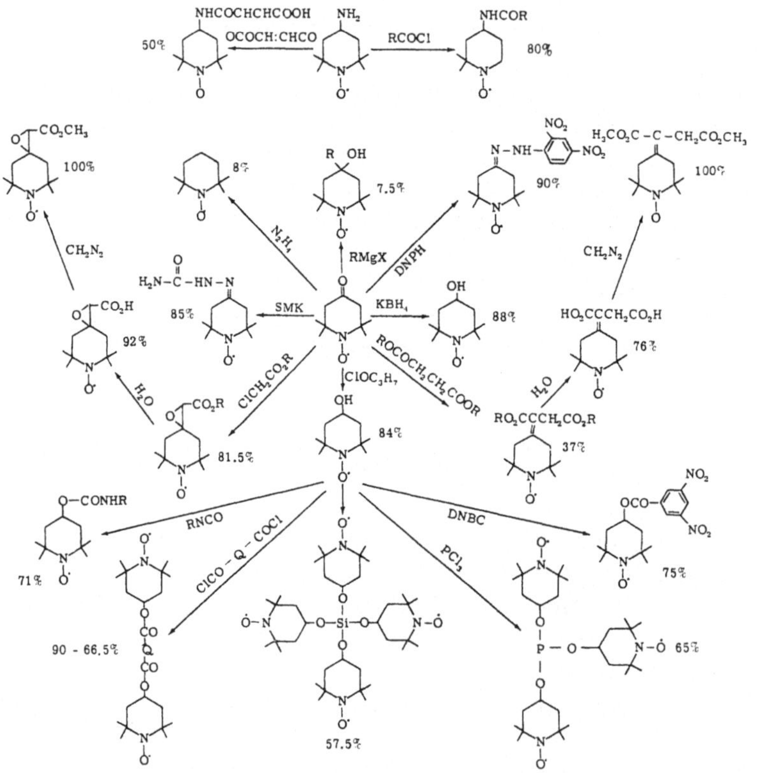

Scheme 1

A series of brilliant investigations recently published by McConnell and his co-workers has opened up a new and promising aspect of the use of such reactions in molecular biology. In accordance with a suggestion due to McConnell, the paramagnetic substances obtained by the reaction of radicals with biomolecules have been named spin-labeled compounds.

An analysis of the EPR spectra of spin-labeled proteins can give valuable information on the molecular symmetry axes, allosteric structural changes, the nature and order of the bonds of the amino acid fragments, the form and chemical structure of the active centers and their relative spatial location, and information on the supermolecular structure of the biomolecules.

The analysis is based on the fact that the HFS of the EPR spectra of nitroxyl radicals depends on the orientation of the orbital of the unpaired electron in a constant magnetic field and on the rate of movement and the exchange interaction of the radicals in solution.

A restriction of the mobility of the nitroxyl radicals leads to changes in the EPR spectra. The components of the HFS broaden nonuniformly, and this is shown in the amplitude of the components of the spectrum.

A criterion of the mobility of the radicals is the correlation time [38] (see Chapter VII), which can be obtained from an analysis of the widths of the components of the EPR spectra by Kivelson's method [39].

In practice, one may use the simplified expression

$$\tau = -8.4\Delta H_{(H)}\alpha \times 10^{-10} \tag{1}$$

where $\Delta H_{(H)}$ is the distance between the maxima of the derived component of the HFS located in the low-field side of the spectrum, and α depends on the ratio of the intensities of the components in low and high fields, $I_{(+1)}$ and $I_{(-1)}$, respectively, $\alpha = 1 - (I_{(+1)}/I_{(-1)})^{1/2}$. The expression given above is correct for those cases where the nonsecular contributions to the width of the EPR signal can be neglected [40].

An analysis of the EPR spectra of nitroxyl free radicals enables us to obtain quantitative characteristics of the degree of their mobility, which may depend fundamentally on the weak intermolecular actions which play an extremely important role in biological systems.

Where there is an arbitrary orientation of the nitroxyl free radicals in a rigid matrix, a qualitatively different kind of EPR spectra arises that is characterized by a marked broadening of the components of the HFS.

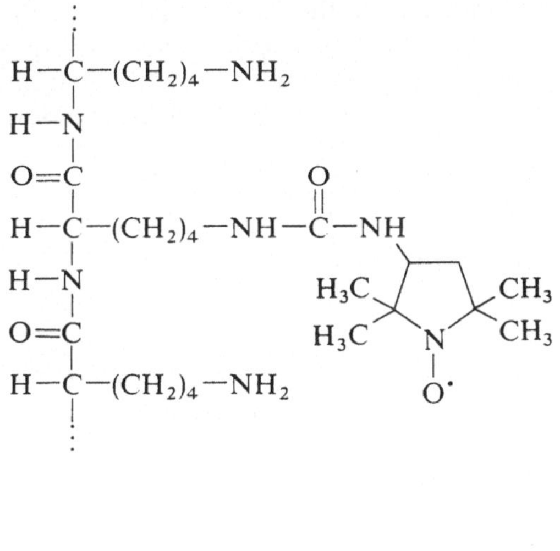

In the case of monocrystals, the anisotropic distribution depends on the orientation of the sample with respect to the external magnetic field. In a study of monocrystals of free nitroxyl radicals, Griffith *et al.* [41] determined the magnitudes of the coefficients of the anisotropic hyperfine interaction with the parallel and the perpendicular orientations of the π-orbital of the unpaired electron of the nitroxyl free radicals in a constant magnetic field and also the components of the $\hat{g}$ tensor, g_x, g_y, and g_z. The reaction of 3-isocyanato-2,2,5,5-tetramethylpyrrolidine-1-oxyl with poly-L-lysine without the involvement of the free valence gave spin-labeled poly-L-lysine [42], the components of whose EPR spectrum were broadened because of the decrease in the mobility of the radical.

In an analysis of the EPR spectra of spin-labeled serum albumen, it was found that the bulk of the free radicals were present in the same bound state as in the case of poly-L-lysine. However, in the spectra studied, it was possible to observe a component of an EPR signal with other parameters which are characteristic of nitroxyl radicals in solid organic glasses.

Since the disordered motion of the molecules of a protein in solution is insufficient for the effective averaging out of the anisotropic interaction, the results obtained show that a small part of the free radicals was fixed to a relatively rigid tertiary structure of the serum albumen. This phenomenon has been given the name "strong-immobilization spin labeling," in contrast to weak immobilization, when the components of the HFS of the EPR spectrum are only slightly broadened.

On the basis of the fact that, on an average, one molecule of protein was attached to between one and two free radicals, the authors assumed that in serum albumen, there are two different amino groups reacting with nitroxyl free radicals. In the case of compounds paramagnetically labeled with an amino group located on the surface of the protein molecule, the weakly-immobilized-labeling EPR spectrum is obtained. The other amino group is located in the depths of the protein globule and the spin labeling connected with it gives the EPR spectrum of strongly-immobilized free radicals. In the latter case, a free radical connected by a covalent bond with the molecule of the proton reacts hydrophobically with the adjacent segments of the polypeptide chain.

When the serum albumen was subjected to acid–alkaline denaturation, and also when the spin-labeled protein was digested with pepsin, the broad components of the EPR spectrum of the strongly-immobilized free radicals disappeared and the parameters of the EPR signal of the system studied approximated to those of the EPR spectrum of spin-labeled poly-L-lysine.

In another investigation [43], 3-maleimido-2,2,5,5-tetramethylpyrroli-dine-1-oxyl was used as the free-radical label. After the reaction in which the

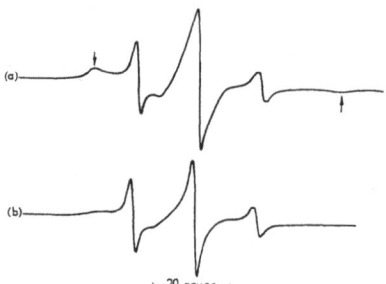

(a)

(b)

⊢—20 gauss—⊣

Fig. 37. EPR spectra of serum albumen spin-labeled with 3-maleimido-2,2,5,5-tetramethyl-pyrrolidine-1-oxyl. In contrast to (a), spectrum (b) relates to a protein the SH groups of which were previously blocked with N-ethylmaleimide [43].

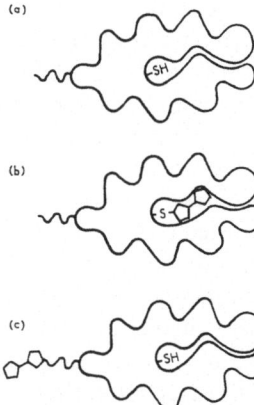

Fig. 38. Schematic illustration (a) of original protein; (b) of protein spin-labeled at the SH groups (immobile spin); (c) of protein with a spin-labeled lysine fragment (mobile spin) [43].

spin label was attached to the serum albumen had been carried out, a super-position of the two EPR signals relating to strong and weakly bound free radicals was observed. If the label was introduced after the preliminary blocking of the SH groups of the proteins by means of a specific reagent, the EPR spectrum lacked the broad components of the strongly-immobilized free radicals (in Fig. 37, these components are marked with arrows).

At pH 2.1, the components of the broad EPR signal were absent, and the amplitude of the narrow EPR signal had increased approximately twofold. On the basis of these facts, it may be concluded that about 50% of the spin labels add specifically to the sulfhydryl groups of the serum albumen, giving the EPR spectrum of the strongly-immobilized free radicals, and the remaining free radicals react with the amino groups of the lysine fragments (Fig. 38). Griffith and McConnell report that the components of the broad EPR signal can be observed in a study of many SH-containing proteins, while in the case of strong immobilization, the paramagnetic label must apparently be located in the depth of the surface of the protein approximately corresponding stereo-chemically to the configuration of the spin label (Fig. 38).

A convenient specific spin label is 4-maleimido-2,2,6,6-tetramethyl-piperidine-1-oxyl [44, 45]: about 90% of this radical passes into the strongly bound state on reaction with serum albumen.

Since the influence of the thermal motion of the molecules of protein in solution on the EPR spectrum of the paramagnetic label can be neglected, it is appropriate to take into account only the contribution of the vibrations of the free radicals around the covalent bond and the mobility of the link of the macromolecule to which the spin label is attached.

Taking into account the ideas mentioned above on the mechanism of the strong immobilization of a paramagnetic label, we may come to the conclusion that even small changes in the tertiary structure of proteins may lead to a substantial change in the mobility of the free radicals. It is known that the tertiary structure of proteins can be changed within certain limits independently of their secondary structure [46] by the action of certain nonpolar solvents, the latter almost selectively modifying the tertiary structure.

On the other hand, proteins lose their secondary structure completely in concentrated solutions of urea, passing into the state of a statistical coil. Thus, with the addition of both urea and dioxane to solutions of serum albumen labeled with nitroxyl free radicals, the initial broad EPR signal of the strongly-immobilized free radicals disappeared, being converted into the EPR spectrum of radicals possessing considerably greater mobility. The action of urea, which destroys the secondary and tertiary structures of the protein, led to a more fundamental increase in the mobility of the nitroxyl radicals. The correlation time calculated for these two cases proved to be 1.09×10^{-9} sec for serum albumen in a $10\,M$ urea solution and 2.04×10^{-9} sec for serum albumen in 50% aqueous dioxane.

It is important to note that during the denaturation of the labeled protein, only the contribution of the narrow EPR signal increased (Fig. 39), while the ratio of the intensities of its components remained constant. These results show that in serum albumen under the action of both urea and dioxane, conformational transitions take place only between two constituents—the native and modified constituents—intermediate stages apparently being absent. It is fundamental that the change in the conformation of serum albumen takes place over a fairly narrow range of concentrations of denaturing agents:

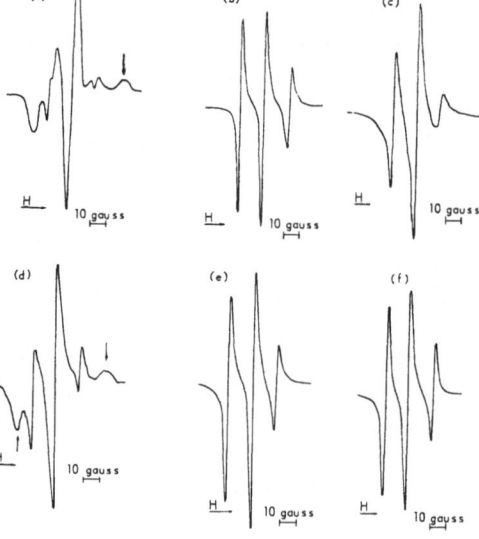

Fig. 39. EPR spectra of proteins labeled with 4-maleimido-2,2,6,6-tetramethylpiperidine-1-oxyl; (a) original serum albumen; (b) serum albumen in $10\,M$ urea solution; (c) 40% dioxane solution of serum albumen; (d) original egg albumen; (e) human gamma globulin with previously reduced disulfide bonds; (f) rabbit gamma globulin treated similarly [44, 45].

On the basis of data on the secondary and tertiary structure of proteins obtained by optical and hydrodynamic methods [47], it is natural to assume that the state of the spin-labeled fragment of a polypeptide chain after treatment of the protein with urea and dioxane differs by the absence or presence of an α-helical structure. The presence of the latter considerably increases the rigidity of the spin-labeled segment of the chain of the macromolecule, thanks to which, the mobility of the paramagnetic label in the presence of dioxane is considerably less than in the presence of urea.

Ogawa and McConnell [48] have studied the conformational changes of hemoglobin during its oxygenation, with the aid of the nitroxyl radical 3-iodoacetamido-2,2,5,5-tetramethylpyrrolidine-1-oxyl.

When the hemoglobin was converted into oxyhemoglobin, there was a transition between two states of the spin label. The absence of intermediate states was shown by the presence of isobestic points when the EPR spectra were superposed. The authors assume that the conformational change due to the addition of oxygen takes place simultaneously in all four subunits of hemoglobin and is cooperative in nature.

Recently, stable nitroxyl radicals have been used in a study of nucleic acids. The best results on RNA were obtained with 3-bromoacetamido-2,2,5,5-tetramethylpyrrolidine-1-oxyl [49]. When this radical was used, the degree of immobilization was greater than in the case of other nitroxyl labels. The treatment of samples of labeled RNA with ribonuclease led to a marked increase in the mobility of the paramagnetic label, and the EPR spectrum of the system studied approximated to the spectra of ordinary nitroxyl radicals in solution.

The addition of a paramagnetic label to DNA led to the superposition of the two EPR signals corresponding to different degrees of immobilization of the free radicals. The results obtained by these workers in a study of the reactions of a spin label with polynucleotides are extremely indicative. In the case of poly(guanylic acid) and poly(adenylic acid) at pH 5.5, the EPR spectrum of strongly-immobilized free radicals was observed, while at pH 11.5, the mobility of the paramagnetic label increased sharply as a result of the destruction of the ordered secondary structure of the polynucleotides. The presence of isobestic points showed that this transition was cooperative in nature.

In a study of poly(cytidylic acid), the authors observed the superposition of two EPR signals, and from this, they concluded that only part of the molecule, and not all segments of this polynucleotide have an ordered secondary structure. The study of poly(uridylic acid) showed that it is present in the state of a statistical coil even at pK values close to neutrality.

In a study by the EPR method of spin-labeled poly(guanidylic acid) and poly(adenylic acid), the pH values of helix-coil transitions were determined, and the values found agreed with the results of optical measurements. In this work, Smith and Yamane [49] first studied the reaction of 3-bromoacetamido-2,2,5,5-tetramethylpyrrolidine-1-oxyl without the involvement of the free radical with five nucleotides and obtained spin-labeled guanine (yield 20%), adenine (3%), cytosine, uracil, and thymine (less than 1%).

Stable nitroxyls may prove useful in a study of enzyme-substrate complexes. Berliner and McConnell [50] used as substrate for reaction with α-chymotrypsin the p-nitrophenyl ester of 3-carboxy-2,2,5,5-tetramethyl-pyrrolidine-1-oxyl. They succeeded in isolating a small amount of paramagnetic acylenzyme. The EPR spectrum of the enzyme–substrate complex showed that the spin label was rigidly fixed in relation to the enzyme molecule. At pH 6.8, the complex decomposed and the EPR signal characteristic for the free nitroxyl label in solution appeared.

In a 10 M solution of urea, the broad EPR signal disappeared, changing into the EPR spectrum of free radicals possessing considerably greater mobility. However, the components of the EPR signal were still broadened, which showed that in the transition of the enzyme into the state of a statistical coil, the substrate was not split off in spite of the fact that its free rotation increased considerably. The paramagnetic label was not removed by dialysis in an acid medium either. Apparently, the acyl group is attached to the enzyme by covalent bonds and is additionally immobilized by van der Waals forces. An analysis of the results obtained in this work leads to the conclusion that the formation of the enzyme–substrate complex involves a fixed orientation of the enzyme molecules and the substrate relative to one another.

The specific configuration of the molecule of an enzyme in the formation of a complex with the substrate ensures a definite mutual orientation of the molecule or substrate and the active groups of the enzyme. Thus, the orientation function of an enzyme apparently makes an extremely large contribution to ensuring its high catalytic activity.

The use of stable nitroxyl radicals in immunological studies has also led to extremely important and interesting results. The substances produced by microorganisms and causing the formation of antibodies are called antigens. The main function of the antibodies consists in binding the antigens. According to the ideas of Ehrlich, Landsteiner, and Pauling, the antigenic determinant (active center) combines with the specific segment of the antibody, since they correspond to one another stereochemically. In other words, the antigenic determinant is the information matrix for the construction of the specific segment of the antibody [51].

In recent years, it has been possible to obtain antibodies to many simple haptens. The dinitrophenyl haptens, the interaction of which with antibodies can be studied by fluorescence methods, have proved to form a very suitable model. Stryer and Griffith [52] have used the 2,4-dinitrophenylhydrazone of

2,2,6,6-tetramethyl-4-oxopiperidine-1-oxyl as an antigen. This hapten can thus be considered as a dual (fluorescence and nitroxyl) label. When a solution of the hapten was added to immune serum, a marked anisotropic broadening of the EPR spectrum, showing a binding of the hapten by antibodies, was observed. The binding of the hapten by antibodies at various ratios of them was studied, and it was shown that the haptens labeled with nitroxyl free radicals can be used to determine the titre of the antibodies.

It was impossible to determine the correlation time from an analysis of the widths of the components of the HFS of the EPR spectrum because of the strong binding of the spin label. However, it was possible to determine an analogous magnitude by the method of polarization luminescence. For this purpose, an aqueous glycerol solution was selected in which the EPR spectra of the dinitrophenyl-nitroxyl hapten were similar to the EPR spectrum of the complex of the hapten with the antibody (90% of glycerol, 5% of H_2O, and 5% of ethanol).

Satisfactory results were also obtained in a study of antigens using 4-maleimido-2,2,6,6-tetramethylpiperidine-1-oxyl. The latter was added to the sulfhydryl groups of the interchain disulfide bonds of the gamma globulins of man and rabbit, that had been reduced with 2-mercaptoethanol [53]. In both cases, the EPR spectra [see Fig. 39(e, f)] correspond to a relatively high mobility of the free radicals (τ approximately 1.1×10^{-9} for human gamma globulins and 7.4×10^{-10} for rabbit gamma globulins). On comparing the correlation time of the nitroxyl free radicals grafted onto the sulfhydryl groups of the cleaved disulfide bonds of the gamma globulins and SH groups of serum albumen treated with urea ($\tau = 1.09 \times 10^{-9}$) and dioxane ($\tau = 2.04 \times 10^{-9}$), it is natural to infer that the fragments of the polypeptide chains of the gamma globulins to which free radicals added do not possess an ordered secondary structure. The mobility of the nitroxyl labels in rabbit gamma globulin is particularly high, even exceeding that for serum albumen denatured with urea. These results are in agreement with data on the extremely low content of α-helical structures in the gamma globulins [54].

The congruence of the EPR spectrum of immune gamma globulin of the rabbit with the retention of activity specific to it opens up routes for the study of conformation and phase transitions in the specific antigen–antibody reaction.

In a study of the precipitation of spin-labeled antibodies with specific (egg albumen) and nonspecific (ammonium sulfate) antibodies, marked differences in the EPR spectra of the precipitates were found.

While the precipitation of antibodies with a specific antigen did not lead to a marked change in the mobility of the paramagnetic label, the precipitation of the gamma globulin with ammonium sulfate caused a strong immobilization of the free radicals.

Thus, these results can be regarded as a direct proof of the alternation theory, according to which, the formation of a precipitate is connected with the immunological multivalence of the antigen and the antibody with respect to one another.

In actual fact, by deliberately excluding the location of the spin label in the active center of the antibody, it is possible to come to the conclusion that the relatively mobile state of the spin labels in the antigen–antibody precipitate can be retained only if there is no marked dehydration of the antibodies through intermolecular interactions. Consequently, in contrast to gamma globulin precipitated with ammonium sulfate, the specific antigen–antibody precipitate has a microcellular structure, in agreement with the alternation theory.

When the antigen–antibody precipitate was stored for a long time in the absence of stabilizers, the degree of mobility of the nitroxyl free radicals decreased sharply. This fact can be explained by the increasing dehydration of the antibodies through the interaction of the protein molecules in the precipitate. Such dehydration is secondary in nature.

Recently, Waggoner et al. [55] have made a successful attempt to use nitroxyl radicals to study the micelles of sodium dodecyl sulfate. The study of such systems is important in order to explain the principles of the organization of biomembranes. The work was carried out with two nitroxyl free radicals: the 2,4-dinitrophenylhydrazone of 2,2,6,6-tetramethyl-4-oxopiperidine-1-oxyl and 2,2,4,4-tetramethyl-1,2,3,4-tetrahydro-γ-carboline-3-oxyl [56]. When the concentration of sodium dodecyl sulfate (NaDS) was increased from 0 to 5% (in 5% solutions, almost all the NaDS molecules are present in micelles), the solubility of both nitroxyl radicals increased by more than two orders of magnitude. Correspondingly, the correlation times of the free radicals in this system increased from 1×10^{-1} (magnitude characteristic for a free radical in solution) to 7×10^{-1} sec. A study of the dependence of the correlation time on the concentration of NaDS enabled the critical micelle concentration to be determined.

As Onishi et al. [57] have shown, it is possible to obtain unique information by studying monocrystals of proteins labeled with nitroxyl radicals. These authors obtained the basic results on horse oxyhemoglobin using maleimide spin labels.

Boeyens and McConnell [58] found that 3-maleimido-2,2,5,5-tetramethylpyrrolidine-1-oxyl reacts in solution without the participation of the unpaired electron both with hemoglobin and with oxyhemoglobin, showing, in the first case, the EPR spectrum of weakly-immobilized free radicals and, in the second case, the EPR spectrum corresponding to strong immobilization. The authors succeeded in showing that in both cases, the spin label reacts with the sulfhydryl group of the ninety-third amino acid residue of the

β-chain of the hemoglobin. Later, Onishi, Boeyens, and McConnell studied monocrystals of oxyhemoglobin labeled in this way.

The EPR study of monocrystals of proteins labeled with stable nitroxyl radicals enabled them to find some symmetry elements arising when the molecule of the protein consisted of two or more subunits. Pair rotational axes must be included in such symmetry elements. Since only the left-hand direction of rotation of the helix is characteristic for proteins, there can be no center of symmetry or plane of mirror reflection in proteins.

Pair rotational axes have also been found in the molecules of hemoglobin and α-chymotrypsin. They may also arise, for example, in the association of virus sheaths made of protein subunits. If these symmetry elements coincide with the crystallographic axes, they can be detected relatively easily by X-ray diffraction. Otherwise, their detection involves considerable difficulties. In both situations, pair rotational axes can be detected by the EPR method considerably more easily and simply than by X-ray structural analysis.

Under favorable conditions, it is also possible to determine by this method even slight changes in the conformation of proteins, such as those due to the replacement of one amino acid residue, or allosteric changes of structure as a result of the interaction of the protein molecule with substrates, activators, inhibitors, and coenzymes.

4. The Construction of Physical Instruments

Recently, so-called nuclear precession magnetometers and gradiometers possessing high sensitivity and accuracy have found ever-wider use for geomagnetic measurements. In contrast to magnetostatic, electromagnetic, induction, and electrodynamic instruments, the operation of a nuclear magnetometer does not depend either on the temperature or on the orientation of the pick-up. In the nuclear magnetometer, the magnetic field is measured by the frequency of the electromagnetic vibrations, which are determined by modern techniques with an accuracy reaching millionth parts of the measured magnitude.

The complete working cycle of a precession nuclear magnetometer includes two successive physical processes: the polarization of the working substance (water or a solution of ethanol in water) and the measurement of the frequency of the nuclear induction signal in a weak magnetic field (the Earth's field).

The use of the phenomenon of the dynamic polarization of protons in paramagnetic solutions has enabled nuclear magnetometers to be substantially improved. Simultaneously, it has been possible to synchronize the processes of polarization and measurement and thus to construct a fast-acting instrument [59, 60]. The creation of a special nuclear generator provides the possibility of obtaining a continuous nuclear induction signal with a frequency proportional to the magnetic field being measured [61].

The presence of free radicals in solution enables the dynamic polarization of the water protons to be effected by the saturation of one or two electronic transitions by means of a powerful high-frequency field, with a frequency of 54 or 56 MHz (constant field of about 0.5 G).

Until recently, the only working substance of the pick-ups of nuclear geomagnetometers with dynamic polarization was an aqueous solution of Fremy's salt (potassium nitrosodisulfonate). When this salt is dissolved in water, it decomposes into the anion radical

$$^{\ominus}O_3S-\overset{\displaystyle .}{N}-SO_3^{\ominus}$$
$$\underset{O^{\ominus}}{|}$$

(I)

which has a well-resolved hyperfine structure in electron paramagnetic resonance spectra. The saturation of any line in the EPR spectrum of the solution leads to a considerable increase in the nuclear magnetization of the solvent. This includes the dynamic polarization effect. A study of this effect in aqueous solutions of potassium nitrodisulfonate as applied to magnetometry has been carried out by Ryzhkov and Stepanov [62]. The great defect of the substance considered is its instability. Thus, in distilled water, the paramagnetic ion-radical decomposes during a few tens of minutes, the decomposition process being autocatalytic in nature and the rate of decomposition increasing with an increase in the concentration of the solution. It can be used in geomagnetometer solutions only when stabilizing substances such as K_2CO_3 are added. A stabilized solution can be kept for a month provided that the temperature does not exceed 40°C.

In view of this, the search for new, highly-stable radicals possessing properties analogous to those of potassium nitrosodisulfonate acquires great importance. Such properties are possessed by a whole series of nitroxyl radicals, of which only 2,2,6,6-tetramethyl-4-oxopiperidine-1-oxyl (I) [63] and 4-hydroxy-2,2,6,6-tetramethylpiperidine-1-oxyl (II) [64] have been studied in their application to the problem under consideration:

(II)   (III)

These individual radicals are readily soluble in many polar and nonpolar proton-containing solvents and possess a very simple and well-resolved HFS of the EPR spectra over the whole range of magnetic fields [65].

In a weak magnetic field, the connection between the electron spin $s = \frac{1}{2}$ and the spin of the nitrogen nucleus $K = 1$ is not broken, and the system of energy levels is characterized by a total spin number F assuming two values, $F = s + K = \frac{3}{2}$ and $F = s - K = \frac{1}{2}$, and by the magnetic quantum number m_F, which assumes $2F + 1$ values from $+F$ to $-F$. The system of energy levels of nitroxyl radicals is shown in Fig. 40.

The selection rule $\Delta F = 0, \pm 1$ and $\Delta m_F = \pm 1$ exists for electronic transitions between these levels. The saturation of one of such transitions by a strong radiofrequency field oriented perpendicularly to the constant magnetic field greatly changes the electronic magnetization of the solution M_z which, in turn, leads to a large increase in the proton magnetization m_z of the molecule of the solvent according to the expression [66]: $m_z - m = fQ(M_0 - M_z)$, where m_z, M_z, and m_0, M_0, are, respectively, the nuclear and electronic magnetizations of the solution with saturation of the electronic transition and at thermal equilibrium, f is a linkage factor, and Q is a coefficient determining the nature of the electronic–nuclear interaction. In solutions of these radicals, a dipole–dipole interaction exists, in the main, for which $Q = \frac{1}{2}$.

The coefficient of the increase of proton polarization, $A = m_z/m_0$, depends on the mechanism of the electron relaxation and on the saturated transition. A study of radical I has shown that only saturation of transitions

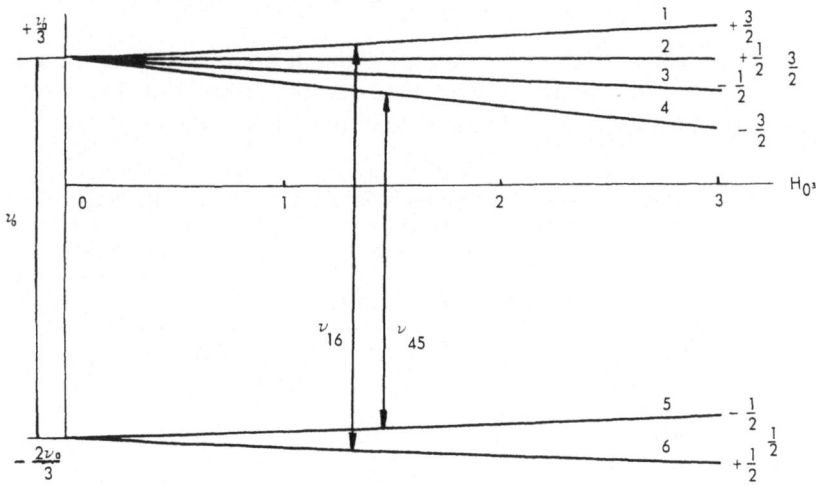

Fig. 40. Scheme of the energy levels of nitroxyl radicals in weak magnetic fields [65].

TABLE 22
Linewidths and Splitting Frequencies in a Zero Field for Solutions of Nitroxyl Radicals

Radical	Solvent	Linewidth, MHz	Splitting in zero field, MHz
I	Water	0.8	54.8
	Benzene	Insoluble	Insoluble
II	Water	1.8	67.6
	Benzene	2.6	61.4
III	Water	2.5	72.0
	Benzene	3.6	65.4

1–6 and 4–5 (Fig. 40) leads to a considerable dynamic polarization of the solvent nuclei.

A similar phenomenon is observed in solutions of radicals II and III. A study of the EPR spectra of solutions of radicals II and III in water and in benzene in the Earth's magnetic field has been carried out on an apparatus [67] enabling comparison between the initial amplitude of the signal of free precession with dynamic polarization and with polarization by a constant magnetic field with a change in the frequency of the saturating generator.

Table 22 gives the linewidths and frequencies v_0 of the splitting in a zero field of all three radicals in 1.6×10^{-3} M aqueous and benzene solutions.

Table 23 gives the frequencies of the electronic transitions and the coefficients of the increase in the proton polarization A in the Earth's magnetic field for the same solutions with saturation of the transitions by a radiofrequency field whose amplitude is equal to the width of the transition line.

It can be seen from the table that the dynamic polarizations in the solutions of nitroxyl radicals are not inferior to that of Fremy's salt. The organic nitroxyls are considerably more stable than the nitrosulfonate anion. Thus,

TABLE 23
Dynamic Polarizations of the Protons in Solutions of Nitroxyl Radicals

Radical	Solvent	Transition 4–5		Transition 1–6	
		A	v_{45}, MHz	A	v_{16}, MHz
I	Water	+820	53.8	−890	55.9
	Benzene		Insoluble		
II	Water	+760	66.6	−850	68.8
	Benzene	+460	60.2	−530	62.6
III	Water	+600	70.8	−720	73.2
	Benzene	+400	64.0	−480	66.6

the characteristics of solutions of nitroxyl radicals do not change over six months. Moreover, in the investigation, the solutions were heated to 90°C. Thus, nitroxyl radicals can be used successfully as working substances for nuclear magnetometers. The nitroxyl radicals are very convenient because they dissolve in various organic solvents. This enables a solvent with a high content of protons and a long proton relaxation time suitable for working under any climatic conditions to be selected.

In the near future, nitroxyl radicals (particularly deuterated ones) will apparently replace the unstable Fremy's salt not only in magnetometers [69–70], but also in maser-type apparatus [71].

References

1. É. G. Rozantsev and V. A. Golubev, *Izv. Akad. Nauk SSSR, Ser. Khim.* **1966**, 891.
2. V. A. Golubev and É. G. Rozantsev, *Izv. Akad. Nauk SSSR, Ser. Khim.* **1965**, 716.
3. É. G. Rozantsev and A. B. Shapiro, *Izv. Akad. Nauk SSSR, Ser. Khim.* **1964**, 1123.
4. É. G. Rozantsev, *Usp. Khim.* **35**, 1549 (1966).
5. É. G. Rozantsev, *Teor. i Éksperim. Khim.* **2**, 127 (1966).
6. É. G. Rozantsev and V. A. Golubev, *Izv. Akad. Nauk SSSR, Ser. Khim.* **1965**, 718.
7. É. G. Rozantsev, V. A. Golubev, M. B. Neiman, and Yu. V. Kokhanov, *Izv. Akad. Nauk SSSR, Ser. Khim.* **1965**, 572.
8. L. A. Krinitskaya, A. L. Buchachenko, and É. G. Rozantsev, *Zh. Organ. Khim.* **2**, 1301 (1966).
9. É. G. Rozantsev and L. A. Krinitskaya, *Izv. Akad. Nauk SSSR, Ser. Khim.* **1967**, 1137.
10. W. Klyne, in: Advances in Stereochemistry [Russian translation]. Goskhimizdat, Moscow (1961), p. 103.
11. A. L. Buchachenko, Stable Radicals, Moscow (1968).
12. A. N. Plyusnin and N. M. Chirkov, *Teor. i Éksperim. Khim.* **1966**, 777.
13. L. A. Volodina, Thesis, Moscow (1968).
14. W. Brackman and C. Gaasbeck, *Rec. Trav. Chim.* **85**, 221 (1966).
15. W. Brackman and C. Gaasbeck, *Rec. Trav. Chim.* **85**, 242 (1966).
16. W. Brackman and C. Gaasbeck, *Rec. Trav. Chim.* **85**, 257 (1965).
17. V. A. Golubev, Thesis, Moscow (1967).
18. L. A. Krinitskaya, Thesis, Moscow (1967).
19. P. Howard-Flanders, *Nature* **180**, 1191 (1957).
20. P. Emmerson and P. Howard-Flanders, *Nature* **204**, 1005 (1964).
21. P. Emmerson and P. Howard-Flanders, *Radiation Res.* **26**, 54 (1965).
22. M. Klimek, *Nature* **209**, 1256 (1966).
23. N. P. Konovalova, G. N. Bogdanov, V. B. Miller, M. B. Neiman, É. G. Rozantsev, and N. M. Émanuél', *Dokl. Akad. Nauk SSSR* **157**, 707 (1964).
24. G. I. Efremova and É. G. Rozantsev, *Genetika* **1965** (2), 63.
25. L. Bretherick, G. Lee, E. Lunt, W. Wragg, and N. Edge, *Nature* **184**, 1707 (1959).
26. S. Cummings, S. Grace, and C. Latimer, *J. Pharm. Exp. Therap.* **141**, 349 (1963).
27. G. Corker, M. Klein, and M. Calvin, *Proc. Natl. Acad. Sci. USA* **56**, 1365 (1966).
28. É. G. Rozantsev and L. A. Krinitskaya, *Tetrahedron* **21**, 491 (1965).
29. E. Weaver and H. Chon, *Science* **153**, 301 (1966).
30. J. Greenstein, *J. Biol. Chem.* **125**, 501 (1938).

31. J. Greenstein and W. Jenrette, *J. Biol. Chem.* **142**, 175 (1942).
32. M. B. Neiman, E. G. Rozantsev, and Yu. G. Mamedova, *Nature* **196**, 472 (1962).
33. É. G. Rozantsev, *Izv. Akad. Nauk SSSR, Ser. Khim.* **1963**, 1669.
34. É. G. Rozantsev and M. B. Neiman, *Tetrahedron* **20**, 131 (1964).
35. É. G. Rozantsev, *Izv. Akad. Nauk SSSR, Ser. Khim.* **1966**, 770.
36. É. G. Rozantsev and L. A. Krinitskaya, *Izv. Akad. Nauk SSSR, Ser. Khim.* **1964**, 1573.
37. É. G. Rozantsev and M. B. Neiman, *Zh. Organ. Khim.* **1**, 1337 (1965).
38. N. Blombergen, E. Purcell, and R. Pound, *Phys. Rev.* **73**, 679 (1948).
39. D. Kivelson, *J. Chem. Phys.* **53**, 1094 (1960).
40. B. I. Sukhorukov, A. M. Vasserman, L. A. Kozlova, and A. L. Buchachenko, *Dokl. Akad. Nauk SSSR* **177**, 454 (1967).
41. O. Griffith, D. Cornell, and H. McConnell, *J. Chem. Phys.* **43**, 2909 (1965).
42. D. Stone, T. Buckman, P. Nordic, and H. McConnell, *Proc. Natl. Acad. Sci. US* **54**, 1010 (1965).
43. O. Griffith and H. McConnell, *Proc. Natl. Acad. Sci. US* **55**, 8 (1966).
44. G. L. Grigorian, V. I. Suskina, É. G. Rozantsev, and A. É. Kalmanson, *Nature* **216**, 927 (1967).
45. G. D. Grigoryan, V. I. Suskina, É. G. Rozantsev, and A. É. Kalmanson, *Molek. Biologiya* **2**, 148 (1968).
46. S. E. Bresler, V. P. Kushner, and S. Ya. Frenkel', *Biokhimiya* **24**, 685 (1959).
47. S. Ya. Frenkel' and P. Orn, *Vysokomolekul. Soedin.* **3**, 541 (1961).
48. S. Ogawa and H. McConnell, *Proc. Natl. Acad. Sci. US* **58**, 19 (1967).
49. I. Smith and T. Yamane, *Proc. Natl. Acad. Sci. US* **58**, 884 (1967).
50. J. Berliner and H. McConnell, *Proc. Natl. Acad. Sci. US* **55**, 708 (1966).
41. D. Pressman, Molecular Structure and Biological Specificity, Washington (1957).
52. L. Stryer and O. Griffith, *Proc. Natl. Acad. Sci. US* **54**, 1785 (1965).
53. G. L. Grigoryan, S. G. Tatarinova, A. Ya. Kul'berg, A.É. Kalmanson, É. G. Rozantsev, and V. I. Suskina, *Dokl. Akad. Nauk SSSR* **178**, 230 (1968).
54. V. N. Okulov, G. V. Troitskii, and Yu. N. Gordeev, *Biokhimiya* **31**, 768 (1966).
55. A. Waggoner, O. Griffith, and H. McConnell, *Proc. Natl. Acad. Sci. US* **57**, 1198 (1967).
56. É. G. Rozantsev and A. B. Shapiro, *Izv. Akad. Nauk SSSR, Ser. Khim.* **1964**, 1123.
57. S. Onishi, J. Boeyens, and H. McConnell, *Proc. Natl. Acad. Sci. US* **56**, 809 (1966).
58. J. Boeyens and H. McConnell, *Proc. Natl. Acad. Sci. US* **56**, 22 (1966).
59. A. Abragam, Phys. Rev. **98**, 1729 (1955).
60. A. I. Filatov, A. P. Stepanov, and V. M. Stotskii, *Pribory i Tekhn. Éksperim.* **1965**(1), 169.
61. A. Abragam, J. Combrisson, and I. Solomon, *Compt. Rend.* **245**, 157 (1957).
62. V. M. Ryzhkov and A. P. Stepanov, *Geofiz. Pribor.* **1962**(12), 35.
63. É. G. Rozantsev, *Izv. Akad. Nauk SSSR, Ser. Khim.* **1964**, 2218.
64. É. G. Rozantsev, *Izv. Akad. Nauk SSSR, Ser. Khim.* **1964**, 2187.
65. É. G. Rozantsev and A. P. Stepanov, *Geofiz. Apparatura* **1966**(29), 35.
66. I. Solomon, *Phys. Rev.* **99**, 559 (1955).
67. A. P. Stepanov, *Pribory i Tekhn. Éksperim.* **1961**(3), 102.
68. J. Combrisson, *J. Phys. Radium* **19**, 840 (1958).
69. R. Besson, H. Lemaire, A. Rassat, A. Salvi, and P. Servoz-Gavin, in: Proc. XIIth Ampère Colloquium, Bordeaux, 17–21 Sept. (1953), p. 327.
70. J. Thomas, *Geophys. Prosp.* **13**, 22 (1965).
71. E. Allais, *Compt. Rend.* **246**, 2123 (1958).

Chapter IX

SYNTHESES OF SOME STABLE RADICALS AND THE MOST IMPORTANT INTER-MEDIATES

I. *2,2,6,6-Tetramethyl-4-oxopiperidine.* A five-liter glass vessel was charged with 2500 g of dry acetone, 800 g of ground anhydrous calcium chloride, and 255 g of liquid ammonia. The reaction mixture was left to stand for four days with periodic stirring. Then, with continuous cooling and stirring, the mixture was carefully decomposed with 1250 ml of 50% caustic soda solution. The organic layer was separated and the aqueous layer was carefully extracted with diethyl ether. The diethyl ether extract was combined with the organic layer. After the combined mixture had been dried with anydrous potassium carbonate, the solvent was distilled off, the residue was subjected to vacuum rectification* through a column with an efficiency of 10 theoretical plates, and the fraction boiling between 84 and 90°C (14 mm Hg) was collected. The distillate was cooled and the deposited crystals of triacetoneamine were collected on a glass filter. The yield of pure substance was 350–380 g (15–17%) after recrystallization from cyclohexane: colorless needles, mp 35.5–36°C; bp 84–85°C (15 mm Hg), 102–105°C (18 mm Hg).

II. *3,5-Dibromo-2,2,6,6-tetramethyl-4-oxopiperidine Hydrobromide.* With ice-water cooling, 17.6 g of triacetoneamine was dissolved in 70 ml of glacial acetic acid and then, with vigorous stirring, a solution of 36 g of bromine in 50 ml of glacial acetic acid was added. The reaction mixture was left to stand for a day and was then pressed out on a Büchner funnel; the residual crystals of 3,5-dibromotriacetoneamine hydrobromide were washed on the filter successively with acetic acid, water, and ether, and were then dried in the air. This gave 35 g (78%) of a white substance with mp 193–195°C, which could be used further without additional purification. Literature data [2]: mp, 203°C.

*The main impurity in this method of preparation is the so-called "acetonin," a tetrahydropyrimidine derivative [1].

III. *2,2,5,5-Tetramethylpyrroline-3-carboxyamide.* With stirring, 35 g of 3,5-dibromo-2,2,6,6-tetramethyl-4-oxopiperidine was slowly added to 350 ml of 25% aqueous ammonia, and the resulting solution was saturated with solid caustic potash. The light crystalline precipitate of the amide was

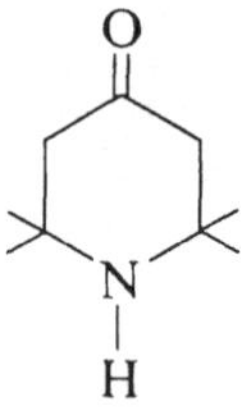

(I) 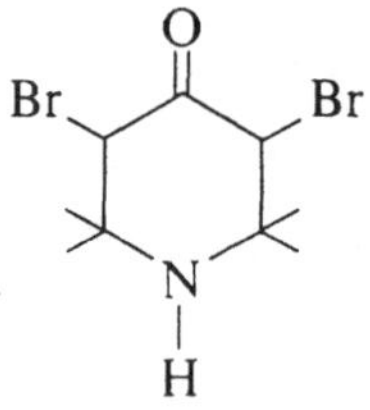   (II) 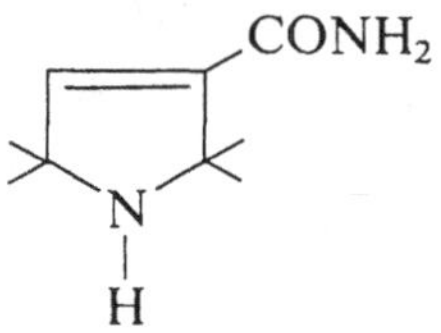   (III)

carefully pressed out on a Büchner funnel and dried in air. The substance was subsequently used as such or was additionally recrystallized from benzene; after purification, 11.8 g (80% of theory) of amide, mp 178–179°C, was obtained. Literature data [3]: mp, 180–181°C.

IV. *2,2,5,5-Tetramethylpyrrolidine-3-carboxyamide.* Thirty grams of 2,2,5,5-tetramethylpyrroline-3-carboxyamide was dissolved in 300 ml of methanol and hydrogenated in a rotating steel autoclave in the presence of Raney nickel (W-6) at room temperature with an initial hydrogen pressure of 180 atm. After the completion of hydrogenation, the ethanolic solution was filtered, the filtrate was evaporated to dryness in vacuum, and the residue was recrystallized from benzene. The yield of 2,2,5,5-tetramethylpyrrolidine-3-carboxyamide was quantitative; mp, 128–129°C (sublimate); literature data [3]: 129–130°C.

V. *2,2,5,5-Tetramethylpyrroline-3-carboxylic Acid Dihydrate.* A mixture of 1.68 g of 2,2,5,5-tetramethylpyrroline-3-carboxyamide and 1.73 g of barium hydroxide (octahydrate) was boiled until the evolution of ammonia ceased (45 hr). Then water was added and the resulting mixture was heated to 60°C with simultaneous addition of carbon dioxide. The barium carbonate was removed by filtration, and carbon dioxide was again passed through the filtrate. When it was certain that no more barium carbonate was being precipitated, the solution was evaporated on a water bath. This gave 1.46 g of the dihydrate of the amino acid (71%), which was used subsequently without purification. When it was melted on a Kofler block, the substance evaporated without melting. Literature data [3]: mp, 300°C (in a sealed capillary).

VI. *2,2,5,5-Tetramethylpyrrolidine-3-carboxylic Acid.* A mixture of 3 g of 2,2,5,5-tetramethylpyrrolidine-3-carboxyamide, 3.5 g of barium hydroxide, and 6 ml of water was boiled for 50 hr, by which time, the evolution of ammonia had ceased. The reaction mixture was saturated with carbon dioxide, and the barium carbonate was removed by filtration. The filtrate was evaporated to dryness on a water bath. This gave 2.5 g (75%) of the monohydrate of the amino acid with mp 219–220°C, which was used subsequently without additional purification. Literature data [3]: mp, 220°C.

VII. *3-Carbamoyl-2,2,5,5-tetramethylpyrroline-1-oxyl* [4]. A mixture of 15 g of 2,2,5,5-tetramethylpyrroline-3-carboxyamide, 150 ml of water, 0.75 g of Trilon B,* 0.75 g of sodium tungstate, and 15 ml of 30% hydrogen peroxide was left in the dark at room temperature for 10 days. The deposited bright yellow crystals (14 g) were filtered off and dried, and the filtrate was acidified with 5% hydrochloric acid and extracted with chloroform. The chloroform extract was dried with sodium sulfate and filtered, and the filtrate was evaporated to dryness in vacuum to give an additional 0.93 g of reaction product. The total yield of the free radical was 14.93 g (91.5%): mp, 203–204°C (decomp.) from ethanol.

VIII. *3-Carbamoyl-2,2,5,5-tetramethylpyrrolidine-1-oxyl* [4, 5]. To a solution of 4.4 g of 2,2,5,5-tetramethylpyrrolidine-3-carboxyamide in a mixture of 50 ml of water and 7 ml of methanol were added 0.25 g of Trilon B, 0.25 g of sodium tungstate, and 6 ml of 30% hydrogen peroxide. The reaction mixture was left to stand at room temperature for a week. The deposited bright yellow crystals (3.41 g) were filtered off and dried over P_2O_5, and the filtrate was acidified with 5% hydrochloric acid and extracted with chloroform. The chloroform extract was dried with sodium sulfate and filtered, and the filtrate was evaporated to dryness in vacuum. This gave an additional

(IV)   (V)   (VI)

0.79 g of reaction product. The total yield of free radical was 4.2 g (88%). The radical melted at 174–174.5°C after recrystallization from a mixture of hexane and dioxane and vacuum sublimation.

*Soviet brand of the sodium salt of ethylenediaminetetraacetic acid, which is also widely known under other names; for example, Complexone-3, Titriplex-III, Irgalone, and Versene.

| (VII) | (VIII) | (IX) |

IX. *3-Carboxy-2,2,5,5-tetramethylpyrroline-1-oxyl* [4]. A mixture of 3.66 g of 3-carbamoyl-2,2,5,5-tetramethylpyrroline-1-oxyl and 40 ml of 10% caustic soda was boiled under reflux until the evolution of ammonia had ceased. After cooling, the reaction mixture was carefully acidified with hydrochloric acid and the deposited precipitate was filtered off with suction on a porcelain filter and dried in air. The acid solution was extracted with ether, and an additional amount of paramagnetic substance was obtained from the extract. The total yield of bright yellow crystalline acid-radical was 3.57 g (97%), mp 210–211°C (dec.), after recrystallization from benzene.

X. *3-Carboxy-2,2,5,5-tetramethylpyrrolidine-1-oxyl* [4]. A mixture of 1.85 g of 3-carbamoyl-2,2,5,5-tetramethylpyrrolidine-1-oxyl, 3.15 g of barium hydroxide octahydrate, and 6 ml of water was boiled under reflux for 20 hr, by which time, the evolution of ammonia had ceased. The reaction mixture was treated with 50 ml of water, heated to 60°C, and saturated with carbon dioxide. The precipitate of barium carbonate was filtered off with suction on a fine glass filter, and the filtrate was heated to 60°C and resaturated with carbon dioxide. When no more precipitate deposited, the solution was acidified with 5% hydrochloric acid to pH 2 and extracted with ether. The ethereal extract was dried with anhydrous sodium sulfate, the desiccant was filtered off, and the filtrate was concentrated in vacuum. This gave 1.62 g (87%) of yellow crystals which, after recrystallization from a mixture of chloroform and hexane, melted at 193°C (decomp.).

XI. *3-Methoxycarbonyl-2,2,5,5-tetramethylpyrroline-1-oxyl* [4]. An excess of an ethereal solution of diazomethane was added to a solution of 1.84 g of 3-carboxy-2,2,5,5-tetramethylpyrroline-1-oxyl in 10 ml of ether until the evolution of nitrogen ceased. The ethereal solution was washed with saturated sodium bicarbonate solution and with water and was then dried with sodium sulfate. The ether was evaporated to give 1.95 g (98.6%) of a yellow crystalline substance possessing a strong odor of flowering lime and melting at 89.3–89.8°C (aqueous methanol). The substance was readily soluble in organic solvents and sparingly soluble in water, and readily sublimed.

XII. *3-Methoxycarbonyl-2,2,5,5-tetramethylpyrrolidine-1-oxyl* [4].
An excess of an ethereal solution of diazomethane was added dropwise to a
solution of 1.27 g of 3-carboxy-2,2,5,5-tetramethylpyrrolidine-1-oxyl in 10 ml
of ether until the evolution of nitrogen ceased. The solvent and the excess of
diazomethane were evaporated under vacuum. The residue (1.41 g) was
dissolved and chromatographed on a column of alumina of activity II (with
benzene as eluant). This yielded an orange-colored oil which was distilled in
vacuum: bp, 120°C (10.5 mm Hg); n_D^{20}, 1.4585; d_4^{20}, 1.0357; MR_D, 52.78.
MR_D, calc. 52.88.

XIII. *3-Cyano-2,2,5,5-tetramethylpyrroline-1-oxyl* [5]. To a solution of 1.53 g of 3-carbamoyl-2,2,5,5-tetramethylpyrroline-1-oxyl in 10 ml
of dry pyridine was added 2.87 g of *p*-toluenesulfonyl chloride, whereupon
the temperature of the mixture rose. It was left at room temperature for two
days and then a solution of 1 g of caustic potash in 25 ml of water was added
and the mixture was heated to 80°C. After cooling, the alkaline solution was
extracted with ether; the extract was carefully washed with 10% hydrochloric
acid, with a saturated solution of sodium carbonate, and with water, and was
dried with sodium sulfate. Evaporation of the ether yielded 0.94 g (57%) of
a yellow-orange paramagnetic substance which was purified by chromatography on alumina of activity II (with benzene as eluant) and subsequent
vacuum sublimation, after which it melted at 62.5–63°C. The substance was
readily soluble in organic solvents and sparingly soluble in water.

XIV. *3-Cyano-2,2,5,5-tetramethylpyrrolidine-1-oxyl* [5]. To a solution of 1.85 g of 3-carbamoyl-2,2,5,5-tetramethylpyrrolidine-1-oxyl in 5 ml
of dry pyridine was added 2.87 g of *p*-toluenesulfonyl chloride. The reaction
mixture was left at room temperature for two days and then a solution of 1 g
of caustic potash in 25 ml of water was added and the mixture was heated to
80°C. After cooling, the alkaline solution was extracted with ether, and the
extract was washed with 10% hydrochloric acid, saturated aqueous sodium
bicarbonate solution, and water, and was then dried with sodium sulfate.
Concentration of the ethereal extract yielded 0.8 g (48%) of a red liquid which,
after chromatography on a column of alumina of activity II (with benzene

(XIII)   (XIV)   (XV)

as eluant) set to orange-colored paramagnetic crystals which melted after vacuum sublimation at 31.5–32.5°C; bp, 99.6–99.8°C (2 mm Hg); n_D^{35}, 1.4576; d_4^{35}, 0.9839; MR_D, 46.34; MR_D calc., 46.43.

XV. *2,2,5,5-Tetramethyl-3-oxopyrrolidine-1-oxyl* [4]. In one portion, 9.15 g of 3-carbamoyl-2,2,5,5-tetramethylpyrroline-1-oxyl was added to a cooled solution of sodium hypobromite obtained from 14 g of caustic soda, 200 ml of water, and 10 g of bromine. The reaction mixture was stirred for 2 hr with ice-water cooling and was then heated to 70°C and kept at this temperature for 1 hr. The resulting mixture was again cooled with ice-water, and 100 g of caustic soda was added with stirring, after which the amide that had not reacted (1.8 g) was filtered off with suction, and the filtrate was extracted with ether. The ether extract was dried over anhydrous sodium sulfate, filtered, and evaporated in vacuum. Chromatography on a column of alumina of activity II (with chloroform as eluant) gave 2.05 g (33%) of a red oil which set to orange-red paramagnetic crystals melting at 41.5–42.5°C (from hexane) which are readily soluble in water and organic solvents.

XVI. *3-(Hydroxyimino)-2,2,5,5-tetramethylpyrrolidine-1-oxyl* [5]. A solution of 0.430 g of sodium acetate in 2 ml of water was added to 0.275 g of hydroxylamine hydrochloride, and then the solution was heated to 60°C and 0.41 g of 2,2,5,5-tetramethyl-3-oxopyrrolidine-1-oxyl was added. The reaction mixture was kept at 60°C with continuous stirring for 30 min, and was then cooled to give 0.38 g (70%) of a light yellow crystalline precipitate. The substance melted at 182.5–183°C after two crystallizations from carbon tetrachloride.

XVII. *3-Amino-2,2,5,5-tetramethylpyrrolidine-1-oxyl* [5]. To a cooled solution of sodium hypobromite prepared from 6 g of caustic soda, 65 ml of water and 5.6 g of bromine was added 4.62 g of 3-carbamoyl-2,2,5,5-tetramethylpyrrolidine-1-oxyl. The reaction mixture was stirred for 2 hr with ice-water cooling, and was then heated to 70°C and kept at this temperature for 1 hr. The resulting mixture was again cooled with ice-water, treated with 50 g of caustic potash with stirring, and extracted with ether. The extract was dried with anhydrous sodium sulfate and, after removal of

(XVI)   (XVII)   (XVIII)

the desiccant, the ether was evaporated in vacuum. The residue consisted of 2.37 g (61%) of a yellow paramagnetic oil which was distilled in vacuum; bp, 75–80°C (1 mm Hg). Picrate; mp, 176–179°C (dec.; acetone–hexane). Benzoate; mp, 148°C (acetone–hexane).

XVIII. *3-Hydroxy-2,2,5,5-tetramethylpyrrolidine-1-oxyl* [6].   Five grams of freshly-prepared aluminum isopropoxide was added to a solution of 1.21 g of 2,2,5,5-tetramethyl-3-oxopyrrolidine-1-oxyl in 30 ml of absolute isopropanol and the mixture was heated in an oil bath with simultaneous removal of isopropanol. After 3 hr, a test for acetone was negative. The mixture was cooled and the isopropanol was evaporated in vacuum. The residue was treated with 25 ml of water and was made alkaline with caustic potash until the aluminum alkoxides had dissolved completely. The alkaline solution was extracted with ether. After drying with sodium sulfate, the extract was concentrated in vacuum to give 0.98 g (80%) of a paramagnetic solid which was recrystallized from hexane; mp, 125.5–126°C. Benzoate: mp 66.5–67°C (hexane).

XIX. *3-Chloroformyl-2,2,5,5-tetramethylpyrroline-1-oxyl* [7].   A suspension of 0.552 g of 3-carboxy-2,2,5,5-tetramethylpyrroline-1-oxyl in 10 ml of dry benzene was treated with 0.3 ml of dry pyridine, and, with cooling and stirring, 0.27 ml of thionyl chloride was added dropwise to the resulting solution. The reaction mixture was stirred at room temperature for 1 hr, the salts which deposited were filtered off and washed on the filter with dry benzene, and the filtrate was evaporated in vacuum at room temperature. The residual dark-yellow crystalline acid chloride was readily hydrolyzed by atmospheric moisture.

XX. *Anhydride of 3-Carboxy-2,2,5,5-tetramethylpyrroline-1-oxyl* [7]. With stirring and cooling, 0.2 ml of dry pyridine and then 0.05 ml of thionyl chloride were added to a suspension of 0.184 g of 3-carboxy-2,2,5,5-tetramethylpyrroline-1-oxyl in 3 ml of dry benzene. After 1 hr, the reaction mixture was poured onto ice, and the organic layer was separated and washed with dilute hydrochloric acid, water, dilute alkali, and water again. After being dried with anhydrous sodium sulfate, the solution was evaporated in

(XIX)   (XX)

vacuum to give 0.131 g (75%) of bright yellow crystals which were recrystallized twice from a mixture of hexane and benzene, mp 141.5–143°C.

XXI. *Anhydride of 3-Carboxy-2,2,5,5-tetramethylpyrrolidine-1-*
oxyl [7]. With stirring and cooling, 0.2 ml of dry pyridine and then 0.05 ml of
thionyl chloride were added to a suspension of 0.186 g of 3-carboxy-2,2,5,5-
tetramethylpyrrolidine-1-oxyl in 3 ml of dry benzene. After 1 hr, the reaction
mixture was poured onto ice, and the organic layer was separated off and
washed with dilute hydrochloric acid, water, dilute alkali, and water again.
After being dried with anhydrous sodium sulfate, the solution was evaporated
in vacuum to give 0.064 g (36%) of bright yellow crystals which were recrystallized twice from anhydrous carbon tetrachloride, mp 179–179.5°C (decomp.).

XXII. *3-Ethoxycarbonyl-2,2,5,5-tetramethylpyrroline-1-oxyl* [7].
With stirring and cooling, 0.2 ml of dry pyridine and then 0.09 ml of thionyl
chloride were added to a suspension of 0.184 g of 3-carboxy-2,2,5,5-tetra-
methylpyrroline-1-oxyl in 3 ml of dry benzene. The reaction mixture was
stirred at room temperature for $\frac{1}{2}$ hr, and then the sodium ethoxide prepared
from 1 ml of ethanol and 0.03 g of sodium was added to it. The salt that
deposited was separated by filtration, and the filtrate was evaporated in
vacuum. The residue was dissolved in hexane and chromatographed on a
column of alumina of activity II (with chloroform as eluant). This yielded
0.1 g (47%) of 3-ethoxycarbonyl-2,2,5,5-tetramethylpyrroline-1-oxyl with
mp 109.5–110°C (hexane).

XXIII. *3-Phenoxycarbonyl-2,2,5,5-tetramethylpyrroline-1-oxyl* [7].
To a solution of 3-chloroformyl-2,2,5,5-tetramethylpyrroline-1-oxyl prepared
from 0.092 g of the acid and 0.05 ml of thionyl chloride in 3 ml of dry benzene
and 0.2 ml of dry pyridine was added 0.05 g of phenol. The mixture was left
for $\frac{1}{2}$ hr at room temperature and was then poured into 5 ml of 5% hydrochloric acid. The organic layer was separated and was washed with water,
10% aqueous caustic potash, and water again. The benzene solution was
dried with fused calcium chloride and the solvent was evaporated in vacuum.
The residue consisted of 0.102 g (78%) of orange-yellow paramagnetic

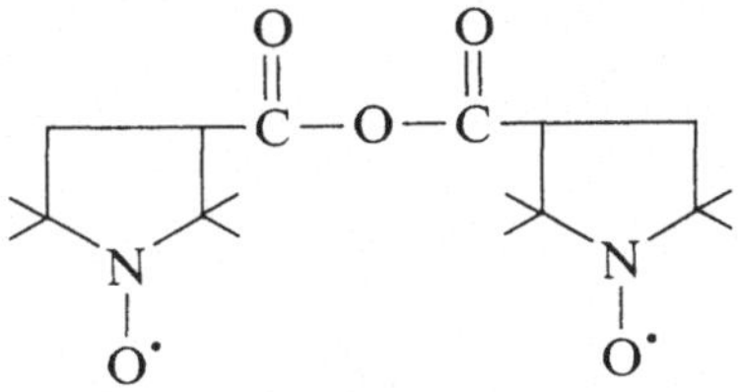

(XXI)

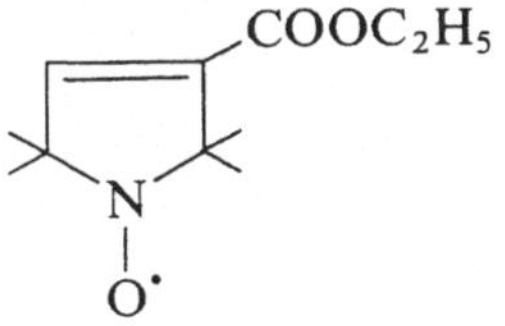

(XXII)

crystals which were purified by chromatography on alumina of activity II (with chloroform as eluant). This yielded 0.084 g (64%) of the radical, which was crystallized from hexane; mp, 113.3–113.7°C.

XXIV. *3-Phenoxycarbonyl-2,2,5,5-tetramethylpyrrolidine-1-oxyl* [7]. To a solution of 3-chloroformyl-2,2,5,5-tetramethylpyrrolidine-1-oxyl obtained from 0.186 g of the corresponding acid and 0.09 ml of thionyl chloride in a mixture of 3 ml of dry benzene and 0.3 ml of dry pyridine was added 0.1 g of phenol. After 30 min, the reaction mixture was poured into 20 ml of water, and the organic layer was separated and was washed with 5% hydrochloric acid, water, 5% aqueous caustic potash, and water again, and was then dried with fused calcium chloride, and the solvent was evaporated in vacuum. The residue consisted of 0.224 g (85.5%) of orange-yellow crystals, which were chromatographed on alumina of activity II (with chloroform as eluant). This yielded 0.149 g (51.5%) of substance, which was purified further by crystallization from hexane and by vacuum sublimation; mp, 88–88.5°C.

XXV. *Piperidide of 3-Carboxy-2,2,5,5-tetramethylpyrroline-1-oxyl* [8]. To a solution of 0.184 g of 3-carboxy-2,2,5,5-tetramethylpyrroline-1-oxyl in a mixture of 3 ml of dry benzene and 0.2 ml of dry pyridine were added 0.09 ml of thionyl chloride and, after 15 min, 0.2 ml of dry piperidine (bp, 104°C at 748 mm Hg). After 30 min, the reaction mixture was poured into 20 ml of water, and the organic layer was separated and was washed with 5% hydrochloric acid, water, 5% aqueous caustic potash solution, and water

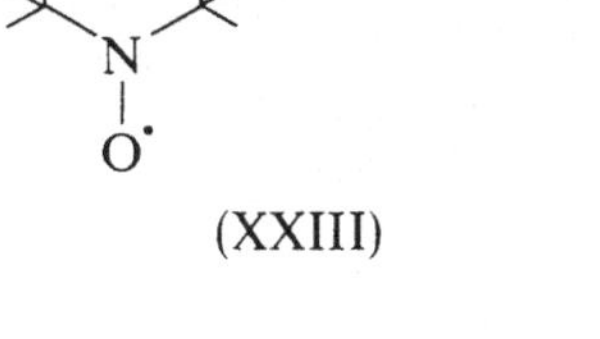

(XXIII)

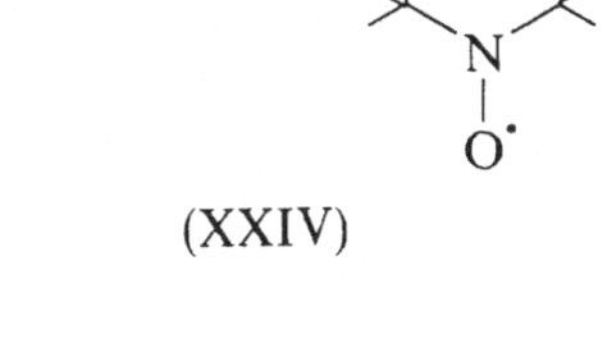

(XXIV)

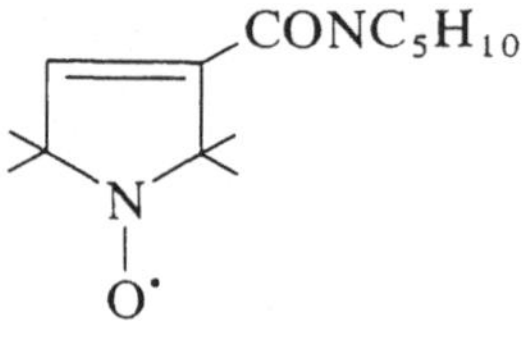

(XXV) 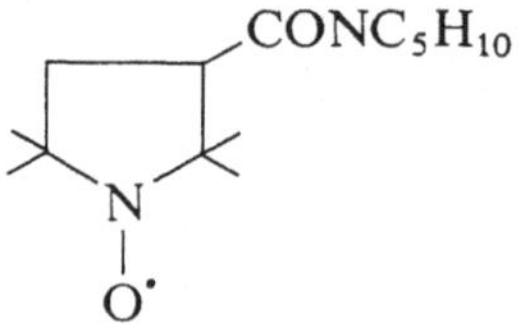   (XXVI)

again, and was then dried over fused calcium chloride and the solvent was evaporated in vacuum. This gave 0.21 g (83%) of a solid which, after crystallization from hexane, melted at 104–104.5°C.

XXVI. *Piperidide of 3-Carboxy-2,2,5,5-tetramethylpyrrolidine-1-oxyl* [8]. To a solution of 0.186 g of 3-carboxy-2,2,5,5-tetramethylpyrrolidine-1-oxyl in a mixture of 3 ml of dry benzene and 0.2 ml of dry pyridine were added successively 0.09 ml of thionyl chloride and, after 15 min, 0.2 ml of dry piperidine. After 30 min, the reaction mixture was poured into 20 ml of water and the organic layer was separated and was washed with 5% hydrochloric acid, water, 5% aqueous caustic potash, and water again, and after drying over fused calcium chloride, the solvent was evaporated in vacuum. This gave 1.81 g (71.5%) of a yellow substance which crystallized well from hexane; mp, 113–114°C.

XXVII. *Catechol Ester of 3-Carboxy-2,2,5,5-tetramethylpyrroline-1-oxyl* [7]. To a solution of 0.184 g of 3-carboxy-2,2,5,5-tetramethylpyrroline-1-oxyl in a mixture of 3 ml of dry benzene and 0.2 ml of dry pyridine were added 0.09 ml of thionyl chloride and, after 15 min, 0.055 g of freshly sublimed catechol. The reaction mixture was left overnight. The salts that had precipitated were filtered off, and the filtrate was chromatographed on alumina of

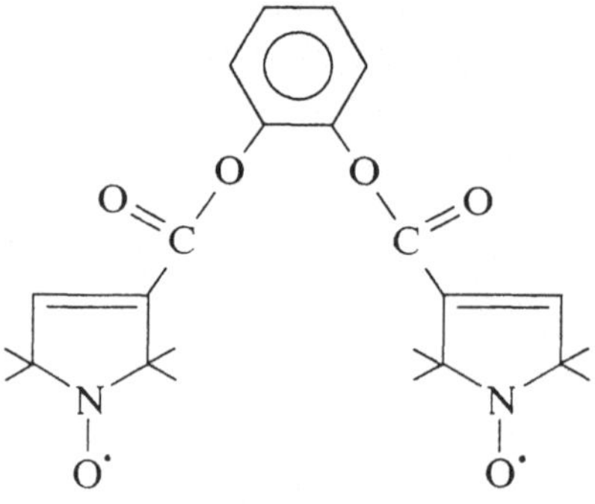

(XXVII)

(XXVIII)

activity II (with chloroform as eluant), giving 0.155 g (70%) of yellow crystals with mp 157–158°C (hexane–benzene).

XXVIII. *Hydroquinone Ester of 3-Carboxy-2,2,5,5-tetramethyl-pyrroline-1-oxyl* [7]. To a solution of 0.184 g of 3-carboxy-2,2,5,5-tetramethylpyrroline-1-oxyl in a mixture of 3 ml of dry benzene and 0.2 ml of dry pyridine were added 0.09 ml of thionyl chloride and, after 15 min, 0.005 g of hydroquinone. The resulting mixture was left overnight. Then it was poured into 20 ml of water, and the yellow crystals were filtered off and dried in air. This gave 0.054 g (24%) of a sparingly soluble substance which was recrystallized from ethyl acetate; mp, 248–250°C.

XXIX. *Resorcinol Ester of 3-Carboxy-2,2,5,5-tetramethylpyrroline-1-oxyl* [7]. To a solution of 0.184 g of 3-carboxy-2,2,5,5-tetramethylpyrroline-1-oxyl in a mixture of 3 ml of dry benzene and 0.2 ml of dry pyridine were added 0.09 ml of thionyl chloride and, after 15 min, 0.055 g of freshly sublimed resorcinol. The reaction mixture was left overnight. Then the precipitated salts were filtered off and the filtrate was chromatographed on alumina of activity II (with chloroform as eluant), giving 0.051 g (23%) of yellow crystals with mp, 183–184.5°C (benzene–heptane).

XXX. *Phloroglucinol Ester of 3-Carboxy-2,2,5,5-tetramethylpyrroline-1-oxyl* [7]. To a solution of 0.276 g of 3-carboxy-2,2,5,5-tetramethylpyrroline-1-oxyl in a mixture of 5 ml of dry benzene and 0.3 ml of dry pyridine were added 0.14 ml of thionyl chloride and, after 15 min, a solution of 0.042 g of freshly sublimed phloroglucinol in 5 ml of absolute ether. The reaction mixture was left overnight. The precipitated salts were filtered off, and the filtrate was chromatographed on alumina of activity II (with chloroform as eluant), to give 0.030 g (14%) of crystals with mp, 225–226°C (from hexane).

XXXI. *2,2,6,6-Tetramethyl-4-oxopiperidine-1-oxyl* [9, 10]. A solution of 155 g of triacetoneamine, 15 g of Trilon B, and 15 g of sodium tungstate in 1000 ml of water was cooled to 3–5°C and 250 ml of 30% hydrogen peroxide was added, after which, by external cooling, the temperature within the reaction flask was kept between 15 and 18°C for 4–6 hr. Then the reaction mixture was left in the dark at room temperature for 8–10 days. The resulting

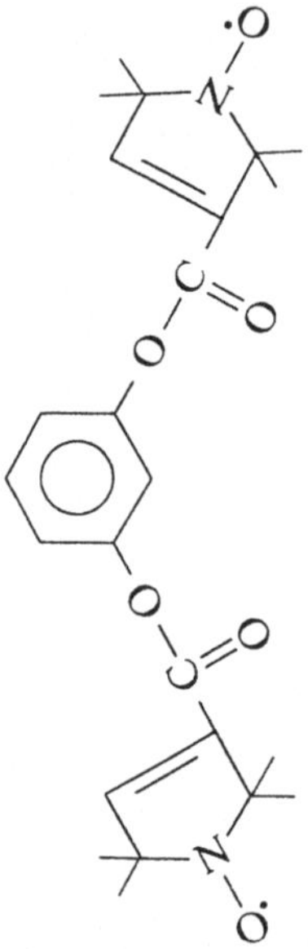

(XXIX) 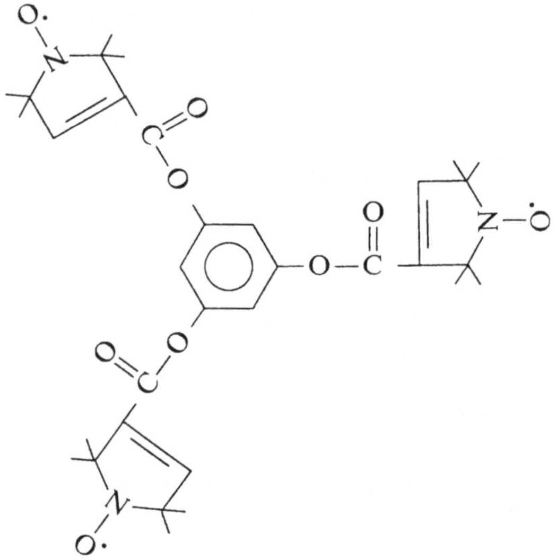   (XXX)

reddish-yellow solution was saturated with potassium carbonate. The radical, which had separated in the form of a dark red oil, crystallized on cooling. It was filtered off with suction, pressed out on the filter paper, and dried in a vacuum desiccator over potassium hydroxide. Yield, 160–170 g. After recrystallization from a minimum amount of hexane, 130–140 g of pure 2,2,6,6-tetramethyl-4-oxopiperidine-1-oxyl was obtained in the form of pink needles, mp, 36°C. Oxime: red crystals with mp, 180°C (from methanol).

XXXII. *4-Hydroxy-2,2,6,6-tetramethylpiperidine-1-oxyl* [11].   A mixture of 30 g of 4-hydroxy-2,2,6,6-tetramethylpiperidine dissolved in 200 ml of water, 40 ml of 30% hydrogen peroxide diluted with 200 ml of water, 2 g of Trilon B, and 2 g of sodium tungstate was left at room temperature for 5 days. The solution was saturated with potassium carbonate and repeatedly extracted with ether. The combined ethereal extracts were dried with fused potassium carbonate, the ether was evaporated, and the residue was recrystallized from a mixture of ether and hexane (2:1). The yield of 4-hydroxy-2,2,6,6-tetramethylpiperidine-1-oxyl with mp, 71.5°C was   278 g (85%). *p*-Toluenesulfonate: pink crystals (from hexane) with mp, 115°C; 3,5-dinitrobenzoate: bright red prisms with mp, 149°C (from ethanol); phenylcarbamate: silvery pink scales with mp, 164°C (from aqueous ethanol); acrylate: bright orange needles with mp, 102.5–103°C (sublimate); methacrylate: lustrous pink plates with mp, 88.7°C (from hexane); cinnamate: lustrous

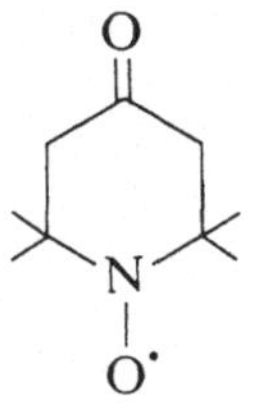

(XXXI)

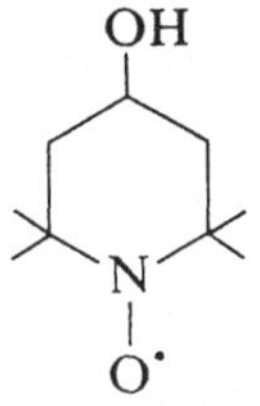

(XXXII)

orange needles (from hexane) with mp, 101°C; butyrate: bright red needles with mp, 28°C, bp, 95–97°C (0.1 mm Hg).

XXXIII. *4-Ethyl-4-hydroxy-2,2,6,6-tetramethylpiperidine* [11]. An ethereal solution of ethylmagnesium bromide was prepared in a $\frac{1}{2}$ liter, three-necked, round-bottomed flask fitted with a dropping funnel, mechanical stirrer, and reflux condenser protected with a calcium tube, from 24 g of magnesium, 119 g of ethyl bromide, and 220 ml of absolute ether. To the cooled reaction mixture was slowly added 46.5 g of anhydrous triacetone-amine in 100 ml of absolute ether. The resulting mixture was heated in a warm water bath for 4 hr and was then left overnight. The solid mass was freed from ether and dissolved in the minimum amount of 15% hydrochloric acid with cooling, and the resulting solution was saturated with solid caustic soda and repeatedly extracted with ether. The combined ethereal extracts were dried with fused potassium carbonate and the ether was evaporated to give 37.5 g of the initial tracetoneamine. The residual, highly alkaline solution was repeatedly extracted with warm *n*-butanol, the combined butanolic extracts were dried with fused potassium carbonate, and the butanol was evaporated in vacuum. The residue was recrystallized from a mixture of ether and hexane to give 7.0 g (76%) of 4-ethyl-4-hydroxy-2,2,6,6-tetramethyl-piperidine with mp, 66.6°C.

XXXIV. *4-Ethyl-4-hydroxy-2,2,6,6-tetramethylpiperidine-1-oxyl* [11]. A solution of 4-ethyl-4-hydroxy-2,2,6,6-tetramethylpiperidine, 0.01 g of Trilon B, 0.1 g of sodium tungstate, and 10 ml of hydrogen peroxide in

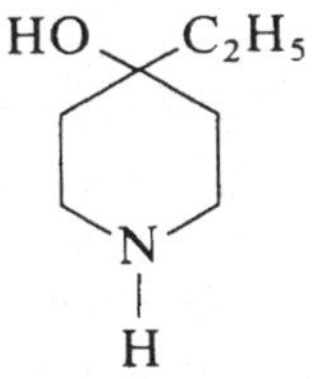

(XXXIII)

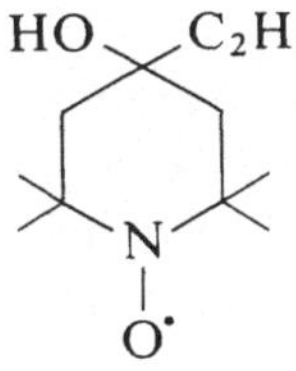

(XXXIV)

OCOC$_6$H$_5$

(XXXV)   (XXXVI)

50 ml of water was left at room temperature for one day. Then it was saturated with potassium carbonate, and the precipitate that deposited was
separated and recrystallized from a mixture of ether and hexane. The yield of
4-ethyl-4-hydroxy-2,2,6,6-tetramethylpiperidine-1-oxyl with mp, 69.5°C was
4.86 g (90%).

XXXV. *4-Benzoyloxy-2,2,6,6-tetramethylpiperidine-1-oxyl* [12].
With stirring and ice cooling, 3.8 ml of benzoyl chloride was added to a solution of 5.2 g of 4-hydroxy-2,2,6,6-tetramethylpiperidine-1-oxyl in 20 ml of
absolute pyridine. The reaction mixture was left at room temperature. After
20 hr, it was poured with vigorous stirring into 150 ml of ice-water. The precipitate that deposited was filtered off with suction, washed with water, dried,
and recrystallized from 20 ml of methanol. The benzoate crystallized in the
form of bright red needles with mp, 105°C, subliming at 95°C. Yield, 7.5 g
(90%).

XXXVI. *Hydrogen Phthalate of 4-Hydroxy-2,2,6,6-tetramethyl-
piperidine-1-oxyl* [12]. A solution of 5 g of 4-hydroxy-2,2,6,6-tetramethyl-
piperidine-1-oxyl and 5 g of phthalic anhydride in 20 ml of absolute pyridine
was heated in the water bath at 80–90°C for 3 hr. After cooling, the solution
was poured with stirring into 200 ml of ice-water. The aqueous solution was
acidified with dilute hydrochloric acid to pH 3 and was extracted with chloroform. The combined extracts were washed with water and were then dried
with anhydrous sodium sulfate. The chloroform was evaporated under
reduced pressure and the residue was recrystallized from a mixture of chloroform and hexane. The yield of the hydrogen phthalate ester was 7.3 g (78%).
Orange crystals with mp, 143°C.

XXXVII. *Hydrogen Diphenate of 4-Hydroxy-2,2,6,6-tetramethyl-
piperidine-1-oxyl*. To a solution of 1.08 g of 4-hydroxy-2,2,6,6-tetramethyl-
piperidine-1-oxyl in 6.5 ml of absolute pyridine was added 1.56 g of diphenic
anhydride, and the mixture was heated with a reflux condenser protected by a
calcium chloride tube for 3 hr. After cooling, the reaction mixture was poured
onto 100 g of crushed ice, and the resulting solution was acidified with dilute

(XXXVII)   (XXXVIII)   (XXXIX)

hydrochloric acid to pH 4. The yellow paramagnetic resin that deposited was extracted with chloroform, and the extract was shaken with dilute sodium carbonate solution. The sodium carbonate extract was acidified with concentrated hydrochloric acid and was reextracted with chloroform. The chloroform extract was washed with water, dried with anhydrous sodium sulfate, and evaporated under reduced pressure. Then the residue was re-crystallized first from a mixture of benzene and cyclohexane and then from a mixture of hexane and acetone. This gave 0.71 g (29.3%) of the radical in the form of orange prisms with mp, 159–160°C.

XXXVIII. *2,2,6,6-Tetramethylpiperidine.* A 1-liter round-bottomed flask fitted with a reflux condenser and heated in an oil bath was charged with 300 ml of diethyleneglycol, 77 g triacetoneamine, 75 ml of 85% hydrazine hydrate, and 100 g of caustic potash. The mixture was heated at 135°C until the evolution of nitrogen ceased, and then the reflux condenser was replaced by a condenser for downward distillation and the bath temperature was gradually raised to 195°C. The distillate was saturated with potassium carbonate and the organic layer was separated and fractionated at ordinary pressure, a fraction boiling at 148–155°C being collected. A second rectification over sodium gave the pure material with bp, 151–152°C (756 mm Hg), n_D^{20}, 1.4458. The yield was 50 g (80.3%). Literature data [14]: bp, 151–152°C (750 mm Hg); n_D^{20}, 1.4455.

XXXIX. *2,2,6,6-Tetramethylpiperidine-1-oxyl* [15]. With cooling, a solution of 141 g of 2,2,6,6-tetramethylpiperidine, 15 g of Trilon B, 800 ml of 45% methanol, 0.3 g sodium of tungstate, and 250 ml of 30% hydrogen peroxide was prepared in a 1.5-liter flat-bottomed flask. The mixture was left at room temperature for 10 days, diluted twofold with water, saturated with potassium carbonate, and extracted with ether. The ethereal extract was dried with anhydrous magnesium sulfate, the ether was evaporated, and the residue was sublimed in vacuum. The yield of radical was 95 g (61%); dark red, transparent prisms (sublimate) with a strong odor of camphor, mp, 37.8–38.1°C. Literature data [16]: mp, 35°C.

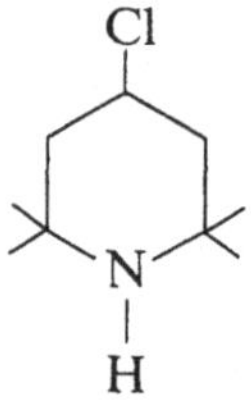

| (XL) | (XLI) | (XLII) | (XLIII) |

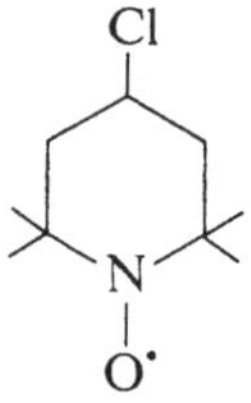

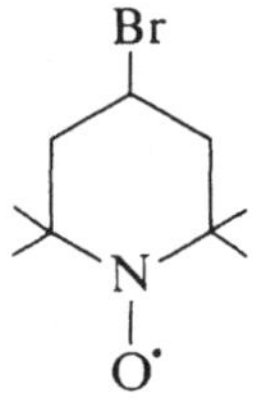

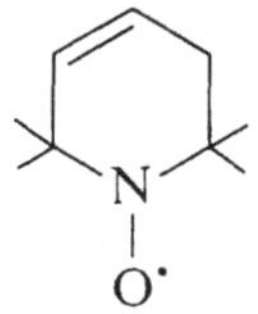

XL. *4-Chloro-2,2,6,6-tetramethylpiperidine* [17]. A mixture of 11.8 g of 4-hydroxy-2,2,6,6-tetramethylpiperidine, 41 g of anhydrous zinc chloride, and 24 ml of concentrated hydrochloric acid saturated with hydrogen chloride at −5°C was heated in a sealed tube at 140°C for 2 h. After cooling, the contents of the tube were poured onto 200 g of crushed ice, and the mixture was made alkaline with dilute caustic potash to a pH of ∼13 and extracted with ether. The ethereal extracts were dried with anhydrous potassium carbonate, the ether was evaporated, and the residue was dried in a desiccator over caustic potash and sublimed in vacuum. The yield of 4-chloro-2,2,6,6-tetramethylpiperidine was 7.8 g (59%); colorless crystals with mp, 37°C.

XLI. *4-Chloro-2,2,6,6-tetramethylpiperidine-1-oxyl* [17]. A small amount of Trilon B, 3.2 ml of 0.5 *M* sodium tungstate, and 8.5 ml of 30% hydrogen peroxide were added successively to a solution of 5.6 g of 4-chloro-2,2,6,6-tetramethylpiperidine in a mixture of 50 ml of methanol and 25 ml of water. The mixture was left to stand at room temperature. After 4 days, the reaction mixture, together with the crystals that had separated, was poured into 300 ml of water. The crystals were filtered off with suction, washed with water, dried, and recrystallized from hexane. The yield of 4-chloro-2,2,6,6-tetramethylpiperidine-1-oxyl was 5.0 g (82%); red needles with a metallic luster, mp, 111.5°C, subliming in the form of needles at 50°C.

XLII. *4-Bromo-2,2,6,6-tetramethylpiperidine-1-oxyl* [17]. The compound was obtained in 36% yield by the catalytic oxidation (cf. XLI) of 4-bromo-2,2,6,6-tetramethylpiperidine. Copper-red needles with a metallic luster, mp, 129°C (from hexane); the substance sublimes in the form of needles at 75°C.

XLIII. *2,2,6,6-Tetramethyl-1,2,5,6-tetrahydropyridine-1-oxyl* [17]. This compound was obtained in 67% yield by the oxidation of 2,2,6,6-tetramethyl-1,2,5,6-tetrahydropyridine by the method used for XXXIX. Bright red crystals with mp, 33°C.

$$HOOC-H_2C-\underset{\|}{C}-COOC_2H_5 \qquad H_3COOC-\underset{\|}{C}-CH_2-COOH$$

(structure XLIV, left; structure XLV, right — 2,2,6,6-tetramethylpiperidine-1-oxyl rings)

(XLIV)   (XLV)

XLIV.   *Monoethyl   2,2,6,6-Tetramethyl-1-oxyl-4-piperidylidene Succinate* [18]. A dry two-necked flask fitted with a reflux condenser and a magnetic stirrer was charged with 45 ml of absolute tert-butanol and 2.34 g of potassium. The mixture was boiled under reflux with stirring until the potassium had dissolved completely. Then the flask was cooled, evacuated at water pump vacuum, and filled with dry argon. To the solution of potassium tert-butoxide was added a mixture of 6.8 g of 2,2,6,6-tetramethyl-4-oxopiperidine-1-oxyl, 12 ml of diethyl succinate, and 15 ml of tert-butanol, and then the reaction mixture was carefully boiled under reflux for 10 min. After cooling with ice and neutralization with dilute hydrochloric acid, the bulk of the alcohol was distilled off under reduced pressure. The residue was poured into 350 ml of ice-water and the resulting mixture was acidified with dilute hydrochloric acid to pH 3 and extracted with methylene chloride. The combined extracts were washed several times with 1% ammonia solution. The ammoniacal extracts were cooled with ice, acidified, and again extracted with methylene chloride. The extract was dried with sodium sulfate. After the methylene chloride had been evaporated, an oily red liquid remained which crystallized on being triturated with hexane. The crystals were pressed out on a porous porcelain plate and were recrystallized from a mixture of ether and hexane. The yield of the monoethyl ester was 9.5 g (64%); yellow prisms with mp, 103°C.

XLV.   *Monomethyl   2,2,6,6-Tetramethyl-1-oxyl-4-piperidylidene Succinate* [18]. With stirring, a mixture of 15.7 ml of dimethyl succinate and a solution of potassium tert-butoxide obtained by dissolving 4.35 g of potassium in 85 ml of absolute tert-butanol was added over 2 hr to 13 g of 2,2,6,6-tetramethyl-4-oxopiperidine-1-oxyl heated to 40°C in a dry two-necked flask fitted with a reflux condenser, a mercury seal, a magnetic stirrer, and a dropping funnel. The reaction mixture was stirred for a further hour at the same temperature and was then cooled and neutralized with dilute hydrochloric acid. The bulk of the alcohol was evaporated under reduced pressure, and the residue was poured into 400 ml of ice-water, acidified to pH 3, and extracted with ether. The ethereal extracts were washed with 1% ammonia

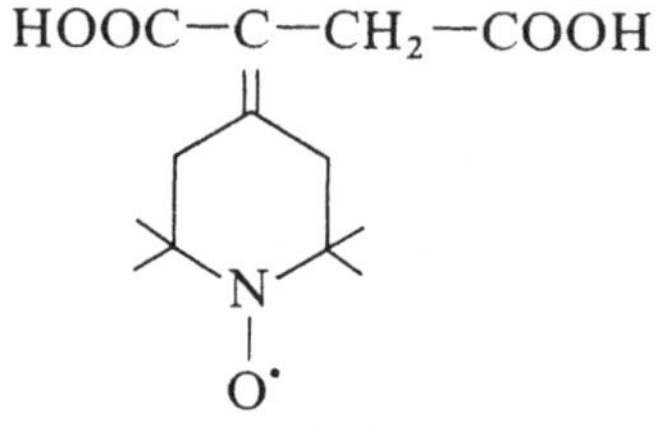

HOOC—C—CH$_2$—COOH   H$_3$COOC—C—CH$_2$—COOCH$_3$

(XLVI) 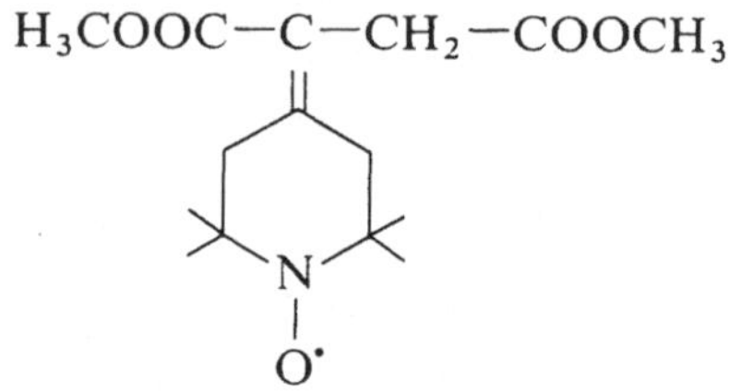   (XLVII)

solution. The ammoniacal extracts were cooled with ice, acidified with dilute hydrochloric acid to pH 3, and again extracted with ether. The combined extract was dried with anhydrous sodium sulfate. The bulk of the ether was distilled off and the residue (30 cm^3), after dilution with an equal amount of heptane, was poured into a watchglass. As the solvent evaporated, the mono-ethyl ether crystallized out; it was filtered off with suction, washed with a mixture of ether and hexane (1:3), and recrystallized from a mixture of ether and hexane; yield, 8 g (37%), yellow crystals with mp, 105.5°C.

XLVI. *2,2,6,6-Tetramethyl-1-oxyl-4-piperidylidenesuccinic Acid* [18]. A solution of 5.9 g of the corresponding monoethyl ester (see XLIV) and 8 g of barium hydroxide in 120 ml of 50% ethanol was boiled under reflux for 7 hr. The hot solution was filtered and left in the refrigerator overnight. The filtrate deposited golden crystals of the barium salt, which were filtered off with suction, washed with ethanol, and dissolved in 200 ml of water. The solution was cooled to 0°C, acidified with dilute sulfuric acid, (1:10), and extracted with ether. The combined ethereal extracts were saturated with sodium sulfate, the bulk of the ether was distilled off, and the residue (30 ml) was poured into a watchglass and hexane was added until an opalescence appeared. As the solvent slowly evaporated, the acid-radical deposited and it was recrystallized from a mixture of acetone and hexane. Yield, 4.1 g (76%); yellow prisms with decomp. pt. 150–153°C.

XLVII. *Dimethyl 2,2,6,6-Tetramethyl-1-oxyl-4-piperidylidene Succinate* [18]. This was obtained in quantitative yield by the reaction of the corresponding acid-radical (see XLVI) with diazomethane. Pink prisms with mp, 47.5°C (from hexane).

XLVIII. *Ethyl Methyl 2,2,6,6-Tetramethyl-1-oxyl-4-piperidylidene Succinate* [18]. Obtained in quantitative yield by the reaction of the mono-ethyl ester (see XLIV) with diazomethane. Pink crystals with mp, 38°C (from hexane).

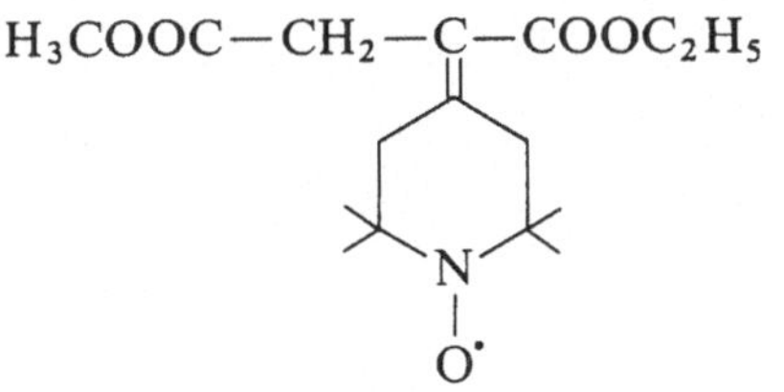

$$H_3COOC-CH_2-\underset{\underset{\text{(XLVIII)}}{}}{\overset{}{C}}-COOC_2H_5$$

(XLVIII)

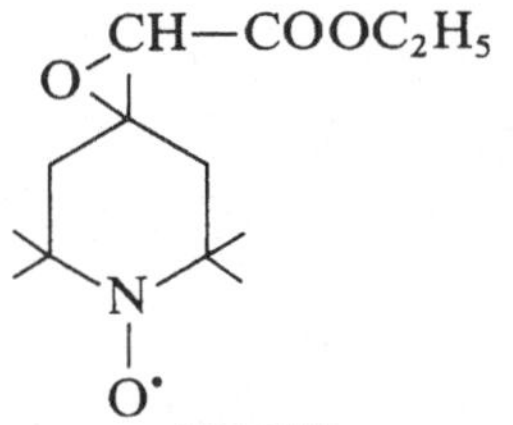

(XLIX)

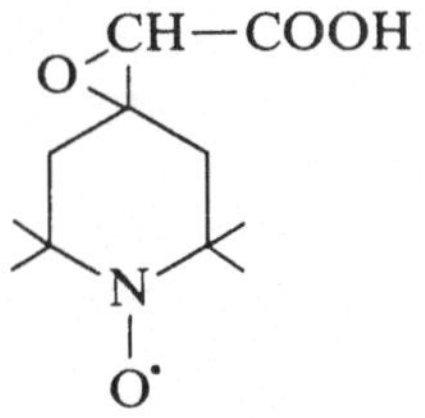

(L)

XLIX. *Ethyl Ester of the Oxide of 2,2,6,6-Tetramethyl-1-oxyl-4-piperidylideneacetic Acid* [19]. In a three-necked flask fitted with a reflux condenser, thermometer, dropping funnel, magnetic stirrer, and gas-inlet tube, a solution of potassium tert-butoxide was prepared from 2.42 g (0.062 g-atom) of potassium and 62 ml of dry tert-butanol. The flask was filled with dry argon, and, with stirring, a mixture of 7 g (0.041 mole) of 2,2,6,6-tetramethyl-4-oxopiperidine-1-oxyl and 7.58 g (0.062 mole) of ethyl chloroacetate was added, the temperature within the flask being kept at 20–25°C. The reaction mixture was left at room temperature for 19 hr and was then neutralized with the equivalent amount of acetic acid (3.55 ml), after which, the bulk of the alcohol was evaporated under reduced pressure and the residue was poured into 500 ml of ice-water. The red oily liquid that separated was extracted with ether. The ethereal extract was washed with sodium bicarbonate solution and water and was dried over sodium sulfate. The residue after the evaporation of the ether was distilled at 0.6–0.7 mm Hg. After the ethyl chloroacetate that had not reacted had distilled off, the glycide ester distilled at 130–140°C. Yield, 8.58 g (81.5%). Viscous, dark red liquid with bp, 135°C (0.7 mm Hg), n_D^{20}, 1.474.

L. *Oxide of 2,2,6,6-Tetramethyl-1-oxyl-4-piperidylideneacetic Acid* [19]. To a solution of sodium ethoxide obtained by dissolving 0.69 g (0.03 g–atom) of sodium in 20 ml of ethanol was added 7.7 g (0.03 mole) of the glycide ester XLIX. The resulting solution was cooled to 0°C, 0.54 ml (0.03 mold) of water was added, and it was left at room temperature. After 17 hr, 150 ml of absolute ether was added. The pink microcrystalline precipitate that deposited was filtered off with suction, washed with absolute ether,

(LI)   (LII)

and dried in a vacuum desiccator. The yield of the sodium salt of glycidic acid was 7.8 g (92%). Six grams of the salt was dissolved in 150 ml of water and, with ice cooling, the solution was acidified with dilute hydrochloric acid to pH 3–4. The aqueous solution was carefully extracted with ether. The ethereal extracts were washed with a small amount of water and were dried with anhydrous sodium sulfate. After the evaporation of the ether, 2.9 g (55%) of acid was obtained. Yellow prisms (from benzene) with decomp. pt. 132–134°C.

LI. *2,2',2,2',6,6',6,6'-Octamethyl-4,4'-bipiperidyl.* A mixture of 33 g of 4-bromo-2,2,6,6-tetramethylpiperidine, 5 g of sodium powder, and 6 drops of acetonitrile in 100 ml of absolute ether was charged into a 0.25-liter two-necked flask with a stirrer and reflux condenser and was boiled with stirring for 30 hr. After cooling, the ethereal solution was filtered and was subjected to distillation, first at atmospheric pressure and then at reduced pressure. The residue, which consisted mainly of the octamethylbipiperidyl and the starting material, was pressed out on a porous plate and separated by fractional vacuum sublimation. The yield of the octamethylbipiperidyl was 0.8 g (3.8%), colorless crystals with mp, 130.5°C, subliming at 70°C.

LII. *2,2',2,2',6,6',6,6'-Octamethyl-4,4'-bipiperidyl-1,1'-dioxyl* [20]. Catalytic amounts of Trilon B and sodium tungstate and 2 ml of 30% hydrogen peroxide were added to a solution of 0.7 g of octamethylbipiperidyl in a mixture of 20 ml of water and 25 ml of methanol. After five days, the solution, together with the crystalline deposit, was treated with 100 ml of water, and the precipitate was filtered off with suction, washed with water, and recrystallized from methanol. The biradical crystallized in the form of small red prisms. Yield, 0.6 g (77.5%), decomp. pt. 236–246°C, subliming at 175°C.

LIII.   *4-Hydroxy-2,2,6,6-tetramethylpiperidine-1-oxyl Ester of Hexamethylenedicarbamic Acid* [21]. A solution of 3.45 g of 4-hydroxy-2,2,6,6-tetramethylpiperidine-1-oxyl in 40 ml of dry benzene in a one-necked flask fitted with a reflux condenser protected with a calcium chloride tube was treated with 1.65 g of freshly distilled hexamethylene diisocyanate in 10 ml of dry benzene. The reaction mixture was heated in the water bath with the benzene gently boiling for 4 hr. The cooled solution was poured into 200 ml

O•—N⟨piperidinyl⟩—O—C(=O)—NH—(CH₂)₆—NH—C(=O)—O—⟨piperidinyl⟩N—•O

$$O^{\bullet}-N \bigcirc -O-\overset{\overset{\textstyle O}{\|}}{C}-NH-(CH_2)_6-NH-\overset{\overset{\textstyle O}{\|}}{C}-O- \bigcirc N-^{\bullet}O$$

(LIII)

of hexane. The bright red resin that separated out was washed several times with hexane, dissolved in 20 ml of nitromethane, and transferred to a chromatographic column (height 430 mm, diameter 26 mm) filled with alumina of activity II. Elution was carried out with nitromethane. The biradical formed the bottom, orange zone, which was collected as eluate. The solvent was distilled off under reduced pressure and the residue was recrystallized from a mixture of ether and hexane. The biradical crystallized in the form of small pink needles with mp, 115–116°C. Yield, 3.3 g (64.5%).

LIV. *4-Hydroxy-2,2,6,6-tetramethylpiperidine-1-oxyl Esters of Dicarboxylic Acids.* With stirring and ice cooling, 0.01 mole of a dicarboxylic acid dichloride was added to a solution of 0.02 mole of 4-hydroxy-2,2,6,6-tetramethylpiperidine-1-oxyl in 15 ml of absolute pyridine. After the addition of the dichloride, the reaction mixure was stirred with cooling for 1.5 hr and then at room temperature for 10–14 hr. The resulting mixture was poured with stirring into 150 ml of ice-water and acidified with dilute hydrochloric acid (1:10) to pH 4. The precipitate that deposited was filtered off with suction, washed with water, dried, and chromatographed on a column (height 400 mm, diameter 25 mm) filled with alumina of activity II. Chloroform was used as eluant. The biradical formed the bottom, brightly-colored zone, which passed into the eluate. Evaporation of the solvent yielded the solid ester, which was recrystallized from hexane or carbon tetrachloride. The constants and yields of the compounds obtained in this way are given in Table 24.

LV. *sym-Bis(tetramethyl-1-oxyl-2,2,6,6-4-piperidyl)urea* [21]. This was obtained by oxidizing the corresponding diamine with a yield of 65%, mp, 198–199°C (from hexane).

LVI. *Tris(2,2,6,6-tetramethyl-1-oxyl-4-piperidyl) Trimesate* [22]. With stirring and ice cooling, 1.37 g of trimesic acid trichloride was added to a solution of 3 g of 4-hydroxy-2,2,6,6-tetramethylpiperidine-1-oxyl in 15 ml of absolute pyridine. The mixture was stirred for 1.5 hr with cooling and was then left at room temperature. After 18 hr, the reaction mixture was poured into 150 ml of ice water and acidified with dilute hydrochloric acid (1:10) to pH 5, after which, the precipitate was filtered off with suction, washed with

TABLE 24

Constants and Yields of 4-Hydroxy-2,2,6,6-tetramethyl-1-oxyl Esters of Dibasic Acids [21]

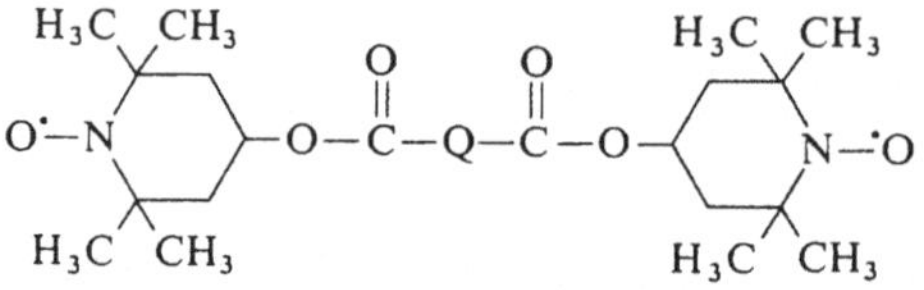

$$H_3C\diagdown CH_3 \qquad \overset{O}{\underset{\parallel}{C}} \quad \overset{O}{\underset{\parallel}{C}} \qquad H_3C\diagdown CH_3$$

O˙—N⟩—O—C—Q—C—O—⟨N—˙O

$$H_3C\diagup CH_3 \qquad\qquad\qquad\qquad H_3C\diagup CH_3$$

Acid component of the ester	Mp, °C	Yield, %	Acid component of the ester	Mp, °C	Yield, %
Carbonic	180	70	trans-Cyclopropanedicarboxylic	121.5	36
Oxalic	179	75	Phthalic	155.5	77
Succinic	141	30	Isophthalic	163	89
Adipic	122	22	Terephthalic	216–217	81
Pimelic	90	68	Diphenic	155.5	51
Sebacic	101	51	Biphenyl-2,3'-dicarboxylic	143	63

water, dried, and chromatographed on a column (length 300 mm, diameter 26 mm) filled with alumina of activity II (with chloroform as eluant). The triradical formed the bottom, orange zone, which passed into the eluate. Evaporation of the chloroform yielded 2.4 g (69%) of the triradical, which was recrystallized from cyclohexane; copper-red needles with mp, 208–209°C.

LVII. *Tris(2,2,6,6-tetramethyl-1-oxyl-4-piperidyl) Phosphite* [23]. A 0.05-liter three-necked flask fitted with a stirrer, reflux condenser with calcium chloride tube, and dropping funnel was charged with 3 g of 4-hydroxy-2,2,6,6-tetramethylpiperidine-1-oxyl, 1.8 g of triethylamine, and 15 ml of absolute benzene. The flask was immersed in an ice bath and, with stirring, 0.8 g of phosphorus trichloride in 10 ml of benzene was added to the reaction mixture. After this, it was stirred for another 2 hr with cooling and was then left at room temperature. After 14 hr, the benzene solution was separated from the triethylamine hydrochloride by filtration, the benzene was evaporated, and the residue was chromatographed on a column (length 400 mm, diameter 26 mm) filled with alumina of activity II (with chloroform as eluant). The phosphite formed the bottom zone, which passed into the eluate.

$$O˙—N⟩—HN—\overset{O}{\underset{\parallel}{C}}—NH—⟨N—˙O$$

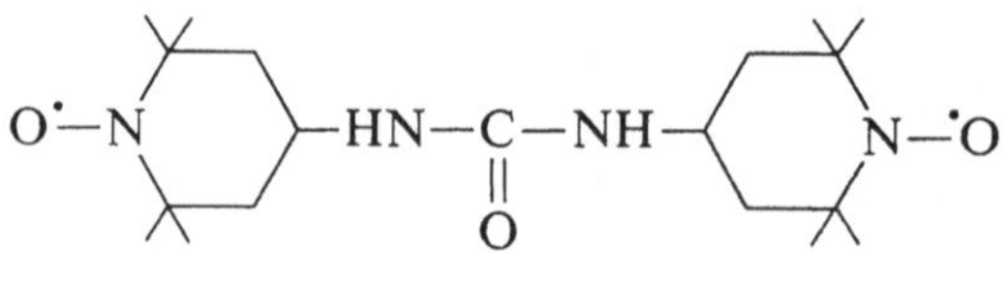

(LV)

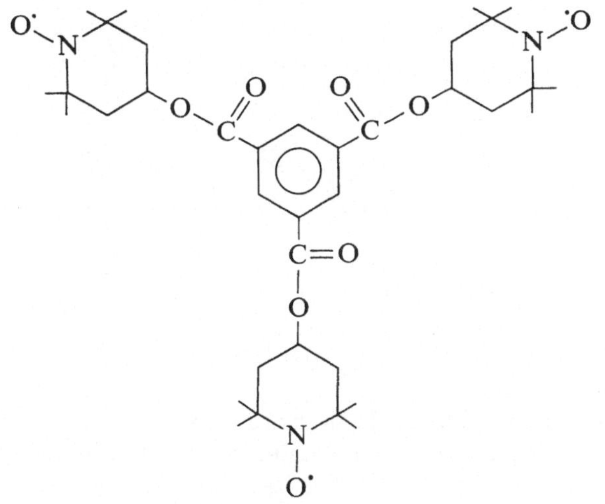

(LVI)

Evaporation of the chloroform yielded 2.1 g (66.5%) of tris(1-oxyl-2,2,6,6-tetramethyl-4-piperidyl) phosphite, which was recrystallized from hexane; red crystals with mp, 143°C.

LVIII. *Tetrakis(2,2,6,6-tetramethyl-1-oxyl-4-piperidyl) Ortho-silicate* [24]. A two-necked, round-bottomed flask fitted with a dropping funnel, reflux condenser with calcium chloride tube, and magnetic stirrer was charged with 3.5 g of 4-hydroxy-2,2,6,6-tetramethylpiperidine-1-oxyl, 3.7 ml of triethylamine, and 15 ml of absolute benzene. The flask was cooled with ice, and, with stirring, a solution of 0.5 ml of silicon tetrachloride in

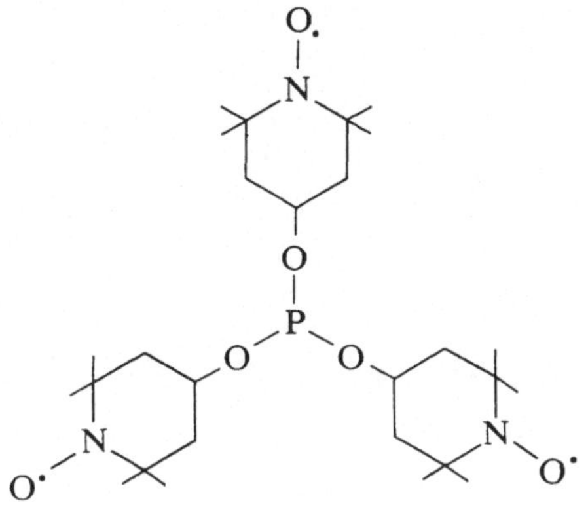

(LVII)

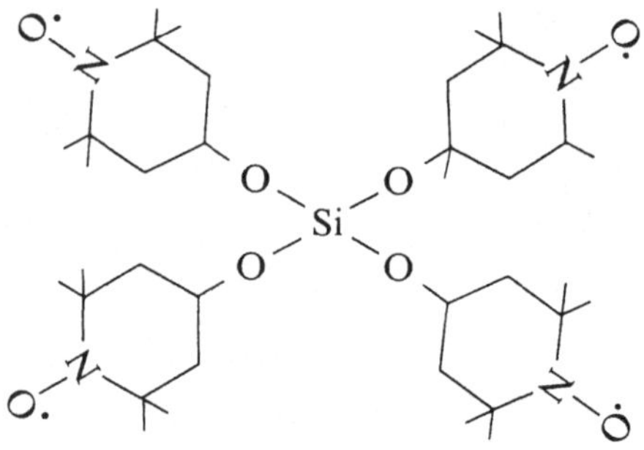

(LVIII)

10 ml of benzene was added. Then the reaction mixture was stirred for 3.5 hr at room temperature. The triethylamine hydrochloride was filtered from the benzene solution and was washed several times with dry benzene. The benzene was evaporated to leave an oily red liquid, which was chromatographed on a column filled with alumina (Brockmann activity grade II). Chloroform was used as the eluate. The tetraradical formed the bottom, pink zone, which passed into the eluate. Evaporation of the chloroform yielded 1.8 g (57.7%) of the orthosilicic tetraester, which was recrystallized from dry hexane; lustrous, golden-red crystals with mp, 160-165°C.

LIX.   *Tetrakis(2,2,6,6-tetramethyl-1-oxyl-4-piperidyl)   Pyromellitate* [24]. A two-necked flask fitted with a stirrer and calcium chloride tube was charged with 4.0 g of 4-hydroxy-2,2,6,6-tetramethyl-1-oxyl and 20 ml of absolute pyridine. With stirring and ice cooling, 1.7 g of pyromellitic acid tetrachloride was added. The reaction mixture was stirred with ice cooling for 30 min and then at room temperature for 15 hr. The reaction

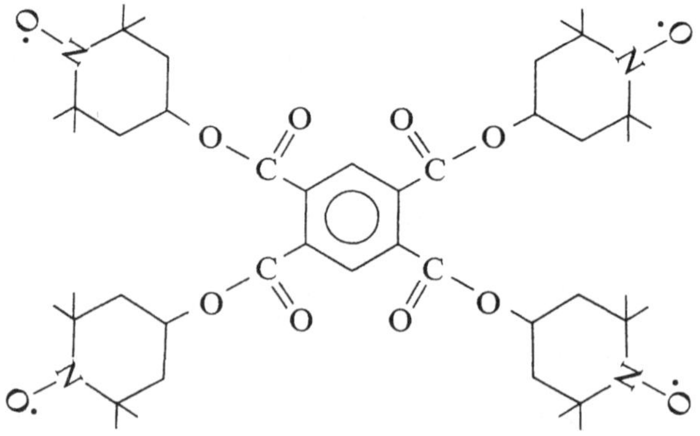

(LIX)

$$\text{HO} \quad \text{C} \equiv \text{CH}$$

(LX)

product was poured into 200 ml of ice-water and acidified with dilute hydrochloric acid to pH 5. The precipitate that deposited was filtered off with suction, washed with water, and dried. This gave 4.05 g (89.5%) of the tetraradical, which was recrystallized from methanol; small pink needles with mp, 225–226°C.

LX.   *4-Ethynyl-4-hydroxy-2,2,6,6-tetramethylpiperidine*. Eight hundred milliliters of pure ammonia (distilled from sodium) was condensed into a two-liter three-necked flask fitted with a Hershberg stirrer, a condenser with solid carbon dioxide, protected by a column of granulated caustic potash, and a gas-inlet tube. To the flask were gradually added 13.8 g (0.062 g-atom) of sodium and 0.01 g of powdered iron nitrate. After the dark blue color of the solution had disappeared, a rapid current of dry acetylene was passed into it for 2 hr. With vigorous stirring, 35.5 g (0.225 mole) of anhydrous triacetonamine was slowly added to the resulting mixture, and acetylene was passed for another 5 hr. The ammonia was evaporated in a current of dry nitrogen and the walls of the flask were rinsed with methanol and then with 100 ml of water. The suspended matter was removed by titration and washed with water. The crude reaction product was dissolved in 10% acetic acid, decolorized with carbon, and precipitated with dilute caustic potash. After washing with water and then drying, 21.6 g (54%) of a white powder with mp, 212–214°C was obtained. Vacuum sublimation gave an analytically pure sample with mp, 215–216°C. Literature data [25]: mp, 214–216°C.

LXI.   *1,4-Bis(4-hydroxy-2,2,6,6-tetramethyl-4-piperidyl)butadiyne*. A half-liter three-necked flask fitted with a mechanical stirrer was charged with 20.1 g (0.112 mole) of 4-ethynyl-4-hydroxy-2,2,6,6-tetramethylpiperidine, 45 g of ammonium chloride, 28 g of copper chloride, 22.5 ml of 6 N hydrochloric acid, and 112 ml of water. The mixture was brought to pH 5.5 with an aqueous solution of ammonia, and then, with stirring, oxygen was passed through it for 1.5 hr. The mixture was made strongly alkaline with aqueous ammonia and the reaction product was filtered off on a porcelain filter with suction. After drying, 18.3 g (92%) of a gray powder with mp,

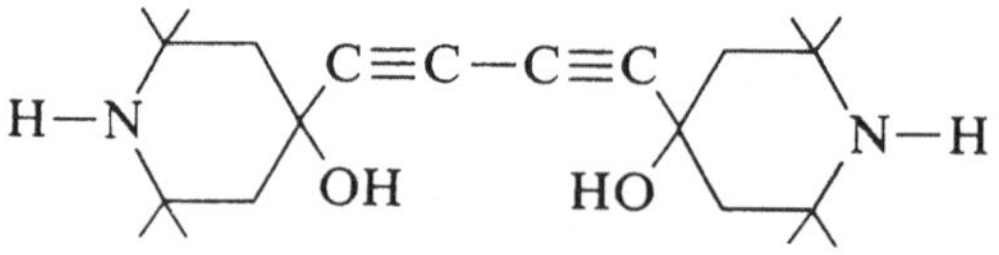

(LXI)

226–236°C was obtained. Dissolution in 10% acetic acid, boiling with carbon, and precipitation with ammonia yielded a white amorphous substance with mp, 231–232°C. Literature data [25]: mp, 231–232°C.

LXII. *1,4-Bis(4-hydroxy-2,2,6,6-tetramethyl-4-piperidyl)butane*. A solution of 1.84 g of the diyne obtained in the preceding experiment (see LXI) in 50 ml of 80% acetic acid was charged to a rotating steel autoclave and hydrogenated in the presence of Adams catalyst at room temperature and an initial hydrogen pressure of 100 atm. After the absorption of the required volume of hydrogen, the reaction mixture was filtered, make alkaline with caustic potash, and extracted with ethyl acetate. The extract was dried with anhydrous magnesium sulfate and the solvent was removed by distillation to give 1.8 g (97%) of product: colorless needles with mp, 128–130°C (hexane). Literature data: mp, 128–129°C (sublimate).

LXIII. *1,4-Bis(4-hydroxy-2,2,6,6-tetramethyl-1-oxyl-4-piperidyl)-butadiyne* [26]. A solution of 7.21 g (0.02 mole) of 1,4-bis(hydroxy-2,2,6,6-tetramethyl-4-piperidyl)butadiyne in the minimum amount of methanol was prepared in a two-liter, flat-bottomed, conical flask, and to this solution were added 0.005 g of sodium tungstate, 0.025 g of Trilon B, and 7.5 ml of 30% hydrogen peroxide. The reaction mixture was left in the dark at room temperature for 1.5 months. Then it was diluted threefold with water and left for the oxidation product to separate out completely, this collecting on the bottom of the flask in the form of golden yellow flakes. After two reprecipitations from methanol with water, 5.75 g (67.2%) of the biradical, decomposing above 220°C, was obtained.

LXIV. *1,4-Bis(4-hydroxy-2,2,6,6-tetramethyl-1-oxyl-4-piperidyl)-butane* [26]. This compound was obtained by oxidizing the diamine in accordance with the method used for LXIII. From 3.6 g (0.01 mole) of 1,4-bis-

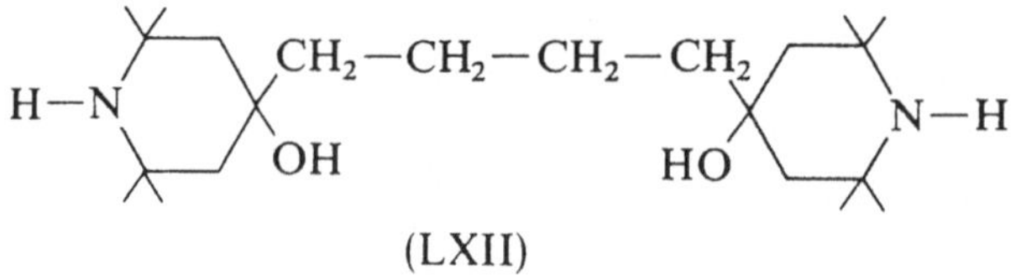

(LXII)

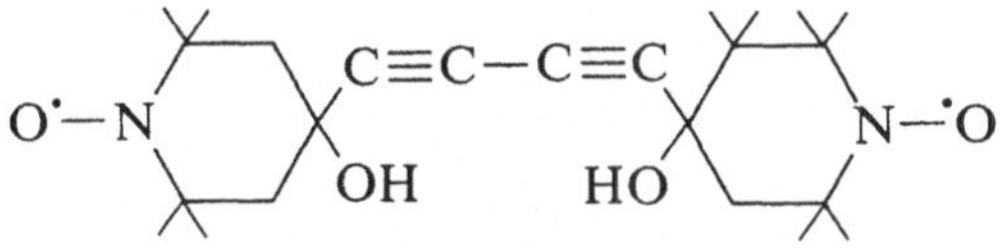

(LXIII)

(4-hydroxy-2,2,6,6-tetramethyl-4-piperidyl)butane, 3.0 g (75.3%) of the bi-radical was obtained; it formed orange needles with mp, 117–118°C (sublimate).

LXV. *Tetrakis(2,2,6,6-tetramethyl-1-oxyl-4-piperidyl) Biphenyl-2,2',4,4'-tetracarboxylate* [27]. A mixture of 3.55 g of biphenyl-2,2',4,4'-tetracarboxylic acid, 9.4 g of phosphorus pentachloride, and 4 ml of phosphorus oxychloride was boiled under reflux for 7 hr. The phosphorus oxychloride was distilled and the residue was added to an ice-cooled solution of 8.54 g of 4-hydroxy-2,2,6,6-tetramethylpiperidine-1-oxyl in 20 ml of absolute pyridine. The reaction mixture was stirred at room temperature for 22 hr, after which it was poured into 200 ml of ice-water and acidified with dilute hydrochloric acid to pH 5. The deposited precipitate was filtered off with suction, washed with water, dried, and chromatographed on a column filled with alumina of activity II, with chloroform as the eluant. Evaporation of the chloroform yielded a red crystalline substance, which was recrystallized several times from aqueous methanol. The yield of the tetra-radical was 2.7 g (37%). Red crystals with a coppery luster, mp 180–185°C.

LXVI. *Tetrakis(2,2,6,6-tetramethyl-1-oxyl-4-piperidyl) Biphenyl-2,2',5,5'-tetracarboxylate* [27]. Obtained in 25% yield by the method described above from 2.58 g of biphenyl-2,2',5,5'-tetracarboxylate, 6.8 g of phosphorus pentachloride, 4 ml of phosphorus oxychloride, and 6 g of 4-hydroxy-2,2,6,6-tetramethylpiperidine-1-oxyl. Transparent red prisms (from methanol) with mp, 202.5–204°C.

LXVII. *4-Amino-2,2,6,6-tetramethylpiperidine* [28]. Thirty three grams of sodium were rapidly added in small pieces to a solution of 27 g of triacetoneamine oxime (mp, 153–153.5°C) in 400 ml of hot amyl alcohol so that the solution boiled vigorously. After the completion of the reaction,

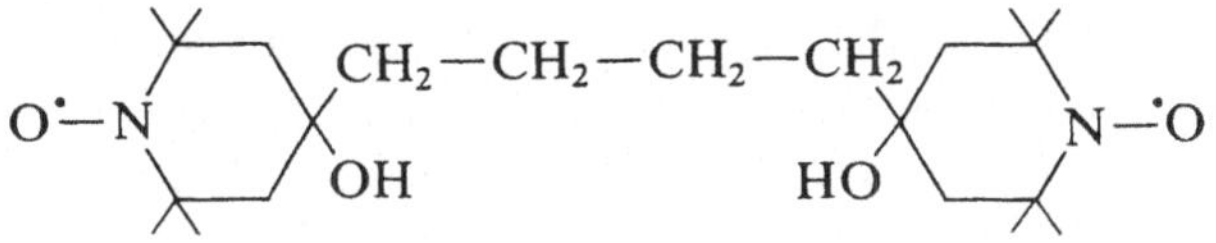

(LXIV)

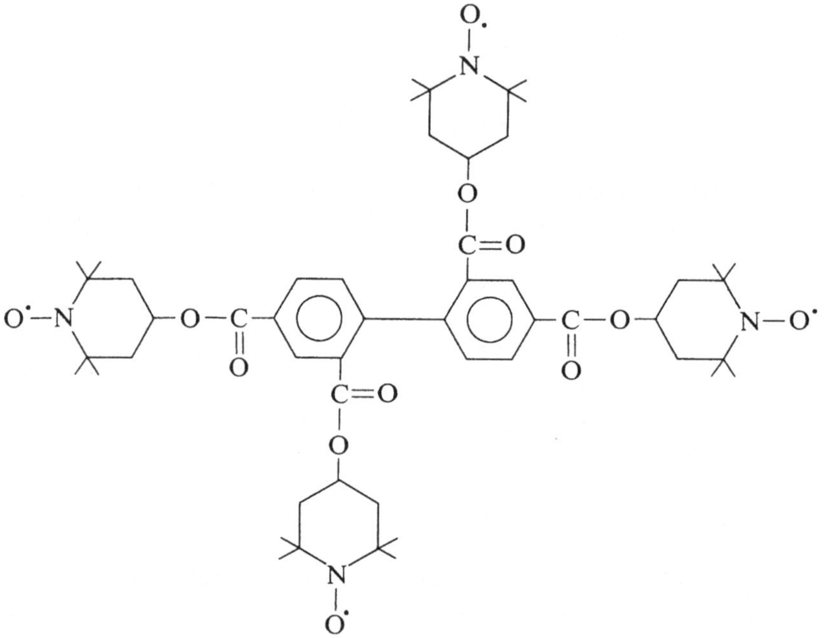

(XLV)

the hot mixture was poured onto ice. The aqueous layer was separated off and the alcoholic layer was extracted four times with equal volumes of 10% hydrochloric acid. The acid extracts were combined with the aqueous layer and the mixture was acidified and evaporated to $\frac{1}{3}$ of its original volume. The concentrate was saturated with solid caustic soda and distilled with steam. The distillate was neutralized with hydrochloric acid and evaporated to dryness. The residual hydrochloride was decomposed with caustic soda and extracted with ether, and the extract was dried with anhydrous potassium carbonate. The ether was evaporated and the residue was distilled in vacuum, a fraction with bp 78–79°C (7 mm Hg) being collected. Yield, 14.8 g (60%); transparent, colorless liquid with $n_D^{20} = 1.4699$, $d_4^{20} = 0.8966$, $MR_D = 48.62$, MR_D(calc.) = 48.58.

LXVIII. *4-Acetylamino-2,2,6,6-tetramethylpiperidine* [28]. An amount of 17.5 g of acetic anhydride was added dropwise to a solution of 8.6 g of 4-amino-2,2,6,6-tetramethylpiperidine in 25 ml of ether with vigorous stirring at such a rate that the temperature of the mixture did not exceed 20°C. After the addition of the acetic anhydride, the reaction mixture was stirred for another 30 min and the precipitate was filtered off, dried in vacuum, and distilled, a fraction with bp 161–163°C (6.5 mm Hg) being collected. The yield of material with mp 120°C was 10.5 g (95%).

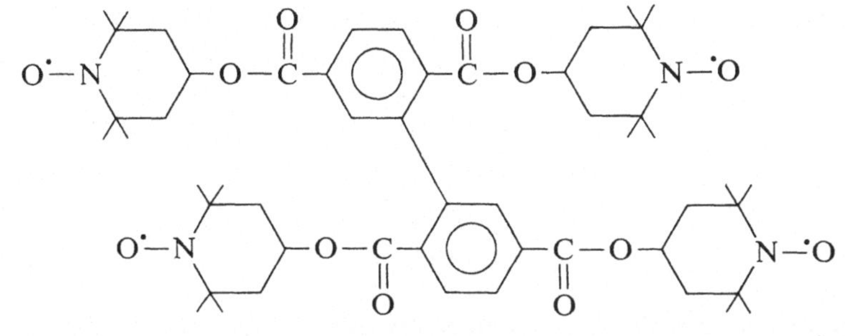

(LXVI)

LXIX. *4-Amino-2,2,6,6-tetramethylpiperidine-1-oxyl* [28]. By the method used for XXXII, 10.5 g of 4-acetylamino-2,2,6,6-tetramethylpiperidine was oxidized to give 11 g of crude 4-acetylamino-2,2,6,6-tetramethylpiperidine-1-oxyl, which was boiled with 25 ml of 15% caustic potash for 12 hr. The mixture was cooled, filtered, saturated with potassium carbonate, and carefully extracted with ether. The combined ethereal extracts were dried with anhydrous potassium carbonate, the ether was evaporated, and the residue was distilled in vacuum, a fraction with bp 97–98°C (4 mm Hg) being collected; yield, 8 g (73%): red hygroscopic needles with mp 34–35°C. The acylated 4-amino-2,2,6,6-tetramethylpiperidine-1-oxyls are listed in Table 25.

LXX. *2,2,7,7-Tetramethyl-5-homopiperazinone* [30]. With vigorous stirring and ice cooling, 8 ml of concentrated sulfuric acid was added dropwise to a solution of 1.55 g (0.01 mole) of anhydrous triacetoneamine in 10 ml of chloroform. Then 1.62 g (0.025 mole) of sodium azide was added in small

TABLE 25
Derivatives of 4-Amino-2,2,6,6-tetramethylpiperidine-1-oxyl

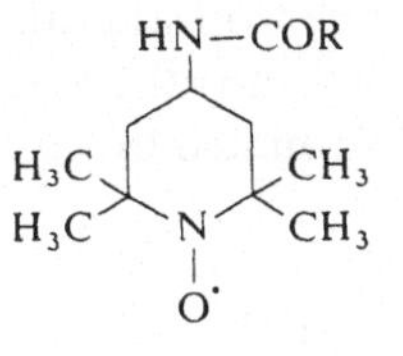

Acyl	Mp, °C	Yield, %	Acyl	Mp, °C	Yield, %
CH_3OCO	128.5	45	CH_2CCH_3CO	150	61
CH_2CHCO	149	57	$CH_3CHCHCO$	156	53
$C_6H_5CHCHCO$	178.5	63	$CH_3CH_2CH_2CO$	84	81
CH_3CH_2CO	101.5	71	$(CH_3)_2NCO$	173	73

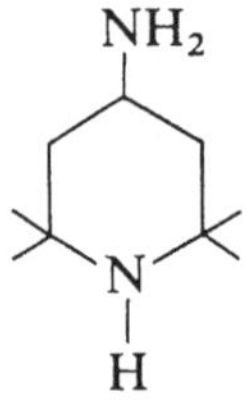

(LXVII) 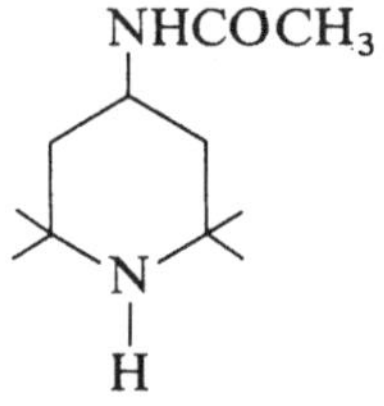   (LXVIII) 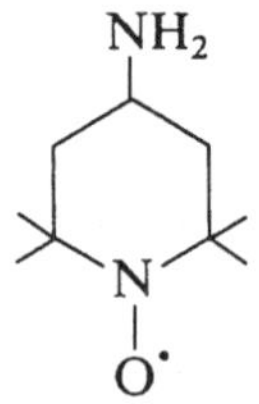   (LXIX)

portions over 30 min. The reaction mixture was stirred for another 15 min and was then poured onto ice (45 g). The resulting solution was saturated with potassium carbonate and repeatedly extracted with ether, and the combined ethereal extracts were dried with fused potassium carbonate. Evaporation of the ether in vacuum and recrystallization of the residue from benzene yielded 1.35 g (79%) of product: white plates with mp, 148–148.5°C (sublimate).

LXXI. *2,2,7,7-Tetramethyl-5-homopiperazinone-1-oxyl* [30]. The oxidation of 2,2,7,7-tetramethyl-5-homopiperazionone by the method used for XXXII gave the corresponding radical in 48% yield: mother-of-pearl-pink flakes melting at 162°C (from a mixture of ether and hexane).

LXXII. *Di-tert-butylnitroxyl*. With stirring and cooling, 34.3 g of tert-nitrobutane in 300 ml of absolute ether was gradually added to a solution of the Grignard reagent obtained from 26 g of magnesium, 112 ml of tert-chlorobutane, and 200 ml of absolute ether. After the mixture had been stirred for 2 hr at room temperature, 120 ml of saturated ammonium chloride solution was added. The dark green ethereal layer was separated and the residual solution was extracted with ether. The ethereal extracts were combined and the ether and the tert-nitrosobutane were evaporated, after which the residue was chromatographed on alumina of activity II. The red zone (with ether as the eluant) passed into the eluate; the ether was evaporated and the residual red oil was fractionated through a Vigreux column, a fraction

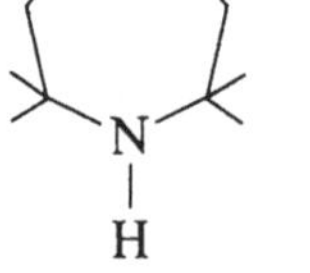

(LXX) 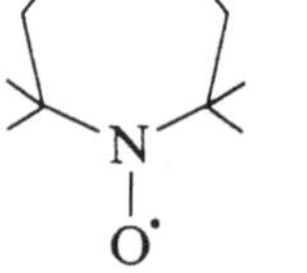   (LXXI) 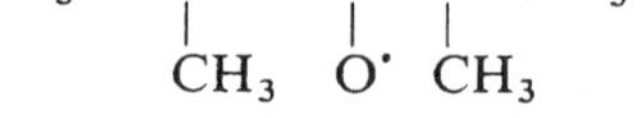   (LXXII)

$$H_3C-\underset{\underset{\displaystyle CH_3}{|}}{\overset{\overset{\displaystyle CH_3}{|}}{C}}\text{---}\underset{\underset{\displaystyle O^{\cdot}}{|}}{N}-\underset{\underset{\displaystyle CH_3}{|}}{\overset{\overset{\displaystyle CH_3}{|}}{C}}-CH_2-CH_3 \qquad HC\equiv C-\underset{\underset{\displaystyle CH_3}{|}}{\overset{\overset{\displaystyle CH_3}{|}}{C}}\text{---}\underset{\underset{\displaystyle H}{|}}{N}-\underset{\underset{\displaystyle CH_3}{|}}{\overset{\overset{\displaystyle CH_3}{|}}{C}}-C\equiv CH$$

(LXXIII)   (LXXIV)

with bp 75–76°C (35 mm Hg) and $d_4^{20} = 0.8600$ being collected. Literature data [31]: bp, 74–75°C (35 mm Hg).

LXXIII. *tert-Butyl-tert-amylnitroxyl* [32]. By the method used for LXXII, 26 g of magnesium, 112 ml of tert-chlorobutane, and 39 g of tert-nitropentane yielded 11 g (17%) of the radical, a red liquid with bp 56–57°C (5 mm Hg), and $d_4^{20} = 0.8630$.

LXXIV. *Bis(1,1-dimethyl-2-propynyl)amine*. By a published method [33], 83 g (1 mole) of 1,1-dimethylpropargylamine, 30 ml of 40% caustic potash, and 153.7 g (1.5 mole) of 1-chloro-1,1-dimethyl-2-propyne in the presence of 0.6 g of copper bronze powder gave 69.5 g (46.5%) of the amine with mp 34.5–35.5°C (sublimate). Literature data [33]: mp 32–35°C.

LXXV. *Di-tert-amylamine*. A solution of 15.7 g (0.1 mole) of di(1,1-dimethyl-2-propynyl)amine in ethanol was hydrogenated in the presence of 1.5 g of Raney nickel at room temperature and an initial hydrogen pressure of 2.5 atm. The usual working up of the reaction mixture gave 8.0 g (51%) of amine with bp 93–95°C (62 mm Hg), $n_D^{25} = 1.4281$. Literature data [33]: bp, 91–95°C (60 mm Hg), n_D^{25}, 1.4279.

LXXVI. *Di-tert-amylnitroxyl* [32]. Obtained in 78% yield by the oxidation of di-tert-amylamine by method XXXIX; after chromatographic separation on alumina and vacuum rectification in a column with an efficiency of 20 theoretical plates, an analytically pure sample was obtained in the form of a bright red liquid with bp 76–77°C (5 mm Hg) and $d_4^{20} = 0.8680$.

$$H_3C-CH_2-\underset{\underset{\displaystyle CH_3}{|}}{\overset{\overset{\displaystyle CH_3}{|}}{C}}\text{---}\underset{\underset{\displaystyle H}{|}}{N}-\underset{\underset{\displaystyle CH_3}{|}}{\overset{\overset{\displaystyle CH_3}{|}}{C}}-CH_2-CH_3$$

(LXXV)

$$\begin{array}{ccccc} & CH_3 & & CH_3 & \\ & | & & | & \\ H_3C-CH_2-C & \!\!\!\!-\!\!\!\!- & N-C-CH_2-CH_3 \\ & | & & | \\ & CH_3 & O^{\cdot} & CH_3 \end{array}$$

(LXXVI)

LXXVII. *tert-Butyl-1,1-dimethyl-2-propynylamine*. A half-liter flask with a gas-inlet tube, stirrer, and dropping funnel was charged with a mixture of 54.3 g (0.75 mole) of tert-butylamine, 50 ml of ether, 25 ml of water, and 0.3 g of catalyst (a mixture of one part by weight of copper powder and one part by weight of copper chloride). With continuous stirring in a current of nitrogen, a solution of 25.5 g (0.25 mole) of 1-chloro-1,1-dimethyl-2-propyne in 50 ml of ether was added. The contents of the flask were stirred for another 2.5 hr and were then worked up in a manner similar to that described in the preceding experiment. This gave 5.6 g (16%) of a substance with bp 135–136°C (750 mm Hg) and mp 24°C; d_4^{25}, 1.7947; n_D^{25}, 1.4305; MR_D, 45.31; MR_D calc., 45.36. Literature data [34]: bp, 135–136°C; mp, 24°C; n_D^{25}, 1.4292.

$$\begin{array}{ccccc} CH_3 & & CH_3 & \\ | & & | & \\ H_3C-C & \!\!\!\!-\!\!\!\!- & N-C-C\equiv CH \\ | & & | & | \\ CH_3 & & H & CH_3 \end{array}$$

(LXXVII)

$$\begin{array}{ccccc} CH_3 & & CH_3 & \\ | & & | & \\ H_3C-CH_2-C & \!\!\!\!-\!\!\!\!- & N-C-C\equiv CH \\ | & & | & | \\ CH_3 & & H & CH_3 \end{array}$$

(LXXVIII)

LXXVIII. *tert-Amyl-1,1-dimethyl-2-propynylamine* [32]. In a similar manner to the preceding experiment (cf. XXVII), 65.3 g (0.75 mole) of tert-amylamine and 25.5 g (0.25 mole) of 1-chloro-1,1-dimethyl-2-propyne gave 1.3 g (30%) of a reaction product with bp 52–53°C (15 mm Hg), $d_4^{20} = 0.8150$, $n_D^{20} = 1.4430$, $MR_D = 49.83$, and MR_D(calc.) = 49.98.

LXXIX. *tert-Butyl-1,1-dimethyl-2-propynylnitroxyl* [32]. A mixture of 3 ml of the corresponding amine (LXXVII), 12 ml of ethanol, 0.25 ml of sodium tungstate, 6 ml of Trilon B, and 10 ml of 30% hydrogen peroxide was left for three days. The solution was diluted with water to give 1.6 g (60%) of golden crystals which, after recrystallization from hexane and vacuum sublimation, melted at 64°C.

LXXX. *tert-Amyl-1,1-dimethyl-2-propynylnitroxyl* [32]. Obtained in 39% yield from 3 g of the corresponding amine (LXXVIII): mp, 20°C (hexane).

$$H_3C-\underset{\underset{CH_3}{|}}{\overset{\overset{CH_3}{|}}{C}}-\underset{\underset{O^{\cdot}}{|}}{N}-\underset{\underset{CH_3}{|}}{\overset{\overset{CH_3}{|}}{C}}-C\equiv CH$$

(LXXIX)

$$H_3C-CH_2-\underset{\underset{CH_3}{|}}{\overset{\overset{CH_3}{|}}{C}}-\underset{\underset{O^{\cdot}}{|}}{N}-\underset{\underset{CH_3}{|}}{\overset{\overset{CH_3}{|}}{C}}-C\equiv CH$$

(LXXX)

LXXXI. *2,2,4,4-Tetramethyl-1,2,3,4-tetrahydro-γ-carboline-3-oxyl* [35]. A solution of 3.8 g of 2,2,4,4-tetramethyl-1,2,3,4-tetrahydro-γ-carboline (mp, 147–148°C) in 125 ml of methanol was treated with 100 mg of sodium tungstate and a mixture of 3.5 ml of 30% hydrogen peroxide, 25 ml of methanol, and 10 mg of Trilon B. The reaction mixture was stirred for 4 hr and left to stand at room temperature for five days. The deposited golden orange crystals were filtered off with suction (2 g). An additional 1.5 g of substance was obtained from the mother liquor. After two recrystallizations from aqueous methanol, 2.7 g (67%) of a paramagnetic substance was obtained in the form of golden orange needles; mp 189–190°C (decomp.).

LXXXII. *Cyanoethylated 2,2,4,4-Tetramethyl-1,2,3,4-tetrahydro-γ-carboline-3-oxyl* [36]. To a solution of sodium ethoxide (0.035 g of Na in 2 ml of absolute ethanol) was added 0.5 g of the salt of Rodionov's catalyst (mp 160–161°C) obtained by heating an equimolar mixture of methyl p-toluenesulfonate and dimethylaniline in 5 ml of absolute ethanol. The sodium p-toluenesulfonate that precipitated was filtered off and the filtrate was used as catalyst. One milliliter of the catalyst was slowly added dropwise to a mixture of 0.5 g of the radical LXXXI and 14 ml of acrylonitrile. (The reaction mixture became hot.) One hour after the end of the reaction, the acrylonitrile was distilled in vacuum and the resinous residue was washed with water and extracted with benzene. The benzene extract was dried with magnesium sulfate and passed through a chromatographic column (with benzene as eluant) filled with alumina of activity II. The benzene was evaporated in vacuum and the residue was twice recrystallized from a mixture of hexane and acetone. The substance was isolated in the form of yellow needles with mp, 158–160°C; yield, 0.2 g (32.8% of theoretical).

(LXXXI)

(LXXXII)

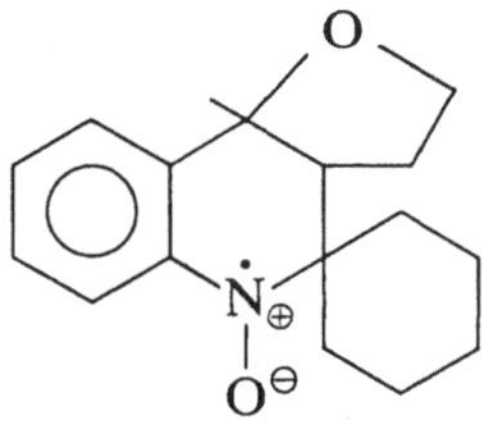

·(LXXXIV) 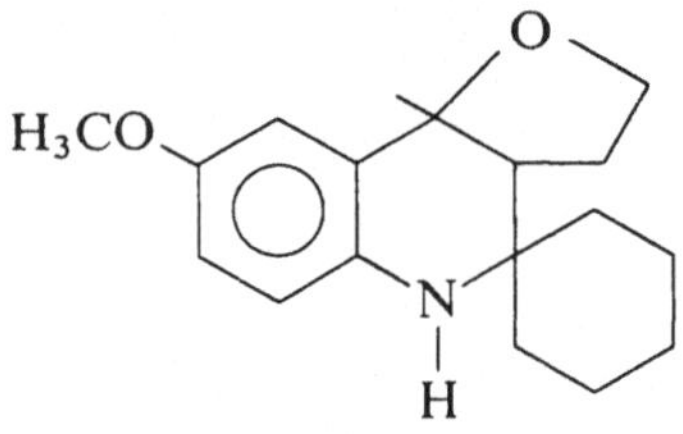   (LXXXV)

LXXXIII. *Oxidation of Derivatives of 2,2,4,4-tetramethyl-1,2,3,4-tetrahydro-γ-carboline.* The initial 1,2,3,4-tetrahydro-γ-carboline derivatives were obtained by the Fischer reaction. The amines were oxidized by the method used for XXXI. 6-Carboxy-9-(β-carboxyethyl)-2,2,4,4-tetramethyl-1,2,3,4-tetrahydro-γ-carboline was oxidized in the form of the dipotassium salt in saturated aqueous solution, and the corresponding oxyl was isolated by careful acidification of the reaction mixture with 10% hydrochloric acid. The properties of the new radicals of the hydrogenated γ-carboline series are given in Table 26.

LXXXIV. *Nitroxide of 4-Methyl-2-spirocyclohexyl-3,4:3',2'-tetra-hydrofurano-1,2,3,4-tetrahydroquinoline* [37]. To a solution of 2.5 g of the corresponding base in 100 ml of methanol were added 0.1 g of sodium tungstate, 0.05 g of Trilon B, and 3.5 ml of 30% hydrogen peroxide. The reaction mixture was left at room temperature for three days, after which

TABLE 26
Free Nitroxyl Radicals of the Hydrogenated γ-Carboline Series [36]

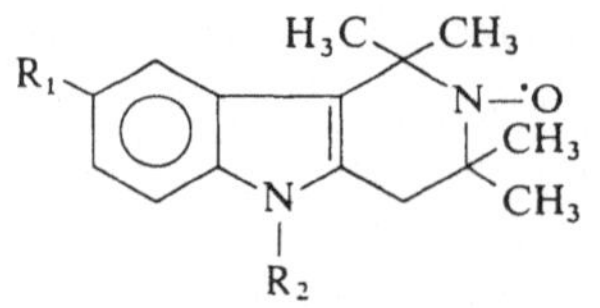

Compound		Yield, %	Mp, °C
R_1	R_2		
H	CH_3	80	154–154.5
CH_3	H	75	213–214 (decomp.)
$C_6H_5CH_2O$	H	77	Decomposes above 220°C
H_2NCO	H	64	Decomposes above 220°C
COOH	CH_2CH_2COOH	45	Decomposes above 220°C

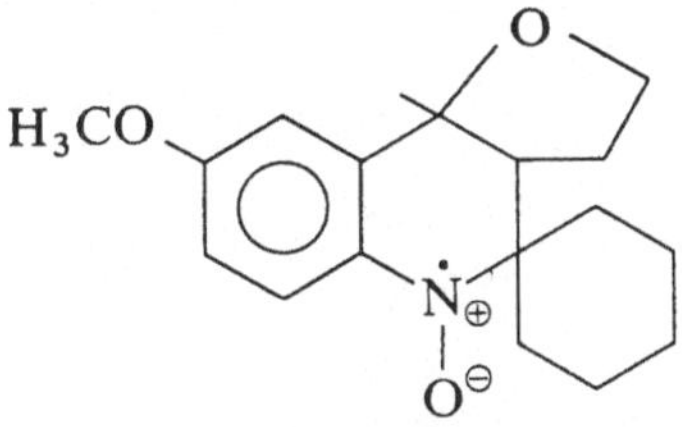

(LXXXVI) 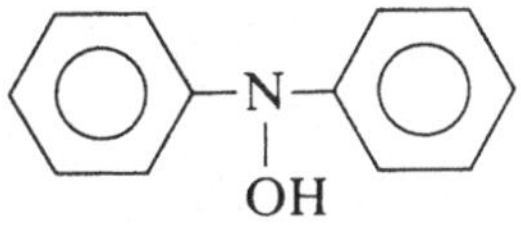   (LXXXVII)

the solvent was distilled in vacuum. The black resinous residue was extracted with benzene and chromatographed on a column of alumina (activity II, column height 400 mm, diameter 20 mm). The mixture was eluted with benzene; the first, bright yellow zone gave, after evaporation of the eluate in vacuum, a red oil. After successive crystallization from ether and hexane, the substance was obtained in the form of bright red prisms; mp, 101–101.5°C; yield, 1.7 g (65% of theoretical).

LXXXV. *6-Methoxy-4-methyl-2-spirocyclohexyl-3,4:3′,2′-tetra-hydrofurano-1,2,3,4-tetrahydroquinoline* [38]. To a solution of 18 g of cyclohexylidene-*p*-anisidine in 50 ml of benzene in the presence of 0.8 ml of boron trifluoride etherate was added 7.7 g of 2-methyl-4,5-dihydrofuran. The reaction mixture was stirred without heating for 3 hr and at 60–70°C for 1 hr and was then washed with 10% sodium hydroxide. When the reaction product was distilled in vacuum, 5.0 g (yield 20% of theoretical) of the amine was obtained with bp 160–165°C (0.1 mm), $n_D^{20} = 1.5640$, mp 83–84°C (from ethanol).

LXXXVI. *Nitroxide of 6-Methoxy-4-methyl-2-spirocyclohexyl-3,4:3′,2′-tetrahydrofurano-1,2,3,4-tetrahydroquinoline* [38]. To a solution of 0.5 g of the amine LXXXV were added 50 mg of sodium tungstate, 20 mg of Trilon B, and 0.8 ml of 30% hydrogen peroxide. The reaction mixture was left at room temperature for two days, after which, the free radical was isolated in a similar manner to that described above (chromatography was carried out on a column 350 mm high and 8 mm in diameter containing alumina of activity II) in the form of bright red prisms with mp 97–98°C, yield 0.3 g (58% of theoretical).

LXXXVII. *Diphenylhydroxylamine.* An ethereal solution of phenyl-magnesium bromide was prepared in a three-necked, round-bottomed flask fitted with a reflux condenser, dropping funnel, and mechanical stirrer, from 9.2 g (0.37 g-atom) of magnesium, 60 g (0.38 mole) of bromobenzene, and 200 ml of anhydrous ether. The solution was cooled to −15° and, with

(LXXXVIII)   (LXXXIX)

continuous stirring, a saturated solution of 18 g (0.17 mole) of nitrobenzene in anhydrous ether was added dropwise. After being stirred at room temperature for 2 hr, the reaction mixture was poured onto ice, and the ethereal layer was separated and dried over calcium chloride at 0°C. The ether was evaporated under reduced pressure in a current of dry hydrogen and the residual oil was mixed with 100 ml of dry hexane, giving crystals of diphenyl-hydroxylamine which were separated, washed with hexane, and recrystallized from a mixture of benzene and hexane. The yield of diphenylhydroxylamine with mp 62°C was 15 g (44.5%). Literature data [39]: mp, 64°C.

LXXXVIII. *Diphenylnitroxide.* A solution of 10 g (0.05 mole) of diphenylhydroxylamine in 60 ml of anhydrous ether was cooled to −5°C and shaken with 25 g (0.1 mole) of dry silver oxide, and the ethereal solution was separated from the silver by filtration. After the evaporation of part of the ether in vacuum, an equal volume of dry hexane was added to the concentrated solution and the mixture was strongly cooled. The crystals that deposited were separated, washed with cold hexane, and recrystallized from a mixture of ether and hexane (1:1). The yield of diphenylnitroxide with mp 64°C was 9 g (90%). Literature data [39]: mp, 64°C.

LXXXIX. *Di-p-anisylnitroxide.* With cooling, 5 ml of concentrated sulfuric acid was added to a solution of 5 g of anisole in 15 ml of glacial acetic acid. Then, with vigorous stirring, 2.5 ml of nitric acid (sp. gr. 1.35) was added, the temperature being kept between 25 and 30°C. The dark violet solution was cooled, mixed with 200 ml of ice-water, and rapidly filtered into 50 ml of 20% perchloric acid solution cooled to 0°C. The orange crystals that deposited on cooling were separated and dried in a vacuum desiccator. The yield of perchlorate with mp 155°C was 0.7 g. The salt was dissolved in 8 ml of cold pyridine, and the nitroxide was precipitated by the slow addition of water; after recrystallization from aqueous methanol, it melted at 150°C with slight decomposition. The yield of purified and dried di-*p*-anisylnitroxide was 0.22 g (45%). Literature data [40]: melting point (from ethyl acetate) 150°C.

XC. *Nitroxide of 2,2,4-trimethyl-1,2,3,4-tetrahydroquinoline* [41]. To a solution of 5 g of 2,2,4-trimethyl-1,2,3,4-tetrahydroquinoline in 75 ml of

(XC)   (XCI)

methanol were added 0.2 g of sodium tungstate, 0.05 g of Trilon B, and 4 ml of 30% hydrogen peroxide. After two days, the methanol was evaporated in vacuum, the residue was dissolved in benzene, and the resulting solution was washed several times with water and dried with magnesium sulfate. After the evaporation of the benzene in vacuum, the residue was chromatographed on a column (length 400 mm, diameter 20 mm) filled with alumina (activity grade II) using chloroform as the eluant. The chloroform was distilled from the first chromatographic fraction and the residue was distilled in vacuum. The distillate (2.5 g) with bp 58°C (6.5×10^{-2} mm Hg) set to a bright red crystalline mass with mp 36°C possessing an EPR signal equivalent to 5.0×10^{23} spins/mole. The paramagnetic crystals obtained resinified completely in 3 hr. Recrystallization of the resinification product from methanol gave colorless needles with mp 42°C. A mixed melting point with an authentic sample of 2,2,4-trimethyl-1,2,3,4-tetrahydroquinoline gave no depression of the melting point.

XCI. *2,2',4,4' - Tetramethoxydiphenylimmonium Oxide Perchlorate.* At 7–8°C, 5 ml (0.06 mole) of nitric acid (sp. gr. 1.35) was added in drops to 10 ml (0.076 mole) of resorcinol dimethyl ether in 30 ml of glacial acetic acid, whereupon the solution became dark green. Then, 120 ml of distilled water cooled to 0°C was added and the mixture was filtered into 100 ml (0.22 mole) of cooled 20% perchloric acid. The yield of purified perchlorate with decomp. pt. 178°C (from acetone) was 7.53 g (49%): violet-brown powder with a metallic luster was readily soluble in chloroform, tetrachloroethane, and acetone, sparingly soluble in water, benzene, and ether, and almost insoluble in petroleum ether.

XCII. *2,2',4,4' - Tetramethoxydiphenylnitroxide* [42]. At −5°C, 6 g of the perchlorate obtained in the preceding experiment, 6 g of sodium iodide, and 30 ml of absolute methanol were ground in an agate mortar for 10–15 min. The resulting red solution was treated with 60 ml of absolute ether and the mixture was strongly cooled. The solution deposited glistening red prisms, which were rapidly filtered off and washed on the filter with cold ether. The yield of purified 2,2',4,4'-tetramethoxydiphenylnitroxide was 4.02 g (91.7%),

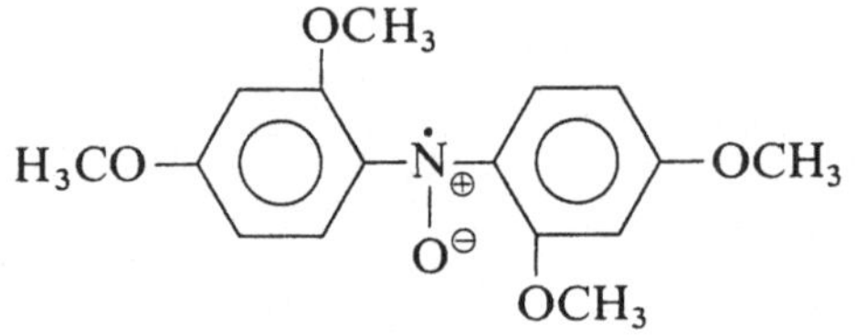

$$\text{H}_3\text{CO}-\underset{}{\bigcirc}-\overset{\cdot}{\underset{\underset{\text{O}^{\ominus}}{|}}{\text{N}_{\oplus}}}-\bigcirc-\text{OCH}_3$$

(XCII)

mp 121.5–122°C (from acetone), readily soluble in chloroform and acetone, less readily in methanol, sparingly soluble in ether and benzene, and insoluble in water and alkalis.

XCIII.   *4,4'-Dimethoxy-2,2',6,6'-tetramethyldiphenylimmonium Oxide Perchlorate.* At 17–18°C, 2.5 ml (0.03 mole) of nitric acid (sp. gr. 1.35) was added dropwise to 5 ml (0.034 mole) of the methyl ether of 3,5-xylenol in 15 ml of glacial acetic acid containing 2 ml of concentrated sulfuric acid. The resulting dark green reaction mixture was treated with 60 ml of ice-water and the mixture was filtered into 50 ml (0.11 mole) of cooled 20% perchloric acid. The yield of dark green perchlorate with a metallic luster was 4.88 g (70.3%): mp, 118°C (from acetone), readily soluble in chloroform and acetone, sparingly soluble in water, benzene, ethanol, and ether, almost insoluble in petroleum ether.

XCIV. *4,4'-Dimethoxy-2,2',6,6'-tetramethyldiphenylnitroxide.* At −5°C, 4.88 g (0.012 mole) of the perchlorate obtained in the preceding experiment and 4.88 g (0.032 mole) of sodium iodide in 40 ml of absolute methanol were ground for 10–15 min. With stirring, the resulting bright red solution was treated with 40 ml of cold absolute ether, which gave 1.90 g (52.6%) of the crystalline radical with mp 163°C (from acetone): ruby-red prisms readily soluble in chloroform, acetone, and ethyl acetate, less readily in methanol, sparingly in ether and benzene, and insoluble in water. Literature data [43]: mp, 163°C.

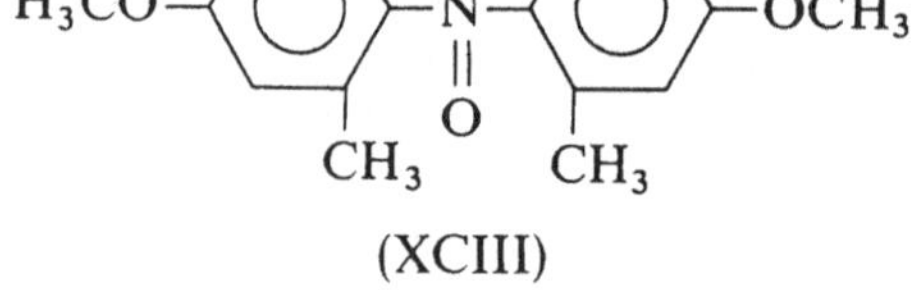

$$\text{H}_3\text{CO}-\bigcirc-\overset{\overset{\text{CH}_3 \quad \text{CH}_3}{\underset{}{|}}}{\underset{\underset{\text{CH}_3 \quad \text{CH}_3}{||}}{\overset{\text{ClO}_4^{\ominus}}{\underset{\oplus}{\text{N}}}}}-\bigcirc-\text{OCH}_3$$

(XCIII)

$$\text{H}_3\text{C}-\text{O}-\overset{\overset{\text{CH}_3}{|}}{\underset{\underset{\text{CH}_3}{|}}{\bigcirc}}-\overset{\overset{\cdot}{\underset{|}{\text{N}_{\oplus}}}}{\underset{\text{O}^{\ominus}}{}}-\overset{\overset{\text{CH}_3}{|}}{\underset{\underset{\text{CH}_3}{|}}{\bigcirc}}-\text{O}-\text{CH}_3$$

(XCIV)

XCV. *2,2',4,4',6,6'-Hexamethoxydiphenylimmonium Oxide Perchlorate.* At 50°C, 1.25 ml (0.015 mole) of nitric acid (sp. gr. 1.35) was added dropwise to a solution of 2.5 g (0.015 mole) of the trimethyl ether of phloroglucinol in 7.5 ml of glacial acetic acid. The dark green reaction mixture was treated with 30 ml of ice-water and the resulting mixture was filtered into 25 ml (0.06 mole) of 20% perchloric acid. The yield of dark brown perchlorate with a metallic luster was 2.97 g (87.3%): decomp. pt., 189°C (from acetone); readily soluble in chloroform and acetone, sparingly soluble in water, benzene, ethanol, and ether, and almost insoluble in petroleum ether. Literature data [40] : mp, 189°C (from acetone).

$$\text{H}_3\text{CO}-\overset{\overset{\text{OCH}_3}{|}}{\underset{\underset{\text{OCH}_3}{|}}{\bigcirc}}-\overset{\overset{\oplus}{\underset{\parallel}{\text{N}}}}{\underset{\text{O}}{}}\overset{\text{ClO}_4^{\ominus}}{}-\overset{\overset{\text{OCH}_3}{|}}{\underset{\underset{\text{OCH}_3}{|}}{\bigcirc}}-\text{OCH}_3$$

(XCV)

XCVI. *2,2',4,4',6,6'-Hexamethoxydiphenylnitroxide.* At 35–40°C, 4.29 g (0.0093 mole) of the corresponding perchlorate, 6.43 g (0.043 mole) of sodium iodide, and 40 ml of absolute methanol were triturated in an agate mortar for 30 min. The resulting red solution was treated with 40 ml of cold absolute ether, and the mixture was strongly cooled. The coarse red prisms were separated, washed with ether, and recrystallized from acetone. The yield of the nitrogen oxide was 3.03 g (78.3%): decomp. pt., 194°C; readily

$$\text{H}_3\text{CO}-\overset{\overset{\text{OCH}_3}{|}}{\underset{\underset{\text{OCH}_3}{|}}{\bigcirc}}-\overset{\overset{\cdot}{\underset{|}{\text{N}_{\oplus}}}}{\underset{\text{O}^{\ominus}}{}}-\overset{\overset{\text{OCH}_3}{|}}{\underset{\underset{\text{OCH}_3}{|}}{\bigcirc}}-\text{OCH}_3$$

(XCVI)

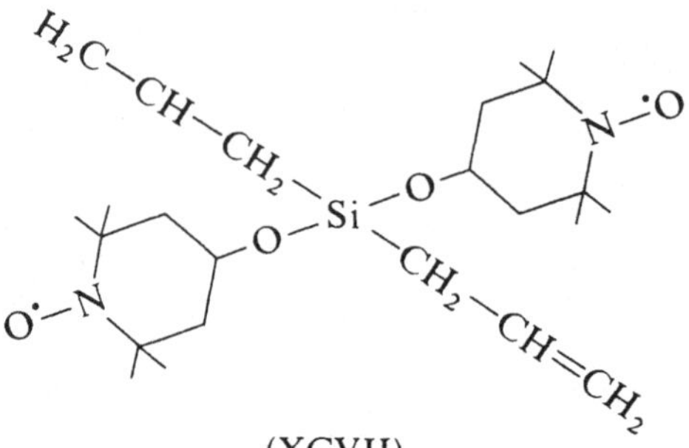

(XCVII)

soluble in chloroform and acetone, sparingly soluble in methanol, slightly soluble in ether and benzene, insoluble in water. Literature data [43]: decomp. pt., 194°C.

XCVII. *Diallylbis(2,2,6,6-tetramethyl-1-oxyl-4-piperidyloxy)silane* [44]. An amount of 3.6 g (0.02 mole) of a solution of diallyldichlorosilane in 25 ml of anhydrous benzene was added dropwise to 6.8 g (0.04 mole) of 4-hydroxy-2,2,6,6-tetramethylpiperidine-1-oxyl dissolved in 18 ml of anhydrous benzene and 5 ml of absolute pyridine. After the mixture had been heated for $\frac{1}{2}$ hr under reflux, the flask was cooled and the solution was separated from the deposited salt. The bright red filtrate was carefully washed with water and dried with anhydrous potassium carbonate. After the solvent had been evaporated under reduced pressure, the solid residue was dissolved in hexane and the solution was filtered and strongly cooled. The red crystalline mass that deposited was filtered off, washed on the filter with cooled hexane, and dried in a vacuum desiccator over paraffin wax. The yield of biradical was 4.6 g (44.2%): brick red plates with mp 57.5–58°C (from hexane).

XCVIII. *Fremy's Salt* [45].   A solution of 84 g (1 mole) of sodium bicarbonate and 70 g (1 mole) of sodium nitrite in 1500 cm³ of distilled water was prepared in a four-liter beaker fitted with a glass mechanical stirrer and a porcelain bubbler. The solution was cooled to 0°C and, with vigorous stirring, purified gaseous sulfur dioxide was passed through until the pH

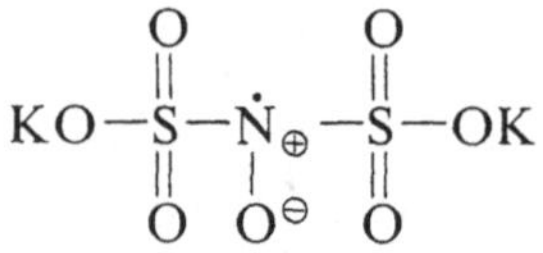

(XCVIII)

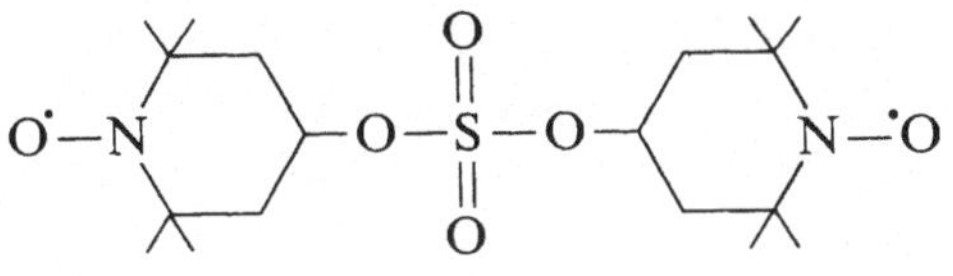

(XCIX)

of the medium had reached 4, after which, the rate of flow was reduced; when the solution had acquired a permanent brown-red color, the passage of gas was stopped (pH about 2; weight of sulfur dioxide absorbed, 124–128 g). To eliminate the excess of sulfur dioxide, nitrogen was bubbled through the solution at 5–10°C until the pH had reached 9–10. With continued stirring, a solution of 43 g of potassium permanganate in 900 cm³ of distilled water was added to the mixture over 20 min. Stirring was continued for another 10–15 min at 5–10°C, and then the mixture was rapidly filtered through a fine glass filter to eliminate the suspended manganese dioxide completely. Six hundred grams of crystalline potassium acetate was dissolved in the clear filtrate and the solution was cooled to 0°C. After 15 min, the orange crystals of Fremy's salt that had precipitated were separated on a glass filter, washed with aqueous potassium acetate solution made slightly alkaline with ammonia (200 g of acetate in 1 liter of water), with 95% ethanol (with the addition of ammonia) and, finally, with pure acetone (with the addition of ammonia). The washing operations each required 200–300 ml of cold liquid. The washed Fremy's salt was well pressed out on the filter and dried under high vacuum. This gave 220 g of anhydrous material free from chloride and sulfate ions.

XCIX.   *Bis(2,2,6,6-tetramethyl-1-oxyl-4-piperidyl) Sulfate* [46]. At 4–5°C, 0.78 g of sulfuryl chloride in 20 ml of benzene was added to a mixture of 20 g of 4-hydroxy-2,2,6,6-tetramethylpiperidine-1-oxyl, 30 ml of benzene, and 1.5 ml of anhydrous pyridine, after which, stirring of the reaction mixture was continued for another hour at room temperature and for 15 min at the boiling point of the solution. The precipitate was filtered off, and the benzene solution was evaporated in vacuum. The residue was dissolved in 20 ml of ethanol, precipitated with water, and sublimed in vacuum. At 60°C (2 mm), 4-chloro-2,2,6,6-tetramethylpiperidine-1-oxyl

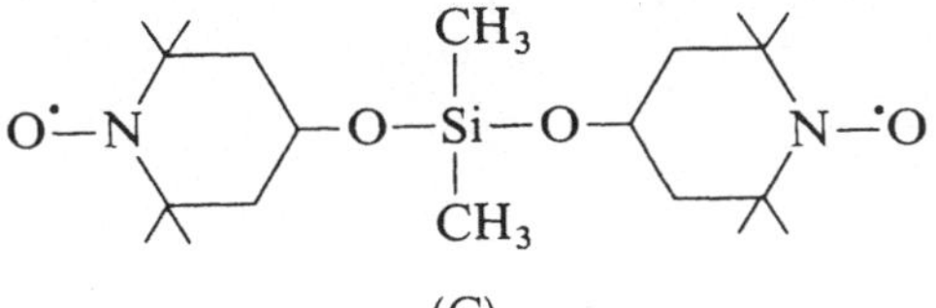

(C)

sublimed, weight 0.05 g (4.5%) mp, 112–113°C (from hexane). The sublimate was separated off and the residue was twice recrystallized from ethanol. The yield of the biradical was 0.1 g (4.5%), mp, 145°C (decomp.).

C. *Dimethylbis(2,2,6,6-tetramethyl-1-oxyl-4-piperidyloxy)silane* [46]. Over 1 hr, with careful stirring, 0.74 g of dichlorodimethylsilane in 20 ml of benzene was added to a mixture of 2.0 g of 2,2,6,6-tetramethyl-4-hydroxypiperidine-1-oxyl, 30 ml of benzene, and 2 ml of triethylamine, after which, the reaction mixture was boiled under reflux for 5 hr. The precipitate that deposited was separated, the benzene was evaporated, and the residue was dissolved in the minimum amount of hexane, the insoluble 4-hydroxy-2,2,6,6-tetramethylpiperidine-1-oxyl being filtered off. The filtrate was diluted with hexane and left in the refrigerator for several hours. This gave 0.67 g (29.0%) of the biradical, with mp, 52–53°C.

CI. *Organosilicon Nitroxyl Polyradicals* [47]. With careful stirring, an equimolar amount of the appropriate organochlorosilane in 20 ml of benzene was added to a mixture of 0.00116 mole of 4-hydroxy-2,2,6,6-tetramethylpiperidine-1-oxyl, 30 ml of absolute benzene, and 2 ml of triethyl-

TABLE 27
Constants and Yields of the Individual Organosilicon Nitroxyl Polyradicals

Formula*	Yield, %	Mp, °C	Mol. wt.		Spins/mole × 10^{24}	
			Calc.	Found	Found	Calc.
$R_2CH_3SiCH_2CH_2SiCH_3R_2$	34.6	123–124	799.2	—	2.4	2.4
$R_2C_6H_5SiCH_2CH_2SiC_6H_5R_2$	38.0	173	923.3	940	2.4	2.4
$R_2C_6H_5SiCH_2CH_2SiCH_3R_2$	26.0	97–98	861.3	—	2.4	2.4
$R_2CH_3SiCH{=}CHSiCH_3R_2$	34.7	149–150	797.2	800	2.3	2.4
$R_2C_6H_5SiCH{=}CHSiC_6H_5R_2$	45.0	165–165.5	921.3	—	2.5	2.4
$R_2CH_3SiCH_2CH_2SiR_3$	58.0	112	955.1	980	2.8	3.0
$R_2C_6H_5SiCH_2CH_2SiR_3$	30.0	98–99	1019.4	—	3.1	3.0
$R_2CH_3SiCH_2CH_2CH_2SiR_3$	21.2	Oil	969.4	—	2.7	3.0
$R_2CH_3SiCH{=}CHSiR_3$	31.8	143–146	952.4	925	2.6	3.0
$R_3SiCH_2SiR_3$	24.0	199	1097.5	—	3.4	3.6
$R_3SiCH_2CH_2SiR_3$	57.0	186	1111.7	1187	3.5	3.6
$R_3SiCH{=}CHSiR_3$	54.2	169–170	1109.6	—	3.5	3.6
$R_2Si(CH_2CH_2SiCH_3R_2)_2$	28.5	Oil	1199.8	1200	3.4	3.6
$R_3SiCH_2CH_2CH_2SiR_3$	22.4	134–135	1127.7	—	3.4	3.6
$\pi\text{-}C_6H_4(SiR_3)_2$	61.5	188–189	1160.8	1140	3.6	3.6
$R_2Si(CH_2{-}CH_2{-}SiR_3)_2$	20.0	Oil	1410	—	4.3	4.8

*Here R represents the 1-oxyl-2,2,6,6-tetramethyl-4-piperidyloxy group.

amine. The reaction mixture was boiled for 5 hr. The precipitate formed was filtered off and the filtrate was evaporated in vacuum. After the usual purification by crystallization or reprecipitation, analytically pure samples were obtained (see Table 27).

References

1. E. Matter, *Helv. Chim. Acta* **30**, 1114 (1947).
2. C. Sandrix and G. Quarisson, *Bull. Soc. Chim. France* **1958**, 345.
3. H. Pauly, *Ann. Chem.* **322**, 113 (1902).
4. É. G. Rozantsev and L. A. Krinitskaya, *Tetrahedron* **21**, 491 (1965).
5. L. A. Krinitskaya, É. G. Rozantsev, and M. B. Neiman, *Izv. Akad. Nauk SSSR, Ser. Khim.* **1965**, 115.
6. É. G. Rozantsev and L. A. Krinitskaya, *Izv. Akad. Nauk SSSR, Ser. Khim.* **1964**, 1573.
7. L. A. Krinitskaya, A. L. Buchachenko, and É. G. Rozantsev, *Zh. Organ. Khim.* **2**, 1301 (1966).
8. L. A. Krinitskaya, Thesis, Moscow (1967).
9. É. G. Rozantsev, USSR Authors' Certificate No. 166032, 27 August, 1962.
10. É. G. Rozantsev, *Izv. Akad. Nauk SSSR, Ser. Khim.* **1964**, 2218.
11. É. G. Rozantsev, *Izv. Akad. Nauk SSSR, Ser. Khim.* **1964**, 2187.
12. É. G. Rozantsev and V. A. Golubev, *Izv. Akad. Nauk SSSR, Sert Khim.* **1966**, 891.
13. É. G. Rozantsev and V. I. Suskina, *Izv. Akad. Nauk SSSR, Ser. Khim.* **1968**, 2106.
14. N. Leonard and E. Nommensen, *J. Am. Chem. Soc. Soc.* **71**, 2808 (1949).
15. Yu. G. Mamedova, Thesis, Moscow (1965).
16. O. L. Lebedev and S. N. Kazarnovskii, *Tr. po Khimii i Khim. Tekhnol.* **1959**, 649.
17. É. G. Rozantsev, V. A. Golubev, and M. B. Neiman, *Izv. Akad. Nauk SSSR, Ser. Khim.* **1965**, 391.
18. V. S. Golubev and É. G. Rozantsev, *Izv. Akad. Nauk SSSR, Ser. Khim.* **1965**, 716.
19. V. A. Golubev, M. B. Neiman, and É. G. Rozantsev, *Izv. Akad. Nauk SSSR, Ser. Khim.* **1966**, 343.
20. É. G. Rozantsev, V. Z. Golubev, and M. B. Neiman, *Izv. Akad. Nauk SSSR, Ser. Khim.* **1965**, 393.
21. É. G. Rozantsev, V. A. Golubev, M. B. Neiman, and Yu. V. Kokhanov, *Izv. Akad. Nauk SSSR, Ser. Khim.* **1965**, 572.
22. É. G. Rozantsev, V. A. Golubev, and M. B. Neiman, USSR Authors' Certificate No. 175504, October 24, 1964.
23. M. B. Neiman, É. G. Rozantsev, and V. A. Golubev, *Izv. Akad. Nauk SSSR, Ser. Khim.* **1965**, 548.
24. É. G. Rozantsev and V. A. Golubev, *Izv. Akad. Nauk SSSR, Ser. Khim.* **1965**, 718.
25. W. Lutz, S. Lazarus, and R. Meltzer, *J. Org. Chem.* **27**, 1695 (1962).
26. É. G. Rozantsev, *Teor. i Éksperim. Khim.* **2**, 415 (1966).
27. É. G. Rozantsev, V. A. Golubev, V. N. Suskina, and M. B. Neiman, *Zh. Organ. Khim.* **2**, 1075 (1966).
28. É. G. Rozantsev and Yu. V. Kokhanov, *Izv. Akad. Nauk SSSR, Ser. Khim.* **1966**, 1477.
29. S. Diskerman and H. Lindwall, *J. Org. Chem.* **14**, 530 (1949).
30. É. G. Rozantsev and R. A. Popko, *Izv. Akad. Nauk SSSR, Ser. Khim.* **1963**, 764.
31. A Hoffman and A. Henderson, *J. Am. Chem. Soc.* **83**, 4671 (1961).
32. É. G. Rozantsev and R. S. Burmistrova, *Dokl. Akad. Nauk SSSR*, **166**, 129 (1966).

33. G. Hennion and C. Digiovanna, *J. Org. Chem.* **30**, 2645 (1965).
34. G. Hennion and R. Hanzell, *J. Am. Chem. Soc.* **82**, 4908 (1960).
35. É. G. Rozantsev and A. B. Shapiro, *Izv. Akad. Nauk SSSR, Ser. Khim.* **1964**, 1123.
36. É. G. Rozantsev, A. B. Shapiro, and N. N. Komzolova, *Izv. Akad. Nauk SSSR, Ser. Khim.* **1965**, 1100.
37. A. B. Shapiro, É. G. Rozantsev, L. S. Povarov, and V. N. Grigos, *Izv. Akad. Nauk SSSR, Ser. Khim.* **1964**, 1725.
38. A. B. Shapiro, É. G. Rozantsev, L. S. Povarov, and V. N. Grigos, *Izv. Akad. Nauk SSSR, Ser. Khim.* **1965**, 1102.
39. H. Wieland and K. Roth, *Ber.* **53**, 210 (1920).
40. K. Meyer and H. Gottlieb-Billroth, *Ber.* **52**, 1476 (1919).
41. A. B. Shapiro, A. A. Medzhidov, and É. G. Rozantsev, *Zh. Organ. Khim.* **2**, 1873 (1966).
42. L. A. Skripko, Thesis, Moscow (1968).
43. K. Meyer and W. Reppe, *Ber.* **54**, 327 (1921).
44. G. F. Pavelko, Thesis, Moscow (1968).
45. G. Harvey and R. Hollingshead, *Chem. Ind.* (*London*) **1953**, 244.
46. É. G. Rozantsev, G. F. Pavelko, A. L. Buchachenko, and M. B. Neiman, *Izv. Akad. Nauk SSSR, Ser. Khim.* **1967**, 2306.
47. G. F. Pavelko and É. G. Rozantsev, *Izv. Akad. Nauk SSSR, Ser. Khim.* **1967**, 2466.

INDEX